Green Synthesis of Nanomaterials

Green Synthesis of Nanomaterials

Biological and Environmental Applications

Edited by

Archana Chakravarty
Jamia Millia Islamia; Central University
New Delhi
India

Preeti Singh
Institute of Chemical Technology
Mumbai
India

Saiqa Ikram
Jamia Millia Islamia; Central University
New Delhi
India

R.N. Yadava
Purnea University Bihar,
Bihar
India

WILEY

Published by John Wiley & Sons, Inc., Hoboken, New Jersey.
Published simultaneously in Canada.

Library of Congress Cataloging-in-Publication Data applied for:

ISBN: HB: 9781119900900, ePDF: 9781119900917, epub: 9781119900924

Cover Design: Wiley
Cover Image(s): © Konstantinos Zouganelis/Getty Images

Set in 9.5/12.5pt STIXTwoText by Straive, Pondicherry, India

Contents

List of Contributors

Rohana Abu
Faculty of Chemical and Process Engineering Technology
Universiti Malaysia Pahang Al-Sultan Abdullah. Persiaran Tun Kalil Yaakob, Gambang Kuantan, Pahang
Malaysia

Khairatun Najwa Mohd Amin
Faculty of Chemical and Process Engineering Technology
Universiti Malaysia Pahang Al-Sultan Abdullah. Persiaran Tun Kalil Yaakob, Gambang Kuantan, Pahang
Malaysia

Isha Arora
Department of Chemistry
Amity Institute of Applied Sciences, Amity University
Noida, UP
India

Deepak Kumar Bhartiya
Department of Zoology
Government Degree College
Dhadha Bujurg-Hata
Kushinagar, UP
India

Archana Chakravarty
Department of Chemistry
Jamia Millia Islamia
New Delhi, Delhi
India

Aayushi Chanderiya
Department of Chemistry
Dr. Harisingh Gour University
Sagar, MP
India

Amrish Chandra
Amity Institute of Pharmacy
Amity University
Noida, UP
India

Vinita Chaturvedi
Biochemistry Division
Central Drug Research Institute
CSIR Lucknow, UP
India

Ritu Rani Chaudhary
Department of Chemistry
B.S.A. College
Mathura, UP
India

Shivani Chaudhary
Department of Chemistry
Dr. Bhimrao Ambedkar University
Agra, UP
India

Neeru Dabas
Department of Chemistry
Amity School of Applied Science
Amity University
Gurugram, HR
India

Ratnesh Das
Department of Chemistry
Dr. Harisingh Gour Central
University
Sagar, MP
India

Rimon Ranjit Das
Department of Physics
Amity Institute of Applied
Sciences, Amity University
Noida, UP
India

Nowsheenah Farooq
Bioinorganic Lab
Department of Chemistry
Jamia Millia Islamia
New Delhi, DL
India

Is Fatimah
Department of Chemistry
Faculty of Mathematics and
Natural Sciences
Universitas Islam Indonesia
Yogyakarta
Indonesia

Manoj Gadewar
Department of Pharmacology
School of Medical and Allied
Sciences, KR Mangalam University
Gurgaon, HR
India

Seema Garg
Department of Chemistry
Amity Institute of Applied
Sciences, Amity University
Noida, UP
India

Mithun Kumar Ghosh
Department of Chemistry
Government College Hatta
Damoh
India

Juhi Gupta
Bioinorganic Research Lab
Department of Chemistry
Jamia Millia Islamia
New Delhi, DL
India

Athar Adil Hashmi
Bioinorganic Lab
Department of Chemistry
Jamia Millia Islamia
New Delhi, DL
India

Afkar Rabbani Hidayatullah Hipeni
Faculty of Chemical and Process
Engineering Technology
Universiti Malaysia Pahang Al-Sultan
Abdullah. Persiaran Tun Kalil
Yaakob, Gambang Kuantan, Pahang
Malaysia

MA Motalib Hossain
Institute of Sustainable Energy
Universiti Tenaga Nasional
Kajang, Selangor
Malaysia

Gautam Jaiswar
Department of Chemistry
Dr. Bhimrao Ambedkar
University
Agra, UP
India

Sarabjeet Kaur
Surface Chemistry and Catalysis:
Characterisation and Application
Team (COK-KAT), Leuven
(Arenberg)
Leuven
Belgium

Nishat Khan
Department of Chemistry
Amity Institute of Applied
Sciences, Amity University
Noida, UP
India

Manoj Kumar
Department of Chemistry
Government Degree College
Dhadha Bujurg-Hata
Kushinagar, UP
India

Vikas Kumar
Department of Chemistry
Government College Khimlasa
Sagar, MP
India

Shivani A. Kumar
Department of Physics
Amity Institute of Applied
Sciences, Amity University
Noida, UP
India

Indu Kumari
CT Group of Institutions
Jalandhar, PB
India

Surbhi Malik
Department of Physics
Amity Institute of Applied
Sciences, Amity University
Noida, UP
India

B.R. Malini
Department of Chemistry
Akshara First Grade College
Bengaluru, KA
India

Sharifah Fathiyah Sy Mohamad
Faculty of Chemical and Process
Engineering Technology
Universiti Malaysia Pahang Al-Sultan
Abdullah. Persiaran Tun Kalil
Yaakob, Gambang Kuantan, Pahang
Malaysia

Baranya Murugan
Nanotechnology & Catalysis
Research Centre
University of Malaya
Kuala Lumpur
Malaysia

Sivasubramanian Murugappan
Department of Biomedical
Engineering
Indian Institute of Technology
Hyderabad
Sangareddy, TS
India

M. Mutthuraju
Department of Chemistry
Sai Vidya Institute of Technology
Affiliated to Visvesvaraya
Technological University
Bengaluru, KA
India

Nahid Nishat
Inorganic Materials Research Lab
Department of Chemistry
Jamia Millia Islamia
New Delhi, DL
India

Sikandar Paswan
Department of Chemistry
Baba Raghav Das PG College
Deoria, UP
India

Monika Pebam
Department of Biomedical
Engineering
Indian Institute of Technology
Hyderabad
Sangareddy, TS
India

G.K. Prashanth
Research and Development Centre
Department of Chemistry
Sir M. Visvesvaraya Institute
of Technology, Affiliated to
Visvesvaraya Technological
University
Bengaluru, KA
India

Benni F. Ramadhoni
National Research and
Innovation Agency
Tangerang Selatan
Indonesia

Srilatha Rao
Department of Chemistry
Nitte Meenakshi Institute of
Technology
Affiliated to Visvesvaraya
Technological University
Bengaluru, KA
India

Aravind Kumar Rengan
Department of Biomedical
Engineering
Indian Institute of Technology
Hyderabad
Sangareddy, TS
India

Annisa Rifathin
National Research and
Innovation Agency
Tangerang Selatan
Indonesia

Atish Roy
Department of Chemistry
Dr. Harisingh Gour University
Sagar, MP
India

Suresh Sagadevan
Nanotechnology & Catalysis
Research Centre
University of Malaya
Kuala Lumpur
Malaysia

Department of Chemistry
Faculty of Mathematics and
Natural Sciences
Universitas Islam Indonesia
Yogyakarta
Indonesia

Yulianti Sampora
National Research and
Innovation Agency
Tangerang Selatan
Indonesia

Sri Amruthaa Sankaranarayanan
Department of Biomedical
Engineering
Indian Institute of Technology
Hyderabad
Sangareddy, TS
India

Athanasia Amanda Septevani
National Research and
Innovation Agency
Tangerang Selatan
Indonesia

Melati Septiyanti
National Research and
Innovation Agency
Tangerang Selatan
Indonesia

Shaily
Inorganic Materials Research Lab
Department of Chemistry
Jamia Millia Islamia
New Delhi, DL
India

Adnan Shahzaib
Inorganic Materials Research Lab
Department of Chemistry
Jamia Millia Islamia
New Delhi, DL
India

Preeti Sharma
Department of Natural Sciences
University of Maryland
Eastern Shore
Princess Anne, MD
USA

K. Shwetha
Department of Chemistry
Nitte Meenakshi Institute of
Technology
Bengaluru, KA
India

A.S. Sowmyashree
Department of Chemistry
Nitte Meenakshi Institute of
Technology
Bengaluru, KA
India

Sudiyarmanto
National Research and
Innovation Agency
Tangerang Selatan
Indonesia

Abu Taha
Bioinorganic Lab
Department of Chemistry
Jamia Millia Islamia
New Delhi, DL
India
India

David Natanael Vicarneltor
National Research and
Innovation Agency
Tangerang Selatan
Indonesia

Fahmina Zafar
Inorganic Materials Research Lab
Department of Chemistry
Jamia Millia Islamia
New Delhi, DL
India

Preface

Nanotechnology and materials science have made remarkable advances in recent years, revolutionizing several industries and creating new opportunities for research and development. Nanomaterials, with their distinct physical and chemical characteristics at the nanoscale, have drawn a lot of interest and are being investigated for a wide range of applications, from electronics and energy to medicine and environmental remediation.

However, despite these promising prospects, there is rising worry regarding the sustainability and environmental impact of the conventional synthesis techniques used to create nanomaterials. Conventional methods frequently employ hazardous chemicals, consume a lot of energy, and produce a lot of waste, which raises severe concerns about their long-term effects and ecological imprint.

To overcome these issues and open the door to the manufacture of sustainable nanomaterials, the idea of "green synthesis" has evolved in this context. Utilizing both ecologically friendly natural resources including plants, microorganisms, and other natural resources, as well as green synthetic techniques, can be used to create nanomaterials.

This book, *Green Synthesis of Nanomaterials: Biological and Environmental Applications*, examines the developing area of "green synthesis of nanomaterials" and its potential biological and environmental pollution remediation applications. It explores the numerous biological sources and fabrication techniques used for the environmentally friendly production of nanomaterials, highlighting their special benefits, constraints, and possible uses.

In addition to highlighting the biological and environmental uses of the synthesized nanomaterials, the goal of this book is to provide a thorough and informative overview of the state-of-the-art methods and developments in green synthesis. The chapters include a wide range of subjects, such as biosynthesis by employing plants and bacteria, as well as the use of natural substances like cellulose and peptide for the green synthesis and biofabrication of nanomaterials and their applications in biomedical as well as environmental pollution remediation.

Readers will obtain a thorough grasp of the concepts driving green synthesis, the characterization methods used for nanomaterial analysis, and the wide range of applications in the biological and environmental domains throughout every chapter of this book. The potential applications of green nanomaterials are numerous and exciting, ranging from pollutant removal to antibacterial agents and targeted medication delivery systems.

This book is a useful resource for students, scientists, engineers, and business executives alike since the contributing authors leading academics and authorities in their respective fields have contributed their wealth of knowledge and expertise. Their combined efforts have produced a thorough compilation that not only illuminates the possibilities of green synthesis but also adds to the continuing discussion about sustainable nanotechnology.

We hope that this book will act as a catalyst for additional study, encouraging scientists to delve more deeply into the field of green synthesis and promoting the creation of brand-new, environmentally friendly nanomaterials. We may work towards a better future where scientific progress and environmental responsibility go hand in hand by harnessing the power of nature and implementing sustainable practices.

We would like to extend our sincere gratitude to everyone who helped with the writing, reviewing, and publishing of this book. We would also like to express our gratitude to the readers for their attention and participation. We can create the conditions for a sustainable and ecologically conscientious future by working together and exchanging knowledge.

1

Introduction to Advanced and Sustainable Green Nanomaterial

Aayushi Chanderiya, Atish Roy, and Ratnesh Das

Department of Chemistry, Dr. Harisingh Gour University, Sagar, MP, India

Abstract

A magical period of scientific observation and science-based regeneration is required to advance human civilization. Technological developments are brimming with deep scientific insight and depth. Today's sustainability is in the midst of a significant crisis. Energy and environmental sustainability are critical for the advancement of human civilization. Sustainable construction is the bedrock of scientific destiny and profound scientific progress. In this chapter, the authors primarily concentrated on the success of green sustainability, synthesis, nanoscience's vast implementation range, and the novel field of specialized and sustainable nanomaterials. The other pillars of this scientific endeavor are expansion efforts. Green and environmental sustainability are today's human forerunners.

Keywords *sustainability; environmental sustainability; scientific progress; nanoscience; green sustainability; green engineering*

1.1 Introduction

Nanotechnology is characterized as the science of the small. It is the manipulation of materials on a microscopic scale. Atoms and molecules behave differently when they are little. These particles have distinct characteristics. It has a wide range of extraordinary and intriguing applications, and research in nanotechnology and nanoscience has exploded across

Green Synthesis of Nanomaterials: Biological and Environmental Applications, First Edition.
Edited by Archana Chakravarty, Preeti Singh, Saiqa Ikram, and R.N. Yadava.

various product areas. It allows for the creation and progress of materials, including medicinal usage, environmental redemption, and so on. Conventional methods may have reached their limits. However, nanotechnology is advancing in a variety of ways. As a result, nanotechnology should not be considered a singular approach that only affects specific research fields; instead, it should be viewed as an exploration of all science disciplines [1a]. Modern green methods are regarded as one of the best aspects to prevent and enhance the environment to achieve sustainable improvement. This concept has piqued the interest of scientists in adopting its principles about indicators of environmental efficiency by demonstrating the real benefit in the stages of planning, design, utilization, and sustainability in multiple vital sectors of the human being [1b, 1c, 2]. Because of their distinct size-dependent qualities, these materials are exceptional and necessary in a wide range of human activities [3]. Nanotechnology covers a wide range of subjects, from adaptations of classical equipment physics to revolutionary tactics centered on molecular self-assembly, from developing products with nanoscale size to determining if we can directly influence things on the atomic scale/level [4] (Figure 1.1).

Several science policy publications show significant potential and value in offering green nano methods that produce nanomaterials and products without pollutants that impair the environment or human health at the management, design, production, and methodology phases. As a result, nanotechnology may help to alleviate issues about safe, sustainable development, such as environmental, human health, and safety issues, as well as assist in a sustainable environment in terms of energy, water, food supply, raw substances, environmental issues, and so on [5].

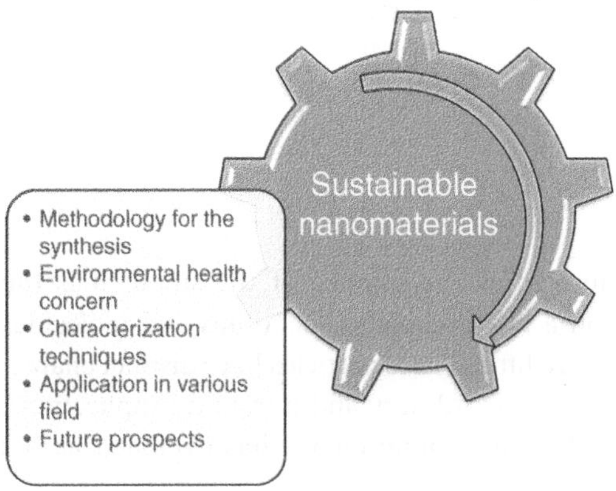

Figure 1.1 Sustainable nanomaterials.

- Advanced nanomaterials are the intelligent materials of today's human society. Advanced nanomaterials may be characterized in a variety of ways. The most extended term refers to any materials that reflect advancements above conventional materials that have been utilized for hundreds, if not thousands, of years. Consider materials early in their product and technology life cycles for a better understanding description of innovative materials.
- What is a sustainable nanomaterial?
  Sustainability is concerned with the needs of the current and coming years' decades. Nanomaterials are at the cutting edge of nanotechnology, which is continually evolving. Because of their distinctive size-dependent qualities, these materials are exceptional and necessary in many human activities.
- Sustainable and green nanomaterial
  Nanoparticles are particles with diameters ranging from 1 to 100 nm [6]. Depending on the shape, nanoparticles can be 0D, 1D, 2D, or 3D [7]. The relevance of these nanoparticles became apparent when researchers discovered that particle size might alter the physiochemical characteristics of substances, such as optical qualities. The nanoparticles are divided into many categories based on their morphology, size, and form (Figures 1.2–1.4).

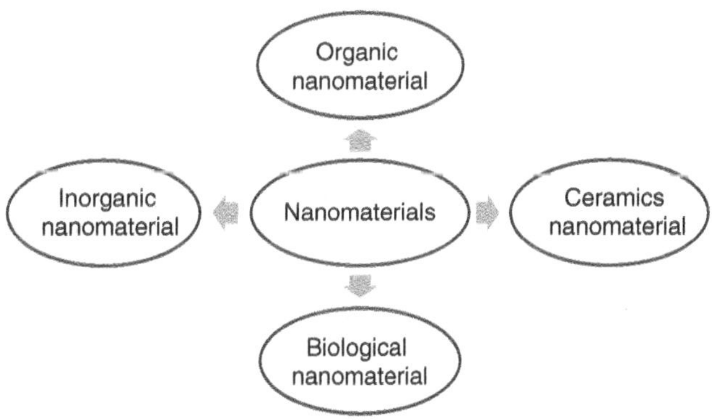

Figure 1.2 Types of nanomaterials.

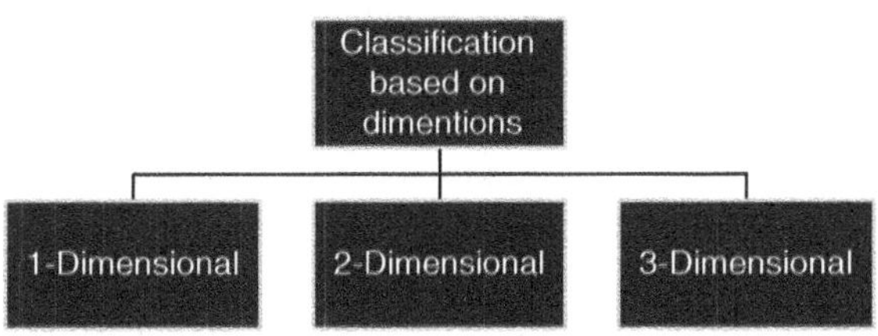

Figure 1.3 Classification based on dimensions of nanomaterials.

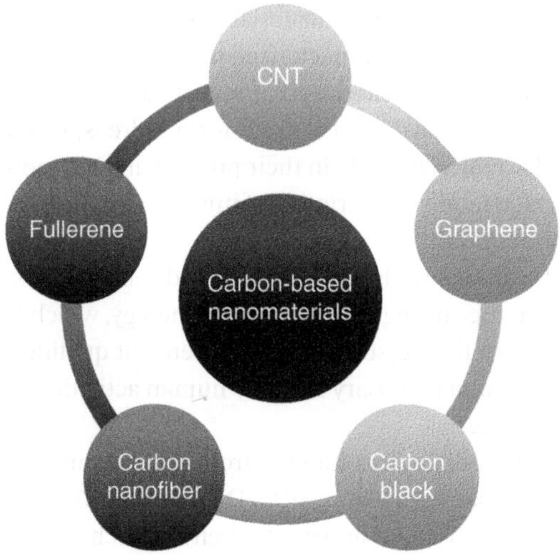

Figure 1.4 Carbon-based nanomaterials.

1.2 Synthesis Methods of Nanomaterials

Metal nanoparticles may be created in a variety of methods. There are two types of conversion methodologies: top-down and bottom-up. Several nanoparticle synthesis processes have been developed, and they are suitable for synthesizing nanoparticles of various sizes and shapes. The top-down technique is destructive, breaking down large molecules into tiny parts before changing them into the desired nanoparticles. Decomposition methods such as chemical vapor deposition (CVD), grinding, and physical vapor deposition (PVD) are used in this procedure. Milling is used to remove nanoparticles from coconut shells, with the size of the crystallites decreasing with time. This method produced iron oxide, carbon, dichalcogenides, and cobalt (III) oxide nanoparticles.

Bottom-up strategy method includes the gradual synthesis of nanoparticles from basic materials. It is least harmful to the environment, more practicable, and less expensive. Typically, the materials used in reduction and sedimentation techniques include green synthesis, biochemical, spin coating, sol–gel, and so on. This method has been used to create titanium dioxide, gold, and bismuth nanoparticles (Figure 1.5).

Nanoparticle synthesis might potentially utilize chemical or biological mechanisms [8]. Some chemical synthesis strategies for nanoparticles include the sol–gel method, wet chemical synthesis, hydrothermal method,

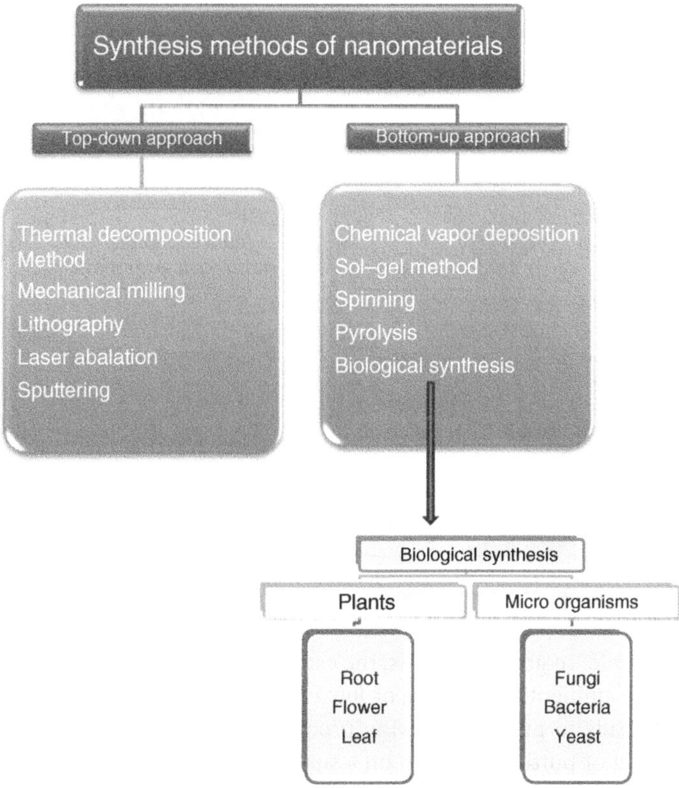

Figure 1.5 Synthesis methods of nanomaterials.

thermal decomposition, microwave method, and so on [9]. In contrast, biological processes contain enzymes, bacteria, plant extracts, and fungi.

1.3 Green Synthesis

Green chemistry and its ideas and environmental efficiency metrics are frequently viewed as fundamental to creating long-term profitability. Green chemistry principles include prevention, atom economy, less hazardous chemical synthesis, safer compound development, energy-efficient design, employing renewable fuel sources, catalysis, and constructing for decay. Green nanomaterial synthesis is an environmentally friendly method of nanomaterial synthesis that employs nontoxic, biodegradable ingredients. This method may be used to mass-produce nanomaterials on a

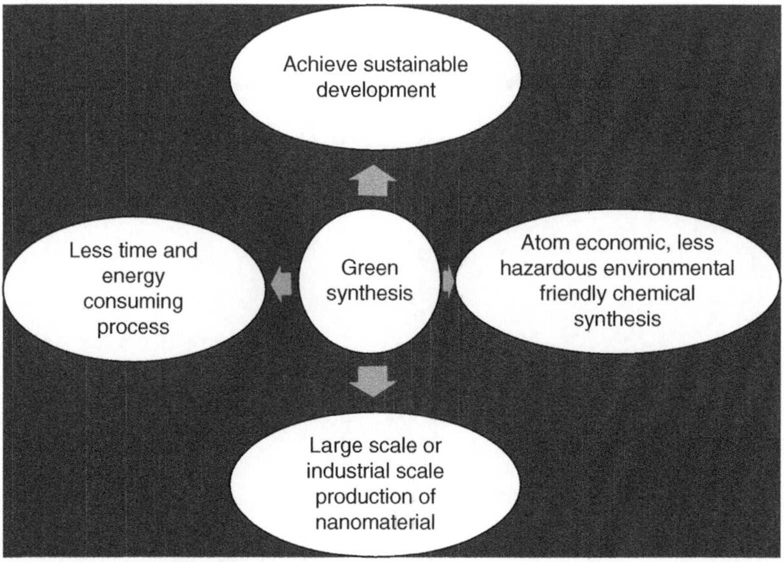

Figure 1.6 Advantages of green synthesis.

considerable scale. In green synthesis, the external experimental conditions should also be ambient, such as nil or low energy requirements and pressure, which leads to an energy-saving procedure. Nanotechnology is an essential aspect of putting the earth on a sustainable path because it combines all of the rewards of the present technology with compact goods that consume minimal energy and resources to run, produce, and integrate the possibility of recycling (Figure 1.6).

1.4 Biosynthesis of Nanoparticles from Plants

Bacteria, fungi, and plants all synthesize different types of nanoparticles [10]. Plants are better suited to the production of nanoparticles (NPs) than bacteria or fungi because metal ion reduction requires less incubation time. Plant tissue culture (PTC) and downstream processing approaches promise to generate metal and oxide NPs on a bigger scale. It has been shown that plants exhibit an innate ability to reduce metals via their particular metabolic pathways [11]. Stampoulis et al. [12] investigated the effects of ZnO, Cu, Si, and Ag NPs on root elongation, seed germination, and biomass production in *Cucurbita pepo* cultivated hydroponically. Compared to the untreated standards, test results showed that root length is reduced by 77% and by 64% when subjected to bulk Cu powder when seedlings are exposed to Cu nanoparticles.

1.5 Characterization of Nanomaterials

Nanoparticles are generally characterized by their size, morphology, and surface charge, using advanced microscopic techniques such as X-ray diffraction (XRD), scanning electron microscopy (SEM), transmission electron microscopy (TEM), atomic force microscopy (AFM), and particle size analyzer (PSA).

1.5.1 X-ray Diffraction (XRD)

XRD is a nondestructive analytical technique that provides crucial information on the lattice pattern of a crystalline component, such as unit cell diameters, bond angles, chemical properties, and crystallographic geometry of pure and produced elements [13]. XRD is predicated on the constructive interference of X-rays, and for the crystalline sample in question, X-rays from a cathode rays tube (CRT), are filtered, collimated, and aimed at the material. The resulting interaction creates constructive interference based on Bragg's equation, which ties the wavelength of entering light to the diffraction angle and lattice spacing.

1.5.2 Scanning Electron Microscope (SEM)

A SEM is a type of electron microscope that images a material by searching it with a high electron beam in a raster scan pattern [13, 14]. Electrons combine with the particles in the substance to generate signals that communicate information about the surface topography, composition, and other properties like thermal conductivity. SEM can provide extremely high-resolution pictures of a sample surface, exposing features as small as 1–5 nm in size. SEM micrographs have a significant depth of field due to the relatively narrow electron beam, resulting in a distinctive three-dimensional appearance beneficial for analyzing the surface structure of a material. A source's electron is accelerated in a field gradient under vacuum conditions. The beam is focused on the specimen after passing via electromagnetic lenses. The representative emits several sorts of electrons as a result of this assault.

1.5.3 Energy Dispersive X-ray (EDX)

Energy scattering X-ray spectroscopy is used for elemental testing or chemical characterization of a substance (Energy Dispersive Spectrometer (EDS) or EDX). X-ray fluorescence spectroscopy includes analyzing X-rays emitted by matter in response to charged particles to examine a sample via interactions among matter and electromagnetic radiation [13, 14].

1.5.4 Thermogravimetric Analysis (TGA)

Thermogravimetric analysis is a method of determining the temperature of a body. TGA is a thermal analysis technique that uses a controlled environment to quantify the weight change in a substance as a function of temperature and time [14].

1.5.5 UV–Visible Spectroscopy (UV–Vis)

Ultraviolet–visible spectroscopy (UV–Vis), sometimes known as UV–Vis, the spectroscopy of photons in the UV–Vis range is known as ultraviolet–visible spectroscopy or ultraviolet–visible spectrometry (UV–Vis). Electronic transitions occur when it makes use of visible light and light in the near-UV and near-infrared (NIR) bands in this area of the electromagnetic spectrum. UV–Vis measures the transmission or absorption of light in transparent or opaque liquids and solids.

1.5.6 Asymmetric Flow Field Fractionation (AF4)

To discover the features of NPs, new combinations of approaches are developed. Asymmetric flow field fractionation (AF4) is a liquid phase size separation technique that can be used with several downstream detectors. The particles are pushed against a semipermeable membrane by the perpendicular cross-flow. Brownian motion is visible in the tiniest particles, which diffuse into the channel's center, where the elution is greatest. The retention time of AF4 can be used to calculate the hydrodynamic diameter. AF4 outperforms microscopy-based approaches. The power of AF4 can be enhanced with the addition of downstream detectors. Inductively coupled plasma mass spectroscopy gives much information about each particle size in a heterogeneous mixture when AF4 couples. Metallic clusters with a diameter of less than 1 nm can be measured [15, 16].

1.5.7 Electrospray Differential Mobility Analysis (ESDMA)

Electrospray differential mobility analysis (ESDMA) is an aerodynamic sizing technique used in nanoparticle characterization that measures the ballistic distance traveled by the particle under an applied voltage to determine the aerodynamic diameter. AF4 can be used in conjunction with ESDMA for reliable analysis of complex nanoparticle synthesis products. For elemental mapping of TEM images, TEM combined with electron energy loss spectroscopy provides a high spatial resolution [17].

1.6 Environmental and Health Concerns

Pollution is an unfavorable alteration wreaking havoc on our primary materials, particularly land and water. At average amounts, soil contains elements, notably metals. The ground uses these elements as micronutrients and macronutrients. Light metals include Mg and Al, metalloids include As and Se, and heavy metals include Cd, Hg, Pb, Cr, Ag, and Sn. Substantial metals are elements with densities higher than 5.0 g/cc and distinct metal characteristics such as conductivity, flexibility, ligand selectivity, and cationic stability. In contrast, light metals are more important for health and the environment [18]. Light elements have densities of more than 5.0 g/cc and distinct metal characteristics such as conductivity, flexibility, ligand selectivity, and cationic stability. Cu, Cr, Zn, Mn, Fe, Co, and Ni are valuable heavy metals required at microscopic levels in metabolism but can be fatal at higher concentrations. In terms of environmental science, NPs have a variety of environmentally friendly applications, such as components that supply clean water from polluted water sources on a huge level and in portable implementations, as well as one that senses and cleans up pollutants (waste and toxic material), identified as remediation [19, 20]. Advanced sustainable materials are now being viewed as a viable option for addressing environmental pollution from its source. Continuous attempts are being made to improve the qualities of these materials to make them more energy-efficient, environmentally friendly, cost-effective, durable, readily available, and recyclable. Their continued research will undoubtedly pave the way toward establishing a clean environment [21]. Nanotechnology improves everyday items' performance and efficiency, making our lives simpler. It helps maintain a balanced environment by providing better air and water and clean, sustainable energy for the long term. Nanotechnology has received a lot of interest, and large institutions, firms, and organizations are investing more in R&D. Nanotechnology has established itself as a cutting-edge field of inquiry, with significant research being performed to bring the concept into action. It is being tested for various applications to improve the object's efficiency and performance while lowering the cost to make it affordable. Nanotechnology has a bright future because of its efficiency and environmental benefits.

1.7 Application

1.7.1 Use as Sensor

Enzymes are proteins that speed up the percentage of processes in which molecules are transformed into near products. Electrochemical biosensors with an enzyme immobilized on the transducer have been created to detect a single molecule alone. Each enzyme contains active regions that bind and react with a certain chemical. In third-generation biosensors, the inserted enzyme continuously exchanges electrons with the electrode, providing the response signal without requiring a mediator. For a range of applications, third-generation biosensors can detect small (e.g. H_2O_2 [22] and O_2 [23]) as well as large molecules (e.g., luteolin [24], medicine [25], and food analysis [26]). Nanostructures utilized as enzyme-binding materials act as channels in direct electrochemistry, transporting the charge from the enzyme to the electrode. These technologies use hemoglobin, laccase, cytochrome c, and horseradish peroxidase enzymes. Reliable electrical contact among enzymes and electrode surfaces is crucial for high-performance third-generation biosensors. The active centers of redox proteins are surrounded by a thick insulating shell, hindering electron transfer to the electrode. Metal NPs and nanopores are useful for improving direct electron transfer (DET) between the enzyme and the electrode transducer because of their good conductivity capabilities and nanoscale size. The enzyme active core is penetrated by NPs and nano-protrusions of porous films, which operate as "wires" of electron transit, connecting protein and electrode [27] (Figure 1.7).

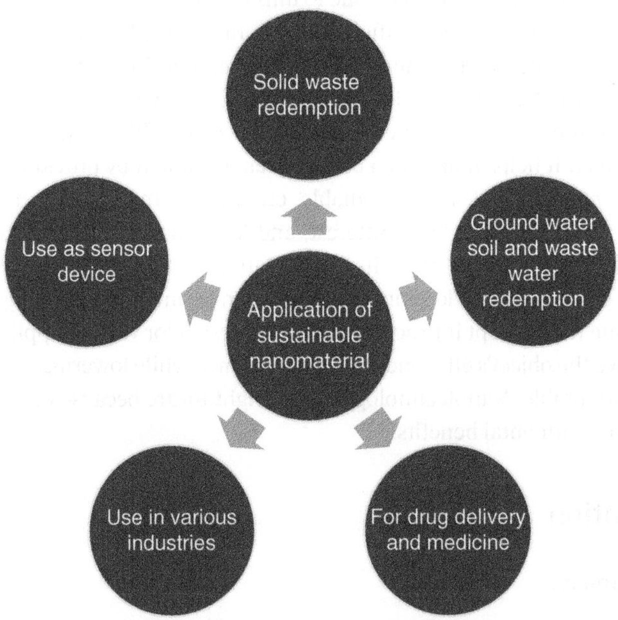

Figure 1.7 Applications of sustainable nanomaterials.

The use of NPs, regardless of the chemical modifiers used, significantly increases the electrochemically active area and improves analyte accessibility to the electrode surface, as seen by improved sensitivity and detection limit. The increased surface available is also advantageous for immobilizing biomolecules and, as a result, for creating biosensors.

Furthermore, the nanostructure's properties can improve sensor selectivity, a significant shortcoming of electrochemical sensors. For instance, Au NPs with varied morphologies and favored crystalline faces can generate selectivity to detect chemically identical analytes. In the case of lactate dehydrogenase (LDHs), the redox-active metal's different standard potential allows for differentiation between oxidizable molecules with one or more hydroxyl groups. Due to their recognition sites, MIPs can be considered the most appealing polymeric modifiers for increasing selectivity [28].

Nanomaterial-based amperometric biosensors present a novel and appealing paradigm in terms of new and enhanced functionality, which may be applied to various analytical applications. Amperometric biosensors based on NPs may have several advantages in enhancing and transcending the capabilities of the present analytical methodology by allowing for quick and precise analysis.

This sector is still impressionable, and several difficulties must be discussed in the innovation of nanomaterial-based amperometric biosensors, including (i) many sophisticated biological processes demand specific physiological environments and a specific degree of biocompatibility, and the biosensor-integrated nanomaterial must meet this necessity; (ii) it is highly desirable to find NPs with enough binding sites for biomolecules, and; (iii) it is incredibly beneficial to find NPs with satisfactory.

Despite significant progress, this sector is still young. Many issues must be resolved in the innovation of nanomaterial-based amperometric biosensors, such as (i) many complexes biological systems involve specific physiological environments and a certain degree of biocompatibility, and the biosensor-integrated nanomaterial must meet this requirement; (ii) it is highly beneficial to find NPs with satisfactory binding sites for biomolecules; and (iii) the possibility of controversies. Future studies could enhance the parameters mentioned above to improve the analytical characteristics of nanomaterial-based amperometric biosensors [29].

Many functional nanoparticles (NPs) based on nanotechnology have provided a novel solid material for very accurate on-site analysis in electrochemical biosensors through signal enhancement. Applicable NPs (carbon nanotubes, graphene, metallic, silica NPs, nanowire, indium tin oxide, and etc.) are frequently used to fabricate highly effective electrode-supportive matrices due to their high electrical conductivity, huge surface area, and other features. Surfaces can be functionalized with various organic groups to provide efficient immobilization (silanes, thiols, and conductive polymers). Labeling techniques are used to increase the sensitivity of electrochemical signals using a range of electroactive nano traces. These technologies may be employed to miniaturize and improve analytical performance via deposition, patterning, and electroactive conveyance in the point-of-care (POC) version of the electrochemical detection platform exemplified by lateral-flow immunoassay and microfluidic devices. Despite electrochemical biosensors being suitable for high-performance analysis in various field applications, matrix interference influencing biomolecular interaction from actual samples (blood, food, etc.) remains one of the most pressing issues that must be addressed to improve analytical performance [30].

1.7.2 Use as Medicine

In drug delivery research, the application of nanotechnology in developing nanocarriers for drug delivery generates a lot of optimism and enthusiasm. Nanoscale drug delivery methods provide several advantages, including higher intracellular absorption than other drug delivery systems [31]. For many causes, formulation scientists are fascinated by nanoscale medicine delivery technologies. The most important reason is that drug delivery techniques increase the ratio of surface atoms or compounds to total atoms or molecules. As a results, the surface area expands, facilitating the binding, adsorption, and transport of various substances such as medicines, probes, and proteins. Drug particles can also be modified to create nanoscale materials [32].

1.7.3 Used for the Removal of Toxicants from Water

Water resources have been extensively contaminated by textile industry effluents, which contain a variety of toxicants, including synthetic and semi-synthetic colors. Consumption of this contaminated water causes a variety of toxins and can lead to severe illnesses in all forms of life. To detect and remove dye from water samples, various synthetic and naturally derived adsorbents have been created. These adsorbents have several advantages, including cost-effectiveness, accessibility, stability, and ease of fabrication. Nanomaterials are used as adsorbents at a minimal cost. Crystal violet dyes can be removed from contaminated water using cellulose nanofibers (CNFs) and a modified CNF microfiltration membrane [33]. The removal of acid yellow 25 (AY25) dye from wastewater is thought to be effective using simple, less expensive, and environmentally friendly procedures employing layered double hydroxides [34]. Cazetta et al. described magnetic-triggered carbon produced from biomass waste as an excellent hazardous dye scavenger [35]. Polyacrylonitrile (PAN) nanofibrous membranes functionalized with cyclodextrin filter reactive dyes from polluted water and make it usable [36]. Similarly, innovative carboxylated functionalized copolymer nanofibers effectively eliminate various colors from wastewater [37].

The number of active radionuclide waste in the environment increases daily, posing a severe threat to all kinds of life. Aside from energy generation, radioactive elements are broadly used in medicine, research, business, agriculture, nuclear weaponry, and other fields. Recent breakthroughs in sustainable technology and practical approaches have supplied new insights into detecting and removing these potentially hazardous contaminants from the environment. Carbonaceous nanofibers generated from bacteria can remove radionuclides [38]. Using multipurpose flexible free-standing sodium titanate nanobelt membranes, effective sorbents, radioactive $90Sr^{2+}$, and $137Cs^{+}$ ions may be efficiently extracted from polluted water and oils [39].

1.7.4 Soil Redemption Use for the Removal of Toxicants from the Soil

The amount of research into the use of NPs in soils and groundwater remediation procedures has expanded dramatically, with encouraging outcomes. Polluted soil remediation using nanotechnologies has emerged as a promising topic with the potential to vastly improve performance over standard remediation technologies [40, 41]. Effective use for contaminants, other inorganic and organic toxins, and emerging contaminants, such as medicinal, beauty products, and hygiene products, in soil contaminants contexts.

1.7.5 In Agriculture and Food Industries

Climate change, growing populations, and the increasing demand for high-quality food and health care necessitate the development of more reliable, environmentally friendly technology. The NPs and technologies have many uses due to their changeable form, size, content, and potential reactivity with organic chemicals. Agricultural products are used in a variety of ways in our life, including food, fuel, furniture, textiles, and feedstock. However, lack of space, diseases, and changes in agroclimatic conditions all pose significant challenges to agricultural productivity. The use of pesticides and fertilizers to increase crop productivity is shown to have serious and even life-threatening consequences. As a result, there is a pressing need to upgrade agricultural practices and methods using next-generation technologies. Several developing nanotechnologies have been demonstrated to have uses in agriculture to increase production. The use of nano-agrochemicals, the development of crop protection technologies, and the proper postharvest management of agricultural goods are all things that need to be addressed. Nanotechnology promotes the least exploitation of natural resources, resulting in better and safer soil, water, and environment. Fertilizer application is very significant in increasing agricultural output. These are usually applied to the soil via surface application and subsurface placement after being mixed with water. Most of these fertilizers do not reach the plants and contribute to environmental damage. As a result, there is a requirement for creative nanotechnological solutions to improve nutrient availability to plants while also reducing the amount of fertilizer used [42]. This has led to scientific advancements in producing smart fertilizers and nanofertilizers. Materials with a size range of 1–100 nm, as well as components modified with nanoscale materials, are included in nanofertilizers. Many NPs have been examined to see how they affect plant growth and production. Because of their ability to promote plant development *in vitro*, NPs have also been described as natural biofertilizers. The properties of several NPs as nanofertilizers are investigated [43].

Nanotechnology has substantially contributed to the food business by enhancing many stages of food production, from the farm to the table. It has important uses in food manufacturing, processing, packaging, safety, and shelf life augmentation, as well as pathogen detection and the development of functional smart food. As a result, the technique can address the majority of customer demands, including nutrient enhancement and organoleptic qualities. Food quality, texture, and nutritional content have all improved due to technological advancements [44]. Nanostructured foodstuffs and food nanosensing devices are two of the technology's significant achievements. All nano additives used in food to assure its quality and functional properties, as well as the packaging material, are included in the former. Nanoscale food additives can potentially increase taste and shelf life while reducing microbial deterioration. Nanotechnology has the potential to completely replace food packaging materials with nanopolymers. Nanosensors help detect pollutants, poisons, and microbiological contamination, ensuring product quality, integrity, and authenticity [45] (Table 1.1).

1.7.6 Use as Photocatalyst

For centuries, fossil fuel-based energy sources such as coal, petroleum, and natural gas have been used to supply the world's energy demands; however, overproduction and overconsumption of these fuels have resulted in a slew

Table 1.1 Different types of water pollutants and their origin.

Pollutant	Origin
Organic	Dyes, pesticides, pharmaceuticals drugs, industrial waste
Inorganic	Metals, metalloids, nitrates, phosphates
Microorganism	Sewage, animal excrement

of acknowledged and unknown issues. In addition to a lack of knowledge regarding mineral fuel sources, including nuclear energy, there is a lack of technology and long-term waste disposal. Fossil fuel-based energy systems have a significant environmental impact. They are often regarded as the primary source of global warming, as well as contamination and pollution of the air, soil, and water. Many countries have been looking for alternative energy sources to replace fossil and mineral-based fuels due to rapid economic development, population increase, environmental and health concerns, and rising demands for clean energy sources [46–48]. These alternative energy sources should really be sustainable, minimize/eliminate issues, and be affordable and accessible to many countries throughout the world. We can design light-harvesting assemblies, devise new ways for producing fuels, and develop tools to synthesize novel functional materials for solar cells, water-splitting units, pollution control devices, and other applications by emulating photoactive green NPs found in nature. Because of their high surface area to volume ratio, photoactive green NPs can be more reactive and potentially more destructive than bulk materials of the same composition. Prior to any photocatalytic applications, the characteristics of these NPs should be investigated.

1.8 Future Scope

The need for ecologically benign and stable NPs that are compatible with biological systems has encouraged scientists to investigate nanoparticle production. This chapter covers the synthesis of sustainable and advanced NPs, as well as their classification, benefits and drawbacks, and several characterization methodologies. Top-down method, bottom-up approach, chemical synthesis, biological method, and mechanical process are all examples of cost-effective and easy manufacturing processes for NPs. Several characterization approaches for NPs are being developed in order to better understand their morphological, structural, optical, size, mechanical, and physiochemical characteristics. Each feature is obtained by the use of various machines and processes. The synthesis and characterization procedures used greatly impact the NPs' properties.

1.9 Ongoing Challenges

Despite nanotechnology's enormous growth and relevance, the uncertainty of a "new" technology persists, particularly because there is a scarcity of information and studies on its impact on health and the environment. This is viewed as a barrier to cross-industry nanotechnology adoption. The use of nanoparticles in various consumer and commercial applications raises concerns about the risks that may arise if people or the environment are exposed to NPs during manufacture, use, or disposal. The environmental, health, and safety risks of NPs vary depending on their chemical composition and physical structure [49, 50]. The absence of information, as well as the potential for negative effects on the environment, human health, safety, and sustainability, remains obstacles [51]. The NP analysis approach is the main issue with NPs. New and unique NPs are gradually developed as nanotechnology advances. The materials, on the other hand, differ in shape and size, which are crucial determinants in determining toxicity. Due to a lack of knowledge and methods for quantifying NPs, existing technology makes detecting NPs in the air for environmental protection extremely difficult. The information on the chemical structure is a critical factor in determining how toxic a non-material is, and minor changes in the chemical function group could drastically change its properties [52]. At all stages of nanotechnology, a full risk evaluation of the safety of human health and environmental consequences is required. The exposure risk and its probability of access, toxicological analysis, transportation risk, persistence risk, transformation risk, and recycling ability should all be considered in the risk assessment. Another element that can be exploited to anticipate environmental impacts is life cycle risk assessment. Material waste can be reduced by good experimental design prior to producing a nanotechnology-based product [53].

1.10 Conclusion

The current chapter focuses mostly on the green synthesis of sustainable advanced materials, their characterization, applications, and future opportunities, as well as addressing ongoing issues in nanotechnology. Nanotechnology is driving progress and innovation in a variety of industries, and it will continue to do so in the future. It is seen as a key enabling technology for a wide range of applications in electronics, health care, chemical products, beauty products, composites, and energy, to name a few, opening up new possibilities for the development of everyday products with improved performance, lower production costs, and less raw material intake. Despite its development, nanotechnology faces some difficulties in getting a greater impact on industry. Despite the efforts made, there is still a shortage of understanding and information regarding the safety and health hazards, as well as the impact on the environment of NPs, and it is necessary to protect all of those involved by supplying precise data.

Abbreviations

AFM	Atomic force microscopy
AY25	Acid yellow 25
CNFs	Cellulose nanofibres
CNT	Carbon nanotubes
ESDMA	Electrospray differential mobility analysis
NPs	Nanoparticles
PSA	Particle size analyzer
PTC	Plant tissue culture
SEM	Scanning electron microscopy
TEM	Transmission electron microscopy
XRD	X-ray diffraction

References

1 (a) Sovan Lal Pal, Utpal Jana, P. K. Manna, G. P. Mohanta, R. Manavalan Nanoparticle: an overview of preparation and characterization. *J. Appl. Pharmaceu. Sci.* 1 (6): 228–234; (b) Michael Fitzpatrick, J., Farren, T.R., and Anastas, P.T. (2002). Green Chemistry: Science and Politics of Change Martyn Poliakoff. *Science* 297; (c) Anastas, P.T. and Warner, J.C. (1998). *Green Chemistry: Theory and Practice*. Oxford: Oxford University Press.

2 Al-Khalaf, A.K.H. (2020). Application of sustainable green nanomaterial for desert cultivation. *Plant Arch.* 20 (2): 2600–2604.

3 Salata, O.V. (2004). Applications of nanoparticles in biology and medicine. *J. Nanobiotechnology* 2: 3.

4 Fakruddin, M., Hossain, Z., and Afroz, H. (2012). Prospects and applications of nanobiotechnology: a medical perspective. *J. Nanobiotechnology* 10: 31.

5 Alaa, K.H., Al-Khalaf, and Hussein, F.H. (2018). *Green and Sustainable Advanced Nonmaterials*. John Wiley and Sons.

6 Laurent, S., Forge, D., Port, M. et al. (2008). Magnetic iron oxide nanoparticles: synthesis, stabilization, vectorization, physicochemical characterizations, and biological applications. *Chem. Rev.* 108 (6): 2064–2110.

7 Tiwari, J.N., Tiwari, R.N., and Kim, K.S. (2012). Zero-dimensional, onedimensional, two-dimensional and three-dimensional nanostructured materials for advanced electrochemical energy devices. *Prog. Mater. Sci.* 57 (4): 724–803.

8 Hasan, S. (2015). A review on nanoparticles: their synthesis and types. *Res. J. Recent Sci.* 4: 3.

9 Wikipedia 2020. "Nanoparticle." https://en.wikipedia.org/w/index.php?title=Nanoparticle&ol did=971565062 (accessed 15 August 2020).

10 Yadav, K.K., Singh, J.K., Gupta, N., and Kumar, V. (2017). A review of nanobioremediation technologies for environmental clean-up: a novel biological approach. *J. Mater. Environ. Sci.* 8 (2): 740–757.

11 Handy, R.D., Owen, R., and Valsami-Jones, E. (2008). The ecotoxicology of nanoparticles and nanomaterials: current status, knowledge gaps, challenges, and future needs. *Ecotoxicology* 17 (5): 315–325.

12 Stampoulis, D., Sinha, S.K., and White, J.C. (2009). Assay-dependent phytotoxicity of nanoparticles to plants. *Environ. Sci. Technol.* 43 (24): 9473–9479.

13 Satyanarayana, T., Ramana Reddy, M.V., and Siva Kumar, J. (2016). Synthesis and characterization of pure and indium doped SnO_2 nanoparticles by sol-gel methods. *Int. J. Sci. Eng. Res.* 7 (12): 310–317.

14 Thodeti, S., Bantikatla, H.B., Kiran Kumar, Y., and Sathish, B. (2017). Synthesis and characterization of ZnO nanostructures by oxidation technique. *Int. J. Adv. Res. Sci. Eng.* 6: 539–544.

15 Hoque, M.E., Khosravi, K., Newman, K., and Metcalfe, C.D. (2012). Detection and characterization of silver nanoparticles in aqueous matrices using asymmetric-flow filed flow fractionation with inductively coupled plasma mass spectrometry. *J. Chromatogr. A* 1233: 109–115.

16 Tsai, D.H., Cho, T.J., Elzey, S.R. et al. (2013). Quantitative analysis of dendron-conjugated cisplatin-complexed gold nanoparticles using scanning particle mobility mass spectrometry. *Nanoscale* 5 (12): 5390–5395.

17 Tsai, D.H., Elzey, S.R., DelRio, F.W. et al. (2012). Tumor necrosis factor interaction with gold nanoparticles. *Nanoscale* 4 (10): 3208–3217.

18 Sanità Di Toppi, L. and Gabbrielli, R. (1999). Response to cadmium in higher plants. *Environ. Exp. Bot.* 42 (2): 105–130.

19 Schrick, B., Hydutsky, B.W., Blough, J.L., and Mallouk, T.E. (2004). Delivery vehicles for zerovalent metal nanoparticles in soil and groundwater. *Chem. Mater.* 16: 2187–2193.

20 Liu, M., Wang, Z., Zong, S. et al. (2014). SERS detection and removal of mercury(II)/silver(I) using oligonucleotide-functionalized core/shell magnetic silica sphere@Au nanoparticles. *ACS Appl. Mater. Interfaces* 6: 7371–7379.

21 Sharma, S., Dave, V., Verma, K., and Dwivedi, J. (2018). *Green and Sustainable Advanced Materials – Environmental Applications*, 125–158. John Wiley and Sons.

22 Kafi, A., Ahmadalinezhad, A., Wang, J. et al. (2010). Direct growth of nanoporous Au and its application in electrochemical biosensing. *Biosens. Bioelectron.* 25 (11): 2458–2463.

23 Feng, J.-J., Zhao, G., Xu, J.-J., and Chen, H.-Y. (2005). Direct electrochemistry and electrocatalysis of heme proteins immobilized on gold nanoparticles stabilized by chitosan. *Anal. Biochem.* 342: 280–286.

24 Jia, J., Wang, B., Wu, A. et al. (2002). A method to construct a third-generation horseradish peroxidase biosensor: self-assembling gold nanoparticles to three-dimensional sol-gel network. *Anal. Chem.* 74: 2217–2223.

25 Brondani, D., Scheeren, C.W., Dupont, J., and Vieira, I.C. (2009). Biosensor based on platinum nanoparticles dispersed in ionic liquid and laccase for determination of adrenalineenaline. *Sens. Actuators B Chem.* 140 (1): 252–259.

26 Rahman, M.A., Noh, H.-B., and Shim, Y.-B. (2008). Direct electrochemistry of laccase immobilized on Au nanoparticles encapsulated-dendrimer bonded conducting polymer: application for a catechin sensor. *Anal. Chem.* 80: 8020–8027.

27 Taurino, I., Sanzò, G., Antiochia, R. et al. (2016). Recent advances in third generation biosensors based on Au and Pt nanostructured electrodes. *Trends Anal. Chem.* 79: 151–159.

28 Tonelli, S.D., Scavetta, E., and Gualand, I. (2019). Electrochemical deposition of nanomaterials for electrochemical. *Sensors* 19: 1186.

29 Hayat, A., Catanante, G., and Marty, J.L. (2014). Current trends in nanomaterial-based amperometric biosensors. *Sensors* 14: 23439–23461.

30 Cho, I.-H., Kim, D.H., and Park, S. (2020). Electrochemical biosensors: perspective on functional nanomaterials for on-site analysis. *Biomater. Res.* 24: 6.

31 Mukherjee, B., Dey, N.S., Maji, R. et al. (ed.) (2014). Current status and future scope for nanomaterials in drug delivery. In: *Application of Nanotechnology in Drug Delivery*, 525–544. IntechOpen.

32 Hadjipanayis, C.G., Machaidze, R., Kaluzova, M. et al. (2010). EGFRvIII antibody-conjugated iron oxide nanoparticles for magnetic resonance imaging-guided convection-enhanced delivery and targeted therapy of glio-blastoma. *Cancer Res.* 70 (15): 6303–6312.

33 Gopakumar, D.A., Pasquini, D., Henrique, M.A. et al. (2017). Meldrum's acid modified cellulose nanofiber-based polyvinylidene fluoride microfiltration membrane for dye water treatment and nanoparticle removal. *ACS Sustain. Chem. Eng.* 5 (2): 2026–2033.

34 Sansuk, S., Srijaranai, S., and Srijaranai, S. (2016). A new approach for removing anionic organic dyes from wastewater based on electrostatically driven assembly. *Environ. Sci. Technol.* 50 (12): 6477–6484.

35 Cazetta, A.L., Pezoti, O., Bedin, K.C. et al. (2016). Magnetic activated carbon derived from biomass waste by concurrent synthesis: efficient adsorbent for toxic dyes. *ACS Sustain. Chem. Eng.* 4 (3): 1058–1068.

36 Foroozmehr, F., Borhani, S., and Hosseini, S.A. (2016). Removal of reactive dyes from wastewater using cyclodextrin functionalized polyacrylonitrile nanofibrous membranes. *J. Text. Polym.* 4: 45–51.

37 Elkady, M.F., El-Aassar, M.R., and Hassan, H.S. (2016). Adsorption profile of basic dye onto novel fabricated carboxylated functionalized co-polymer nanofibers. *Polymers* 8 (5): 177.

38 Sun, Y., Wu, Z.Y., Wang, X. et al. (2016). Macroscopic and microscopic investigation of U(VI) and Eu(III) adsorption on carbonaceous nanofibers. *Environ. Sci. Technol.* 50 (8): 4459–4467.

39 Wen, T., Zhao, Z., Shen, C. et al. (2016). Multifunctional flexible free-standing titanate nanobelt membranes as efficient sorbents for the removal of radioactive $90Sr^{2+}$, and $137Cs^{+}$ ions and oils. *Sci. Rep.* 6: 20920.

40 Liao, C., Xu, W., Lu, G. et al. (2015). Accumulation of hydrocarbons by maize (*Zea mays* L.) in remediation of soils contaminated with crude oil. *Int. J. Phytoremediation* 17 (7): 693–700.

41 Mojiri, A. (2011). The potential of corn (*Zea mays*) for phytoremediation of soil contaminated with cadmium and lead. *J. Biol. Environ. Sci.* 5 (13): 17–22.

42 Tilman, D., Cassman, K.G., Matson, P.A. et al. (2002). Agricultural sustainability and intensive production practices. *Nature* 418: 671–677.

43 Dimkpa, C.O., White, J.C., Elmer, W.H., and Gardea-Torresdey, J. (2017). Nanoparticle and ionic Zn promote nutrient loading of sorghum grain under low NPK fertilization. *J. Agric. Food Chem.* 65: 8552–8559.

44 Singh, T., Shukla, S., Kumar, P. et al. (2017). Application of nanotechnology in food science: perception and overview. *Front. Microbiol.* 8: 1501.

45 Ezhilarasi, P.N., Karthik, P., Chhanwal, N., and Anandharamakrishnan, C. (2013). Nanoencapsulation techniques for food bioactive components: a review. *Food Bioprocess Technol.* 6: 628–647.

46 Rosa, A.V. (2012). *Fundamentals of Renewable Energy Processes*, 3e. New York: Academic Press.

47 Nuraje, N., Asmatulu, R., and Kudaibergenov, S. (2012). Metal oxide-based functional materials for solar energy conversion: a review. *Curr. Inorg. Chem.* 2: 124–146.

48 Lei, Y., Asmatulu, R., and Nuraje, N. (2015). Photo-active metal oxide nanomaterials for water. Splitting. *ScienceJet* 4: 169–173.

49 Persons, T.M., Droitcour, J.A., Larson, E.M. et al. (2014). *Nanomanufacturing: Emergence and Implications for U.S. Competitiveness, the Environment, and Human Health.* U. S. G. A. Office, Ed.

50 Nanoscale Science, Engineering and Technology (NSET)/Nanotechnology Environmental and Health Implication (NEHI) (2011). *Introduction to 2011 NNI Environmental, Health and Safety Research Strategy*, 1–8.

51 Santos, C.S.C., Gabriel, B., Blanchy, M. et al. (2015). Industrial applications of nanoparticles – a prospective overview. *Mater. Today Proc.* 2: 456–465.

52 Dreher, K.L. (2003). Health and environmental impact of nanotechnology: toxicological assessment of manufactured nanoparticles. *Science* 5: 403–412.

53 Bhatt, I. and Tripathi, B.N. (2011). Interaction of engineered nanoparticles with various components of the environment and possible strategies for their risk assessment. *Chemosphere* 82: 308–317.

2

Medicinal Plant-Mediated Nanomaterials

Abu Taha, Nowsheenah Farooq, and Athar Adil Hashmi

Bioinorganic Lab, Department of Chemistry, Jamia Millia Islamia, New Delhi, DL, India

Abstract

The use of plants for medicinal purposes is from ancient times such as in Greek civilization, Indus Valley civilization, etc. Plants of nearly all categories (bacteria, algae, fungi, bryophyte, pteridophyta, gymnosperm, and angiosperm) are found to be useful as medicine for various diseases. Several methods for synthesizing nanomaterials such as sol–gel, polyols, microemulsion, hydrothermal synthesis, and green synthesis. Out of these methods, green synthesis is considered to be best because it involves natural and renewable resources for the synthesis of nanomaterials. Green synthesis of nanomaterials involves using bacteria, fungi, algae, and plants. Plants of different categories are used in the synthesis of nanomaterials, and those plants that have some medicinal value are found to be more useful for green synthesis of nanomaterials. Medicinal plant-mediated nanomaterials are found very useful for the diagnosis of various diseases. In this chapter, we will explore various medicinal plants used for synthesizing nanomaterials.

Keywords *medicinal plants; nanomaterials; green synthesis; phytochemicals; anticancer; antimicrobial*

Green Synthesis of Nanomaterials: Biological and Environmental Applications, First Edition.
Edited by Archana Chakravarty, Preeti Singh, Saiqa Ikram, and R.N. Yadava.

2.1 Introduction

Plants are well known to humans from ancient times for providing nutrition, either in the form of fruit or some other part used as a nutrition source. Not only as a nutrition source, they are also used as medicine for the diagnosis of several diseases such as stomachache, malaria, diarrhea, dysentery, cough, etc. [1]. Those plants having the ability to cure disease are known to contain some medicinal property in them are called medicinal plants. Plants from the very lowest category such as algae fungi to the highest category plants like angiosperms and gymnosperms are known to have medicinal properties [2, 3]. In ancient times of Indian Ayurvedic Medicine, Bhasma (Swarna Bhasma, Muktashukti Bhasma, Abhrak Bhasma, Tamra Bhasma, and Louha Bhasma) having nanoscale range was used to treat different kind of disease which were metallic preparation with medicinally important herbs. These bhasmas were thought to be free from toxicity with low side effects [4]. Nanotechnology is a branch of science that deals with material sizes in the range of 1–100 nm. The emergence of nanotechnology has opened a vast range of applications in different fields such as medicine, pharmacy, catalyst, material, space, agriculture, sensors, and optics. The ability of nanomaterials to show extraordinary properties depends upon their high volume-to-surface ratio value and the shape and size of nanomaterials [5]. There are various methods for nanomaterial synthesis, which two approaches can achieve, that is, top-down and bottom-up. In the top-down method, bulk material is reduced to the nano range by using different physical and chemical methods [6], while in the bottom-up approach, materials are assembled in the form of atoms, ions, or clusters to the nano range [7]. Some of the synthesis methods of nanomaterial are sol–gel, hydrothermal, ball mill, microemulsion, radiation, plant extract, etc. [8].

Any synthesis that involves less harmful toxic free solvent and reusable resources that do not produce harmful waste and pollute environment are called green synthesis [9]. Green synthesis involves the use of radiation, which is called photoinduced synthesis, and extract of plants are also used in green synthesis of nanomaterial [10]. Plant extract has several reducing substances such as sugar, protein, ketone, phenol, amines, etc. which reduce metal salt to metal NPs having the desired size and shape [11]. This green synthesis of nanomaterial is called plant-assisted synthesis of nanomaterial. The use of plants for medicinal purposes is well known from ancient times. Plants of Neem (*Azadiracta indica*), Tulsi (*Ocimum tenuiflorum*), Ashwagandha (*Withania somnifera*), Garlic (*Allium sativum*), and Ginger (*Zingiber officinale*) have been used for their medicinal purpose to all civilizations from ancient times. With the emergence of green synthesis of

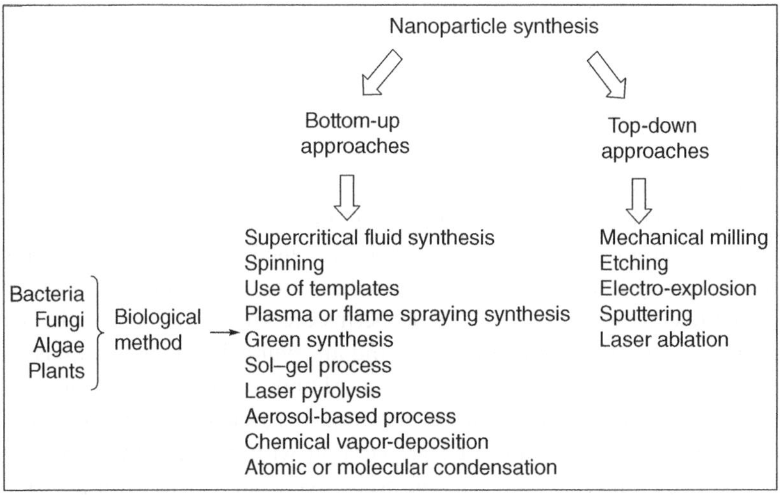

Figure 2.1 Flowchart showing different methods for nanomaterial synthesis.

nanomaterials by using plant extracts, scientists have gotten their attention toward plants with some medicinal properties.

Plants of nearly all categories like bacteria, algae, fungi, and higher division plants have been found to contain reducing material or their byproduct having reducing properties for the synthesis of nanomaterials. *Plectonema boryanum,* a cyanobacteria, was found to be useful in the synthesis of Au, Ag, and Pt NPs [12]. Nearly 6400 bioactive substances have been found in filamentous fungi and some other species of fungi. These are widely used as reduction and stabilizing agents for nanoparticles. The fungi also have a high tolerance for heavy metals and accumulate them in very high concentrations. They are also called "nanofactories," producing metal NPs of controlled size and desired morphology [13]. Blue-green algae (*Cyanophyta*), brown algae (*Phaeophyta*), red algae (*Rhodophyta*), green algae (*Chlorophyta*), and diatoms (*Bacillariophyta*) are found to be useful in the synthesis of silver nanoparticles with different kind of production method either intracellular of extracellular process [14]. Higher plants like herb, shrub, and tree extracts also have reducing properties and are used in nanomaterial synthesis (Figure 2.1) [15].

2.2 Synthesis of Gold Nanoparticles

Gold NPs find their application in ancient paintings due to their optical properties. With the advancement of nanotechnology, different gold NPs could be synthesized with controlled shapes and sizes, having diverse biological applications [16]. Green synthesis of gold NPs could be done using bacteria or plant-mediated, but plant-mediated synthesis has some advantages including less time-consuming, cell culture maintenance, and large-scale production of NPs [17]. The mechanistic approach of gold NP synthesis from plant extract is the reduction by secondary metabolic polyphenols. Polyphenols encourage oxidation and conversion into quinine by reducing gold ions into gold NPs [18]. During the plant-mediated synthesis of NP-reducing agents and other factors including temperature, pH, and concentration of reducing biomass.

While considering plant-mediated synthesis of gold NPs, herb extracts from medicinal plants have gotten more attention. In this method, the extract from medicinal plants mainly polyphenols, which are used for large-scale pharmaceutical purposes. Also, the bioavailability and therapeutic efficiency of such nanomaterials will be better than that of pristine gold nanoparticles and drugs alone [19]. In recent studies, several plants with medicinal properties have been used successfully in the synthesis of gold NPs such as *Jatropa curcas* (Barbados nuts), *Tridax procumbens, Solanum melongena, Calotropis gigantea, Carica papaya* and *Datura metel* [20]. *Ocimum sanctum* (*Tulsi*) is a very famous medicinal plant that is used for the synthesis of gold NPs [21]. Organic compounds extracted from *Tulsi* are cirsimaritin, rosmarinic acid, apigenin, estragole, linalool, urosolic acid, and carvacrol were found to have pharmaceutical applications and ligands of these compounds were found to have reduced ability [22]. Dr. Sharad Gupta and the team have synthesized gold nanoparticles using *Pimenta dioica* plant extract and evaluated their biomedical applications. The leaf extract of this plant has a major constituent of eugenol, a polyphenol responsible for reducing gold capping. *In vitro*, cytotoxicity of synthesized gold nanoparticles was found safe for human cervical cancer cell lines and human embryonic kidney cell lines. The gold nanoparticle was also used for plasmonic sensing of analyte molecules and a photoacoustic signal generator [23]. Xuelin Duan et al. synthesized gold nanoparticles using plant leaf extract of *Curcumae kwangsiensis* Folium and explored its antioxidant, cytotoxicity, and antihuman ovarian activities. MTT assay was performed on ovarian cancer cell lines, that is, SW-626, SK-OV-3, and PA-1, to examine cytotoxicity and antiovarian cancer effect of gold nanoparticles, gold chloride, and *Curcumae kwangsiensis* Folium leaf extract. The best result of cytotoxicity and antiovarian cancer properties was found for gold nanoparticles (AuNPs) having very low cell viability and high antiovarian cancer activity dose-dependent against SW-626, SK-OV-3, and PA-1 cell lines without affecting normal cell lines (HUVEC). Antioxidant properties of gold chloride, *Curcumae kwangsiensis* Folium leaf extract, and gold nanoparticle, DPPH test was used in the presence of butylated hydroxytoluene as positive control, and the best result was found for gold nanoparticle [24]. *Chaetomium globosum*, an endophytic fungal strain, the extract was used by Arivalagan Pugazhendhi and coworkers to synthesize gold nanoparticles (AuNPs) and evaluate its anti-inflammatory activity. FTIR spectroscopy was used to identify the presence of various functional groups in mycosynthesized Cg-AuNPs. FTIR also demonstrated the presence of amine, flavonoid, and amide linkage, which reduced Au^+ to Au^0 nanoparticles. Cg-AuNPs also exhibit good cytotoxicity against the HeLa cell line in a dose-dependent manner and also significant anti-inflammatory activity [25]. Ahmad A. Hussein et al. used *Helichrysum foetidum*, a medicinal plant in South Africa used as folk medicine, from which two chalcones helchysetin and helichrysin was extracted and used for biosynthesis of gold nanoparticle. Biosynthesized gold nanoparticle exhibits potential glucose enhancement in mammalian kidney cells and inhibition of carbohydrate hydrolyzing enzyme. This also provides AuNPs/chalcones conjugate in developing antidiabetes drugs derived from *Helichrysum foetidum* and its nanoparticle (Figure 2.2) [26].

Moringa oleifera plant extract contains important bioactive compounds that exert neuroactive modulation against seizures. A team of Pankaj Kalita used *Moringa oleifera* plant extract to prepare gold nanoparticles, characterized it by various techniques and evaluated it is *in vitro* biological and photocatalytic activity. Green synthesized gold nanoparticle has less cytotoxicity, the ability to regenerate neuronal cells in animal model, and photocatalytic ability against dyes [27]. B. A. Varghese et al. used rhizome extract of plant *Kaempferia parviflora* (black ginger, BG) for green preparation of gold nanoparticle (BG-AuNPs) and evaluated their antimicrobial, antioxidant property, and catalytic activity. The average size of gold nanoparticle was determined to be about 20–60 nm and was spherical in shape. BG-AuNPs showed good antibacterial properties against gram (−) and gram (+) and provided high antioxidant properties. The BG-AuNPs also

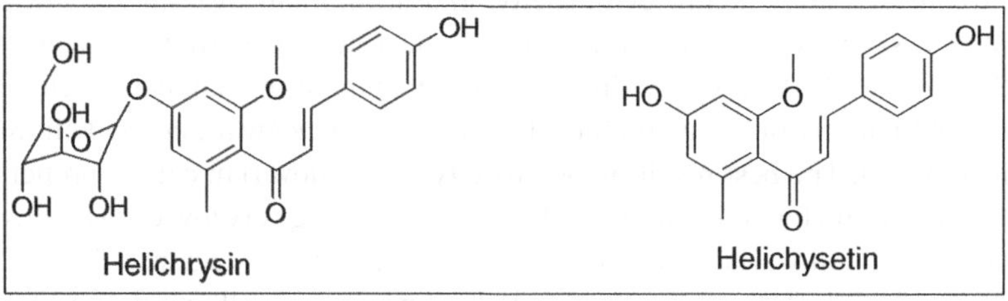

Figure 2.2 Bioactive compound present in *H. foetidum*.

exhibit good catalytic property for dye degradation such as methyl orange using sodium borohydride. Overall results show the potential of BG-AuNPs in biological application and environment remediation [28]. Fahimeh Mobaraki et al. used flower extract of *Achillea biebersteinii* plant for the green preparation of Ab-AuNPs. Ab-AuNPs were in the range of 2–30 nm, having a size of 8 nm. Ab-AuNPs were evaluated for their antioxidant, cytotoxic effect on human embryonic carcinoma stem cell lines (NTERA-2). Ab-AuNPs show dose-dependent cell viability on cancer cells, and the IC50 value was found 10 μg/ml. Cytotoxic results of Ab-AuNPs on NTERA-2 cells show that they can be used in designing novel anticancer agents [29]. N. Muniyappan et al. used curcumin-mediated synthesis of gold nanoparticles (Cur-AuNPs) using *Curcuma pseudomontana* plant. Curcumin is a diphenolic naturally occurring bright yellow color compound having curative attributes such as antioxidant, antiarthritic, anticancer, anti-inflammatory, neuroprotective, radioprotective, and cardioprotective activities. Gold nanoparticles (Cur-Au-NPs) were evaluated for their antibacterial activity on many strains of bacteria such as *Pseudomonas aeruginosa, E. coli, S. aureus,* and *Bacillus subtilis,* exhibited potential antibacterial property. Antioxidant and radical scavenging property of Cur-AuNPs were also very significant (Figure 2.3) [30].

Yali Chang and team have used leaf extract of the plant *Cannabis sativa* for green synthesis of gold nanoparticles (AuNPs). *Cannabis sativa* has been used as a psychoactive drug in folk tradition as a medicine ingredient and a source of fiber for the textile industry for a long time. This plant contains many substances, but Delta-9-tetrahydrocannabinol greatly affects the central nervous system and peripheral parts of an organism. Synthesized gold nanoparticles were evaluated for their antileukemia effect. The best result for antiacute leukemia property of gold nanoparticles was found for the Clone E6-1 cell line. This study confirms the excellent role of gold nanoparticles as a new chemotherapeutic drug for diagnosis of various types of leukemia (Figure 2.4; Table 2.1) [31].

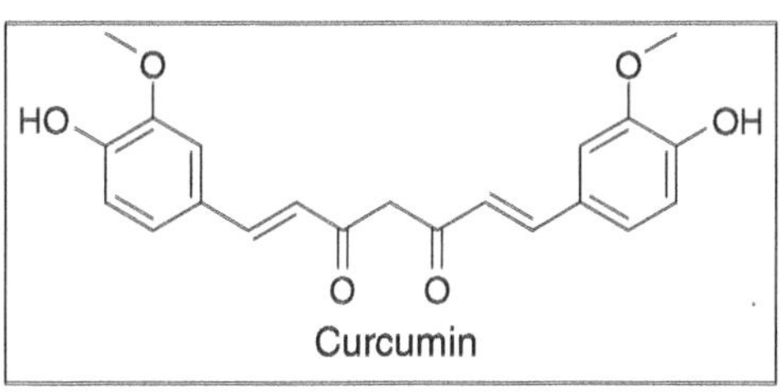

Figure 2.3 Curcumin found in *C. pseudomontana*.

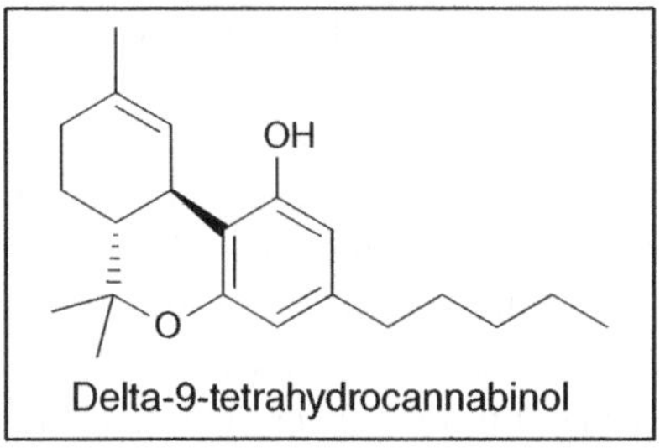

Figure 2.4 Structure of delta-9-tetrahydrocannabinol.

Table 2.1 Different medicinal plants used for gold nanoparticle synthesis.

S. no.	Plant	Size of nanoparticle (nm)	Applications
1	*Pimenta dioica*	13–17	Anticancer, plasmonic
2	*Curcumae kwangsiensis*	8–25	Anticancer, antioxidant
3	*Chaetomium globosum*	23	Anticancer, anti-inflammatory
4	*Helichrysum foetidum*	2–12	Antidiabetic drug
5	*Moringa oleifera*	15–20	Neural cell regeneration, photocatalysis
6	*Kaempferia parviflora*	20–60	Antimicrobial, photocatalysis
7	*Achillea biebersteinii*	8	Antioxidant, anticancer
8	*Curcuma pseudomontana*	20	Antimicrobial, antioxidant, radical scavenging
9	*Cannabis sativa*	18.6	Antileukemia

2.3 Synthesis of Silver Nanoparticles

Silver NPs are one of the important members of noble metal NPs. These NPs find applications in catalysis, electronics, medicines, drug delivery, bio-imaging, bio-sensing, optics, and information storage [32]. But silver NPs find much more application in medicinal fields due to their antimicrobial activity and application in wound dressing, food packaging, ointments, and antimicrobial coatings [33]. Silver nanoparticles are used as broad-spectrum antimicrobials against drug-resistant strains of bacteria [34]. Currently, silver nanoparticles are being used as antiviral agents [35]. Vikrant Sharma et al. prepared silver nanoparticles from a plant extract of

Andrographis paniculata, *Phyllanthus niruri*, and *Tinospora cordifolia*. They studied the cytotoxic effect of these silver NPs on Vero cells, and maximum nontoxic doses were determined before determining antiviral activity. The maximum nontoxic dose was found for silver NPs synthesized using *T. cordifolia*, and the minimum was found for *A. paniculata*. From cytotoxicity assay, it was found that silver NPs synthesized using *A. paniculata* were highly toxic to Vero cells [36]. Henry F. Aritonang et al. have also prepared silver NPs using plant extract of *Impatiens balsamina* and *Lantana camara* and studied their activity against microbes. The formation of silver NPs in plant extract was indicated by color change. These silver NPs antimicrobial activity was examined against *S. aureus* and *E. coli*, and silver NPs of *L. camara* extract showed the best antimicrobial activity [37]. Deepika Tripathi and her coworker used the endangered medicinal plant species *Withania coagulans* for silver NP synthesis and studied their biomedical applications. The silver nanoparticles were spherical in shape and 14 nm in size. Silver NPs have shown antibacterial activity against gram (−) (*E. coli* and *K. pneumoniae*) and gram (+) (*S. aureus*) bacteria strains, and it was found that it is more effective against gram (+) strains of bacteria. Synthesized silver NPs also showed great antioxidant activity. The toxicity of silver nanoparticles was evaluated against the cervical carcinoma cell line (SiHa cell line), and 50% minimum inhibitory concentration was found to be 13.74 μg/ml [38]. Aneta Salayova and coworker have synthesized silver NPs using plant extract of medicinal plants from their different parts *Brassica nigra, Capsell bursa-pastoris, Origanum vulgare, Lavandula angustifolia*, and *Berberis vulgaris* [39].

Aparajita Verma and coworkers have used Neem (*Azadirachta indica*) leave extract for the preparation of silver NPs and studied their antimicrobial behavior. The formation of silver NPs was confirmed by color change from pale yellow to brow color and finally dark brown. Neem leaf extract is suggested to contain reducing agents like terpenoids and flavanones, reducing silver ions. pH was an important factor in the synthesis of silver NPs. The effect of pH was observed on the shape and size of silver NPs. pH 13 was the most favorable pH for the synthesis of silver NPs [40]. Siddhant Jain and his coworker have used leave extract from *Ocimum sanctum* (Tulsi) plant and neat quercetin to synthesize silver NPs under different physiological conditions and studied their antibacterial activity. NP synthesis occurs in three steps: reduction of metal ion, clustering, and further growth of NPs. The probable mechanism of reduction of silver ion to silver is due to presence of −OH group in flavonoids. The keto-enol tautomerism is responsible for the release of active hydrogen, which reduces silver ions to silver NPs (Figure 2.5) [41].

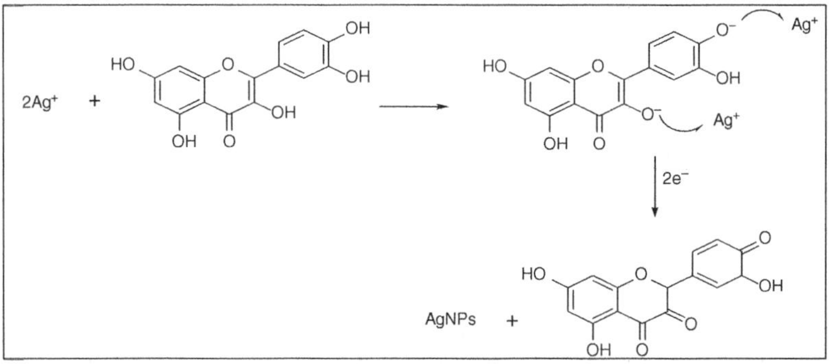

Figure 2.5 Reduction of Ag^+ to Ag by bioactive compound.

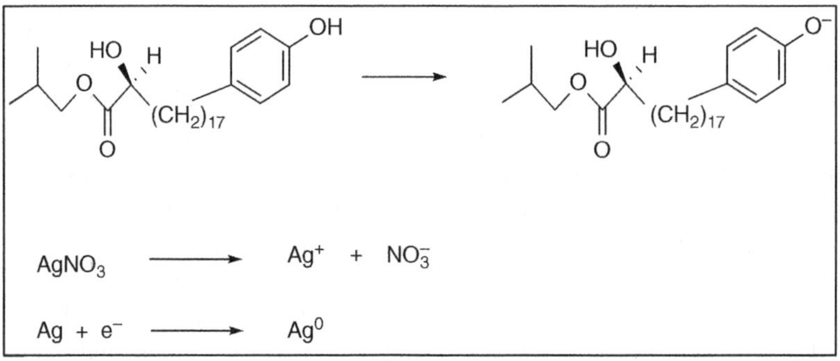

Figure 2.6 Reduction of Ag^+ to Ag by bioactive compound.

M. J. Ahmed and coworkers synthesized silver NPs using extract of the plant *Skimmia laureola* and studied their antimicrobial activity. Synthesized silver nanoparticles were spherical in shape. Antimicrobial study of *Skimmia laureola* leaves extract and silver NPs was done on five pathogenic bacteria: *E. coli, K. pneumoniae, P. aerugenosa, P. vulgaris,* and *S. aureus.* Bioactive compound present in *Skimmia laureola* leaf extract was responsible for silver nanoparticle formation (Figure 2.6) [42].

S. Balu and coworkers utilized medicinal grass (*Saccharum spontaneum*) for the synthesis of silver NPs. These synthesized NPs have a heterogeneous shape having a size of 30–40 nm with antimicrobial properties and cytotoxic effect on the MG-63 cell line [43].

M. Karuppiah et al. synthesized silver NPs in very controlled size (in the range of 13–57 nm) from *Ixora coccinea* leaf extract [44]. In another work, A. Sasikala and coworkers used aqueous stem bark extract of *Cochlospermum religiosum* and studied their antibacterial and antifungal activity. The combined effect of silver NPs and medicinal biomolecules present in NPs has inhibited activity on the array of bacteria and fungi (Table 2.2) [45].

Table 2.2 Different medicinal plants used for silver nanoparticle synthesis.

S. no.	Plant	Size of nanoparticle (nm)	Applications
1	*Tinospora cordifolia*	4–54	Cytotoxic
2	*Andrographis paniculata*	68.06	Cytotoxic
3	*Lantana camara*	10–12	Antibacterial
4	*Withania coagulans*	14	Antioxidant, antibacterial, anticancer
5	*Azadirachta indica*	—	Antimicrobial
6	*Ocimum sanctum*	—	Antibacterial
7	*Skimmia laureola*	—	Antimicrobial
8	*Saccharum spontaneum*	30–40	Antimicrobial, cytotoxic
9	*Ixora coccinea*	13–57	—
10	*Cochlospermum religiosum*	—	Antibacterial, antifungal

2.4 Synthesis of Zinc Oxide Nanoparticles

Zinc oxide is a kind of semiconductor because it has a large value of band gap energy (3.37 eV) and finds application in electronics, optics, and biomedical fields [46, 47]; their biomedical application includes drug delivery, anticancer, antibacterial, antidiabetic, and antifungal properties [48, 49]. United States Food and Drug Administration has recognized zinc oxide as a safe metal oxide. Zinc oxide also can filter UV light, so they are extensively used in sunscreen lotion [50]. K. Elumalai et al. used Neem (*Azadirachta indica*) extract to synthesize zinc oxide nanoparticles and studied their antimicrobial behavior. Neem leaf extract contains phenolic acid and flavonoid, which are responsible for bio-reduction reaction. XRD study confirms the hexagonal phase of ZnO (wurtzite structure). The antimicrobial property of the ZnO nanoparticle formed by the green method was found to be more potent than ZnO and Neem leaf extract alone [51]. Khursheed Ali and their coworker have used *Aloe vera* extract to synthesize functionalized ZnO-NPs as nano-antibiotics against multi-drug-resistant bacteria. Spectral analysis showed that phenolic compounds, terpenoids, and proteins present in *Aloe vera* extract played an important role in nucleation and stabilization of ZnO-NPs [52]. P.C. Nagajtothi and coworkers have synthesized ZnO-NPs using plant extract of *Coptidis rhizoma* and were analyzed by SEM, TEM, UV–visible spectroscopy, etc. The size of the nanoparticle was 2.90–25.20 nm. Synthesized NPs moderately affect gram (+) and gram (−) bacteria and have good free radical depletion properties. These NPs were also cytotoxic against

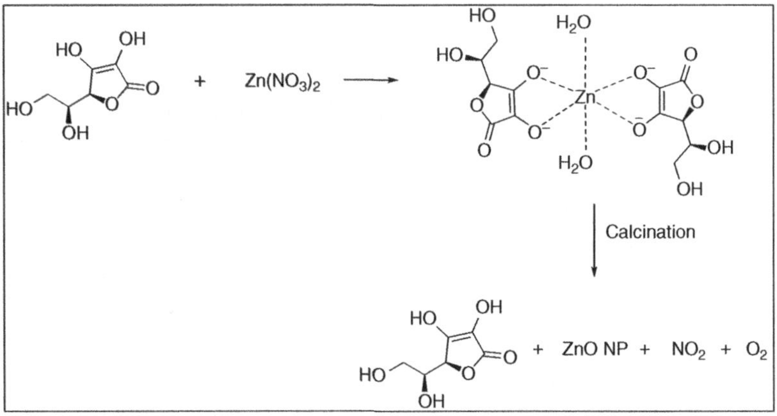

Figure 2.7 Formation of ZnO-NP from bioactive compound.

the RAW 264.7 cell line [53]. S. Jafarirad et al. used *Rosa canina* fruit extract to synthesize ZnO-NPs and evaluated the toxic potential of nanoparticles against bacteria. They also proposed a mechanism for the synthesis of ZnO nanoparticles (Figure 2.7) [54].

Renata Debrucka and Jolanta Dlugaszewska used fruit extract of *Trifolium pratense* for ZnO-NPs synthesis. The NPs have a size of 60–70 nm and a spherical shape. They also evaluated antimicrobial properties of nanoparticles against some gram (+) and gram (−) bacteria [55]. Shagufta Irshad and coworker synthesized ZnO-NPs using green tea leaves (*Camellia sinensis*) extract and evaluated their antimicrobial activity. This plant contains phenolic content and high antioxidant activity. This plant belongs to the theaceae family and is enriched with phytochemicals with antiseptic, anticancer, and antimicrobial properties that are used as potent drugs. The medicinal properties of these materials are enhanced with ZnO-NPs, where they are locked on the surface of NPs [56]. Shagufta Irshad and a coworker in another work synthesized ZnO-NPs using an extract of *Ocimum basilicum* and explored their antimicrobial activity toward pathogenic strains *S. aureus, E. coli,* and *A. niger*. They found that synthesized ZnO-NPs showed better antimicrobial properties than standard antibiotics [57]. P.C. Nethervathi et al. prepared ZnO-NPs from a plant extract of *Garcinia xanthochymus* and studied its photoluminescence, photocatalytic, and antioxidant activity. This plant contains phytochemicals such as xanthone, garciniaxanthone, xanthochymol, isoxanthochymol, volkensi flavone, 1,5-dihydroxyanthone, and 1,7-dihydroxyxanthone which are widely used in folk medicine [58]. Ahmed Abdelkhalek and Abdulaziz A. Al-Askar synthesized ZnO-NPs using plant extract of *Mentha spicata* and studied their antiviral property toward tobacco mosaic virus (TMV). It was the first time someone synthesized ZnO-NPs for antiviral activity, and they opened a gate for the green synthesis of nanoparticles showing antiviral properties [59]. A. T. Khalil et al. prepared ZnO-NPs using *Sageratia thea* (Osbeck) extract and explored the biological activity of these NPs. Medicinal uses of *S. thea* are well known and used in treating jaundice, hepatitis, cardiovascular, and circulatory diseases. Bioactive compound present in *S. thea* are myricentrin, glucopyranoside, syringic acid, quercetin, daucosterol, and *beta*-sitosterol [60]. Ashwini Jayachandran et al. used *Cayratia pedata* (Birdfoot Grapewine) leaf extract to synthesize ZnO-NPs. This plant is indigenous endangered medicinal plant found in south India. The plant is rich in alkaloid, tannins, phenolics, flavonoids, and terpenoids. This plant was used preparation of ZnO nanoparticles which was stable due to presence of capping agents (Figure 2.8; Table 2.3) [61].

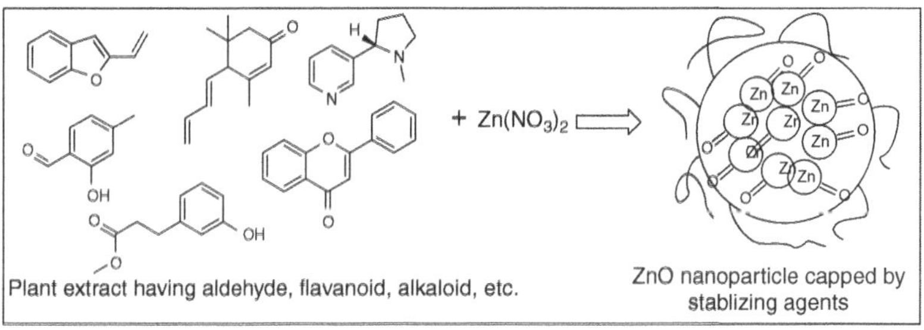

Figure 2.8 Synthesis of ZnO-NPs from bioactive compound and stabilization.

Table 2.3 Different medicinal plants used for ZnO-nanoparticle synthesis.

S. no.	Plant	Size of nanoparticle (nm)	Applications
1	*Aloe vera*	15	Nano antibiotics
2	*Coptidis rhizoma*	2.90–25.20	Antibacterial, antioxidant
3	*Rosa canina*	50	Antibacterial
4	*Trifolium pratense*	60–70	Antimicrobial
5	*Camellia sinensis*	30–40	Antiseptic, anticancer
6	*Ocimum basilicum*	30–40	Antimicrobial
7	*Garcinia xanthochymus*	4–8	Photocatalysis, antioxidant
8	*Mentha spicata*	11–88	Antiviral
9	*Sageratia thea*	20	Antibacterial, antifungal, enzyme inhibition
10	*Cayratia pedata*	52.24	Enzyme immobilization

2.5 Synthesis of Titanium Oxide Nanoparticles

Flower like titanium oxide (TiO_2) NPs have find application in various field such as energy storage device material [62], photocatalysis [63], antibacterial product [64], and electrode material for lithium ion batteries [65]. Very few examples for TiO_2NPs by green synthesis have been reported in literature by using plants. P.N.K. Reddy et al. used *Ocimum tenuiflorum* and *Calotropis gigantean* plant extract to synthesize TiO_2NPs. They evaluated the morphology, optical properties, and electrochemical properties of nanoparticles [66]. Waseem Ahmad et al. utilized *Mentha arvensis* leaf extract to synthesize TiO_2NPs and evaluated its antimicrobial properties against bacteria and fungi [67]. Ghulam Nabi et al. synthesized spherical TiO_2 nanoparticles with a size range of 80–100 nm using a citrus plant (*Citrus limetta*) and studied their photocatalytic activity to decontaminate RhB dye. TiO_2NPs are a good candidate for water purification. Citric acid present in plant extract works as a reducing as well as a capping agent for nanoparticle synthesis [68]. D. Achudhan et al. used multiple plant extracts (*Azadirachta indica, Ficus benghalensis*, and *Syzygium aromaticum*) for the synthesis of TiO_2 nanoparticles having crystalline and tetragonal structure and studied their antibacterial, antibiofilm, antifogging, and mosquitocidal activities [69]. E. R. Silva-Osuna and coworker synthesized TiO_2NPs using *Salvia rosmarinus* and evaluated their photocatalytic and optical property. The TiO_2 nanoparticles were made in the anatase phase, as revealed by XRD. Band gap energy was affected by the increasing concentration of plant extract during synthesis. Synthesized TiO_2 showed great catalytic degradation of dyes [70]. S. M. Roopan and coworker synthesized rutile TiO_2 nanoparticles having 23–25 nm size range using peel extract of *Annona squamosal*. In this synthesis, for the first time, rutile TiO_2 was synthesized at low temperatures using agricultural waste material [71]. Afzal Ansari et al. used *Acorus calamus*, a well-known medicinally important plant leaf extract, to synthesize TiO_2NPs and evaluated their photocatalytic and *in vitro* antimicrobial activity. The XRD study showed that TiO_2NPs were in the anatase phase, while the DLS study predicted nanoparticle size of 37.8–58.8 nm [72]. In some other work, Harpreet Kaur and coworkers prepared TiO_2 nanoparticles using leaves of *Carica papaya* and evaluated their photocatalytic properties. Size TiO_2 was 15.6 nm and exhibited excellent photocatalytic efficiency (91.19%) on the degradation of Reactive Orange-4 dye [73]. S. Subhapriya and P. Gomathipriya synthesized TiO_2NPs using plant extract of *Trigonella foenum-graecum* synthesized nanoparticles and were further characterized by different techniques. Nanoparticles have 20–90 nm in size, spherical in shape, and showed good antimicrobial properties against all tested microorganisms [74].

Table 2.4 Different medicinal plants used for TiO_2 nanoparticle synthesis.

S. no.	Plant	Size of nanoparticle (nm)	Applications
1	*Azadirachta indica*	10–33	Antimicrobial, mosquitocidal
2	*Mentha arvensis*	20–70	Antimicrobial
3	*Citrus limetta*	80–100	Photocatalysis
4	*Salvia rosmarinus*	—	Photocatalysis
5	*Annona squamosal*	23–25	—
6	*Acorus calamus*	37.8–58.8	Antimicrobial, photocatalysis
7	*Carica papaya*	15.6	Photocatalysis
8	*Trigonella foenum-graecum*	20–90	Antimicrobial
9	*Pouteria campechiana*	73–140	Antimosquito
10	*Jatropha curcas*	10–120	Photocatalysis

M. Narayanan et al. also synthesized TiO_2NPs using a green route with leave extract of *Pouteria campechiana*. SEM study confirmed the spherical shape of NPs, and further characterization was also done. They studied these NPs' toxic activity on *Aedes aegypti*, indicating that these NPs can be a potential mosquito control agent. This was the first time that green synthesized TiO_2NPs were used for antimosquito activity [75]. S.P. Gautam et al. prepared TiO_2 nanoparticles using plant leaf extract of *Jatropha curcas*. Nanoparticles were in spherical shape, anatase phase, and further characterized by various analytical techniques. They used these NPs for photocatalytic activity and used for *in-situ* remediation of tannin wastewater (TWW) from which chemical oxygen demand 82.26% and chromium 76.48% was removed. Overall, TiO_2 nanoparticles showed outstanding potential for a clean-green water treatment solution (Table 2.4) [76].

2.6 Synthesis of Iron Oxide Nanoparticles

Iron oxide NPs have attracted scientists due to their different oxidation states, magnetic properties, less expensive, and crystal morphology. Iron oxide NPs find applications in photocatalysis, poisonous gas sensing, batteries, and pharmaceuticals [77–79]. Waseem Ahmad et al. formed iron oxide (Fe_3O_4)

nanoparticles using plant extract of *Euphorbia herita* belonging to the family Euphorbiaceae. Milky juice plants have good antifungal, antibacterial, antioxidant, and anti-inflammatory properties. The main constituents of plant extract are alkaloids, flavonoids, saponins, carbohydrates, and polyphenols. Synthesized NPs were evaluated for antimicrobial activity on different pathogenic microorganisms such as *Staphylococcus aureus, E. coli,* and *Pseudomonas* [80]. In other work, Zahra Izadiyan and a coworker synthesized iron oxide NPs using husk extract of *Juglans regia*. Green synthesis was done using the co-precipitation method and further analyzed by different techniques. Extract of *Juglans regia* was found to be capped on iron oxide NPs. The NPs were evaluated for their *in vitro* cytotoxic properties on normal and cancerous cells. Its toxicity is dose-dependent, making it a very useful candidate for medical uses [81]. Silvia Groiss et al. used *Cyanometra ramiflora* leaf juice for the synthesis of iron oxide NPs and evaluated its antibacterial and catalytic activity. Synthesized iron oxide NPs were analyzed by various techniques and confirmed spherical shape. Iron oxide NPs were used to degrade Rhodamine-B dye and screened against *E. coli* and *S. epidermidis* as antibacterial drug [82]. Arivalagan Pugazhendhi et al. prepared iron oxide NPs using leaf juice of *Ruellia tuberosa* and studied their antimicrobial as well as photocatalytic activity. Further, iron oxide NPs were analyzed by different spectroscopic methods such as UV–visible, FTIR, DLS, XRD, etc. Antimicrobial activity was observed on cotton fabric for gram (+) and gram (−) bacteria, while degradation was performed on crystal violet dye [83]. Valentin V. Makarov et al. used an aqueous extract of *Hordeum vulgare* and *Rumex acetosa* for iron oxide NP preparation. Iron oxide NPs synthesized by *H. vulgare* were not stable, and buffer solution (pH 3.0) can stabilize these nanoparticles, whereas iron oxide NPs synthesized by *R. acetosa* were stable in a very acidic medium (pH 3.7), which suggest that plant having low molecular weight organic molecule have the ability to make stable iron oxide NPs [84]. *Aloe vera* and flaxseed extract (*Linum usitatissimum*) were used by Mohsen Gharanfoli and coworkers to prepare superparamagnetic iron oxide nanoparticles. XRD study suggests iron oxide (Fe_3O_4) pure spinal structure, which remains unchanged upon herbal coating from plant extract. Spherical iron oxide nanoparticles coated with herbal extract have a size of about <50 nm. Synthesized iron oxide NPs were also evaluated for their cytotoxicity assessment on cancer cell line (MCF-7) and reported dose-dependent toxicity [85].

Mina Sarani and coworker biosynthesized iron oxide (Fe_2O_3) NPs using an aqueous solution of *taranjabin* (camel thorn, also called *Alhagi maurorum*) and used it for biomedical and water cleaning applications. Synthesized nanoparticles were characterized using different analytical techniques. Synthesized iron oxide nanoparticles were found to be more useful for removing lead from wastewater (WW) [86]. L. Rokhum et al. prepared magnetic iron oxide (Fe_3O_4) NPs taking leaf extract of *Mikania mikrantha* and characterized it through different techniques such as XRD, UV–Vis spectroscopy, SEM, TEM, and FTIR. TEM images showed rhomboidal NPs having an average mean size of about 20.27 nm, while FTIR showed that the Fe_3O_4NPs are associated with phytochemicals like steroid, terpenoids, phenolic acid, flavonoids, and proteins. The antimicrobial activity of nanoparticles was tested on five strains of bacteria (*B. cereus, Acinebactor johnsonii, P. aerugenosa, Achromobacter spanius,* and *Chromobaterium pseudoviolaceum*) and four strains of fungi (*A. niger, P. citirinum, Fusarium oxyporium,* and *C. albicans*) which proved to be used as medicine [87]. C. G. Watz and coworker biosynthesized iron oxide NPs using plant extract of *Artemisia absinthium*. Using various analytical techniques, they confirmed that synthesized iron oxide NPs were nonmagnetic and reddish-brown color confirmed hematite (Fe_2O_3) as a major phase. Biosynthesized iron oxide NPs were tested *in vitro* cytotoxicity on healthy (HaCaT) cell line, which exhibited good viability (above 80%) and lacked apoptotic features while tumorigenic (A375) cell lines experienced a higher sensitivity and exposed significant apoptotic markers [88]. Coconut pulp (*Cocos nucifera*) was used for the biogenic preparation of iron oxide nanoparticles by A. Rajendran and coworkers. Based on their characterization result, the synthesized iron oxide (Fe_2O_3) NPs are in a husked rice structure and have a rhombohedral crystal structure. They studied the antibacterial, anticancer, and photocatalytic activity of iron oxide nanoparticles. Iron oxide NPs showed good photocatalytic degradation efficiency (92%) for Rhodamine-B dye, while antibacterial and anticancer activity was shown on gram (−) (*E. coli*), gram (+) (*S. aureus*) bacteria, and HepG2 cancer cell lines, showing the good potential of iron oxide (Fe_2O_3) NPs for environment remediation and biomedical application (Table 2.5) [89].

2.7 Conclusion and Future Perspective

Though there are several chemical and physical methods for the synthesis of nanoparticles, green synthesis is considered the best method as it involves natural and nonhazardous chemicals that can be obtained from medicinal or nonmedicinal plants. Thousands of plant species contain bioactive compounds, causing the reduction of metal ions to metal nanoparticles. These plants give enormous possibilities for nanoparticle synthesis by green methods having diverse applications such as antimicrobial medicine, anticancer drugs, and photocatalysts. Also, specific chemicals in plant extract impart specific properties to synthesized nanoparticles and act as capping or stabilizing agents. Nanoparticles surrounded by

Table 2.5 Different medicinal plant used for iron oxide nanoparticle synthesis.

S. no.	Plant	Size of nanoparticle (nm)	Applications
1	*Juglans regia*	5–7	Anticancer
2	*Cyanometra ramiflora*	—	Antimicrobial, photocatalysis
3	*Ruellia tuberosa*	52.78	Antimicrobial, photocatalysis
4	*Hordeum vulgare*	30	—
5	*Aloe vera*	48	Anticancer
6	*Alhagi maurorum*	20–60	Lead removal from WW
7	*Mikania mikrantha*	20.27	Antimicrobial
8	*Artemisia absinthium*	4–5	Anticancer
9	*Cocos nucifera*	—	Photocatalysis, antimicrobial, anticancer

bioactive molecules have better biomedical properties than nanoparticles alone. In the future, more bioactive molecules will be recognized and discovered from plant extract, which will impart excellent biomedical properties and will be used for the green synthesis of nanoparticles.

References

1 Rakotoarivelo, N.H., Rakotoarivony, F., Ramarosandratana, A.V. et al. (2015). Medicinal plants used to treat the most frequent diseases encountered in Ambalabe rural community, Eastern Madagascar. *J. Ethnobiol. Ethnomed.* 11 (1): 68.

2 Bamola, N., Verma, P., and Negi, C. (2018). A review on some traditional medicinal plants. *Int. J. Life-Sci. Sci. Res.* 4 (1): 1550–1556.

3 Ajith, T.A. and Janardhanan, K.K. (2007). Indian medicinal mushrooms as a source of antioxidant and antitumor agents. *J. Clin. Biochem. Nutr.* 40 (3): 157–162.

4 Chandra, H., Kumari, P., Bontempi, E., and Yadav, S. (2020). Medicinal plants: treasure trove for green synthesis of metallic nanoparticles and their biomedical applications. *Biocatal. Agric. Biotechnol.* 24: 101518.

5 Akhtar, M.S., Panwar, J., and Yun, Y.-S. (2013). Biogenic synthesis of metallic nanoparticles by plant extracts. *ACS Sustainable Chem. Eng.* 1 (6): 591–602.

6 Mittal, A.K., Chisti, Y., and Banerjee, U.C. (2013). Synthesis of metallic nanoparticles using plant extracts. *Biotechnol. Adv.* 31 (2): 346–356.

7 Thakkar, K.N., Mhatre, S.S., and Parikh, R.Y. (2010). Biological synthesis of metallic nanoparticles. *Nanomed. Nanotechnol. Biol. Med.* 6 (2): 257–262.

8 Su, S.S. and Chang, I. (2018). Review of production routes of nanomaterials. In: *Commercialization of Nanotechnologies–A Case Study Approach*, 15–29. Cham: Springer International Publishing.

9 Devatha, C.P. and Thalla, A.K. (2018). Green synthesis of nanomaterials. In: *Synthesis of Inorganic Nanomaterials* (ed. S.M. Bhagyaraj, O.S. Oluwafemi, N. Kalarikkal, and S. Thomas), 169–184. Elsevier.

10 Kumar, V., Gundampati, R.K., Singh, D.K. et al. (2016). Photoinduced green synthesis of silver nanoparticles with highly effective antibacterial and hydrogen peroxide sensing properties. *J. Photochem. Photobiol. B* 162: 374–385.

11 Mohamad, N.A.N., Arham, N.A., Jai, J., and Hadi, A. (2013). Plant extract as reducing agent in synthesis of metallic nanoparticles: a review. *Adv. Mater. Res.* 832: 350–355.

12 Zinicovscaia, I. and Cepoi, L. (2016). Nanoparticle biosynthesis based on the protective mechanism of cyanobacteria. In: *Cyanobacteria for Bioremediation of Wastewaters* (ed. I. Zinicovscaia and L. Cepoi), 113–121. Cham: Springer International Publishing.

13 Guilger-Casagrande, M. and de Lima, R. (2019). Synthesis of silver nanoparticles mediated by fungi: a review. *Front. Bioeng. Biotechnol.* 7: 287.

14 Dahoumane, S.A., Mechouet, M., Wijesekera, K. et al. (2017). Algae-mediated biosynthesis of inorganic nanomaterials as a promising route in nanobiotechnology – a review. *Green Chem.* 19 (3): 552–587.

15 Iravani, S. (2011). Green synthesis of metal nanoparticles using plants. *Green Chem.* 13 (10): 2638.

16 Jennings, T. and Strouse, G. (2007). Past, present, and future of gold nanoparticles. *Adv. Exp. Med. Biol.* 620: 34–47.

17 Kumar, V. and Yadav, S.K. (2009). Plant-mediated synthesis of silver and gold nanoparticles and their applications. *J. Chem. Technol. Biotechnol.* 84(2): 151–157.

18 Wang, W., Chen, Q., Jiang, C. et al. (2007). One-step synthesis of biocompatible gold nanoparticles using gallic acid in the presence of poly-(N-vinyl-2-pyrrolidone). *Colloids Surfaces A Physicochem. Eng. Asp.* 301 (1–3): 73–79.

19 Connor, E.E., Mwamuka, J., Gole, A. et al. (2005). Gold nanoparticles are taken up by human cells but do not cause acute cytotoxicity. *Small* 1 (3): 325–327.

20 Rajasekharreddy, P., Usha Rani, P., and Sreedhar, B. (2010). Qualitative assessment of silver and gold nanoparticle synthesis in various plants: a photobiological approach. *J. Nanopart. Res.* 12 (5): 1711–1721.

21 Philip, D. and Unni, C. (2011). Extracellular biosynthesis of gold and silver nanoparticles using Krishna tulsi (*Ocimum sanctum*) leaf. *Phys. E Low-dimensional Syst. Nanostruct.* 43 (7): 1318–1322.

22 Panchal, P. and Parvez, N. (2019). Phytochemical analysis of medicinal herb (*Ocimum sanctum*). *Int. J. Nanomater. Nanotechnol. Nanomed.* 5(2): 008–011.

23 Kharey, P., Dutta, S.B., Gorey, A. et al. (2020). *Pimenta dioica* mediated biosynthesis of gold nanoparticles and evaluation of its potential for theranostic applications. *ChemistrySelect* 5 (26): 7901–7908.

24 Chen, J., Li, Y., Fang, G. et al. (2021). Green synthesis, characterization, cytotoxicity, antioxidant, and anti-human ovarian cancer activities of *Curcumae kwangsiensis* leaf aqueous extract green-synthesized gold nanoparticles. *Arabian J. Chem.* 14 (3): 103000.

25 Ningaraju, S., Munawer, U., Raghavendra, V.B. et al. (2021). *Chaetomium globosum* extract mediated gold nanoparticle synthesis and potent anti-inflammatory activity. *Anal. Biochem.* 612: 113970.

26 Omolaja, A.A., Pearce, B., Omoruyi, S.I. et al. (2021). The potential of chalcone-capped gold nanoparticles for the management of diabetes mellitus. *Surf. Interfaces* 25: 101251.

27 Boruah, J.S., Devi, C., Hazarika, U. et al. (2021). Green synthesis of gold nanoparticles using an antiepileptic plant extract: in vitro biological and photo-catalytic activities. *RSC Adv.* 11 (45): 28029–28041.

28 Varghese, B.A., Nair, R.V.R., Jude, S. et al. (2021). Green synthesis of gold nanoparticles using *Kaempferia parviflora* rhizome extract and their characterization and application as an antimicrobial, antioxidant and catalytic degradation agent. *J. Taiwan Inst. Chem. Eng.* 126: 166–172.

29 Mobaraki, F., Momeni, M., Taghavizadeh Yazdi, M.E. et al. (2021). Plant-derived synthesis and characterization of gold nanoparticles: investigation of its antioxidant and anticancer activity against human testicular embryonic carcinoma stem cells. *Process Biochem.* 111: 167–177.

30 Muniyappan, N., Pandeeswaran, M., and Amalraj, A. (2021). Green synthesis of gold nanoparticles using *Curcuma pseudomontana* isolated curcumin: its characterization, antimicrobial, antioxidant and anti-inflammatory activities. *Environ. Chem. Ecotoxicol.* 3: 117–124.

31 Chang, Y., Zheng, C., Chinnathambi, A. et al. (2021). Cytotoxicity, anti-acute leukemia, and antioxidant properties of gold nanoparticles green-synthesized using *Cannabis sativa* L leaf aqueous extract. *Arabian J. Chem.* 14 (4): 103060.

32 Ahmad, N., Sharma, S., Alam, M.K. et al. (2010). Rapid synthesis of silver nanoparticles using dried medicinal plant of basil. *Colloids Surf., B* 81 (1): 81–86.

33 Kim, J.S., Kuk, E., Yu, K.N. et al. (2007). Antimicrobial effects of silver nanoparticles. *Nanomed. Nanotechnol. Biol. Med.* 3 (1): 95–101.

34 Rai, M.K., Deshmukh, S.D., Ingle, A.P., and Gade, A.K. (2012). Silver nanoparticles: the powerful nanoweapon against multidrug-resistant bacteria. *J. Appl. Microbiol.* 112 (5): 841–852.

35 Lara, H.H., Ayala-Nuñez, N.V., Ixtepan-Turrent, L., and Rodriguez-Padilla, C. (2010). Mode of antiviral action of silver nanoparticles against HIV-1. *J. Nanobiotechnol.* 8 (1): 1.

36 Sharma, V., Kaushik, S., Pandit, P. et al. (2019). Green synthesis of silver nanoparticles from medicinal plants and evaluation of their antiviral potential against chikungunya virus. *Appl. Microbiol. Biotechnol.* 103 (2): 881–891.

37 Aritonang, H.F., Koleangan, H., and Wuntu, A.D. (2019). Synthesis of silver nanoparticles using aqueous extract of medicinal plants' (*Impatiens balsamina* and *Lantana camara*) fresh leaves and analysis of antimicrobial activity. *Int. J. Microbiol.* 2019: 1–8.

38 Tripathi, D., Modi, A., Narayan, G., and Rai, S.P. (2019). Green and cost effective synthesis of silver nanoparticles from endangered medicinal plant *Withania coagulans* and their potential biomedical properties. *Mater. Sci. Eng., C* 100: 152–164.

39 Salayová, A., Bedlovičová, Z., Daneu, N. et al. (2021). Green synthesis of silver nanoparticles with antibacterial activity using various medicinal plant extracts: morphology and antibacterial efficacy. *Nanomaterials* 11 (4): 1005.

40 Verma, A. and Mehata, M.S. (2016). Controllable synthesis of silver nanoparticles using Neem leaves and their antimicrobial activity. *J. Radiat. Res. Appl. Sci.* 9 (1): 109–115.

41 Jain, S. and Mehata, M.S. (2017). Medicinal plant leaf extract and pure flavonoid mediated green synthesis of silver nanoparticles and their enhanced antibacterial property. *Sci. Rep.* 7 (1): 15867.

42 Ahmed, M.J., Murtaza, G., Mehmood, A., and Bhatti, T.M. (2015). Green synthesis of silver nanoparticles using leaves extract of *Skimmia laureola*: characterization and antibacterial activity. *Mater. Lett.* 153: 10–13.

43 Balu, S., Andra, S., Kannan, S. et al. (2020). Facile synthesis of silver nanoparticles with medicinal grass and its biological assessment. *Mater. Lett.* 259: 126900.

44 Karuppiah, M. and Rajmohan, R. (2013). Green synthesis of silver nanoparticles using *Ixora coccinea* leaves extract. *Mater. Lett.* 97: 141–143.

45 Sasikala, A., Linga Rao, M., Savithramma, N., and Prasad, T.N.V.K.V. (2015). Synthesis of silver nanoparticles from stem bark of *Cochlospermum religiosum* (L.) Alston: an important medicinal plant and evaluation of their antimicrobial efficacy. *Appl. Nanosci.* 5 (7): 827–835.

46 Sundrarajan, M., Ambika, S., and Bharathi, K. (2015). Plant-extract mediated synthesis of ZnO nanoparticles using *Pongamia pinnata* and their activity against pathogenic bacteria. *Adv. Powder Technol.* 26 (5): 1294–1299.

47 Vanathi, P., Rajiv, P., Narendhran, S. et al. (2014). Biosynthesis and characterization of phyto mediated zinc oxide nanoparticles: a green chemistry approach. *Mater. Lett.* 134: 13–15.

48 Hameed, A.S.H., Karthikeyan, C., Ahamed, A.P. et al. (2016). In vitro antibacterial activity of ZnO and Nd doped ZnO nanoparticles against ESBL producing *Escherichia coli* and *Klebsiella pneumoniae. Sci. Rep.* 6 (1): 24312.

49 Martínková, L., Uhnáková, B., Pátek, M. et al. (2009). Biodegradation potential of the genus Rhodococcus. *Environ. Int.* 35 (1): 162–177.

50 Wodka, D., Bielańska, E., Socha, R.P. et al. (2010). Photocatalytic activity of titanium dioxide modified by silver nanoparticles. *ACS Appl. Mater. Interfaces* 2 (7): 1945–1953.

51 Elumalai, K. and Velmurugan, S. (2015). Green synthesis, characterization and antimicrobial activities of zinc oxide nanoparticles from the leaf extract of *Azadirachta indica* (L.). *Appl. Surf. Sci.* 345: 329–336.

52 Ali, K., Dwivedi, S., Azam, A. et al. (2016). *Aloe vera* extract functionalized zinc oxide nanoparticles as nanoantibiotics against multi-drug resistant clinical bacterial isolates. *J. Colloid Interface Sci.* 472: 145–156.

53 Nagajyothi, P.C., Sreekanth, T.V.M., Tettey, C.O. et al. (2014). Characterization, antibacterial, antioxidant, and cytotoxic activities of ZnO nanoparticles using Coptidis Rhizoma. *Bioorg. Med. Chem. Lett.* 24 (17): 4298–4303.

54 Jafarirad, S., Mehrabi, M., Divband, B., and Kosari-Nasab, M. (2016). Biofabrication of zinc oxide nanoparticles using fruit extract of *Rosa canina* and their toxic potential against bacteria: a mechanistic approach. *Mater. Sci. Eng., C* 59: 296–302.

55 Dobrucka, R. and Długaszewska, J. (2016). Biosynthesis and antibacterial activity of ZnO nanoparticles using *Trifolium pratense* flower extract. *Saudi J. Biol. Sci.* 23 (4): 517–523.

56 Irshad, S., Salamat, A., Anjum, A.A. et al. (2018). Green tea leaves mediated ZnO nanoparticles and its antimicrobial activity. Yang J, editor. *Cogent Chem.* 4 (1): 1469207.

57 Irshad, S., Riaz, M., Anjum, A.A. et al. (2020). Biosynthesis of ZnO nanoparticles using *Ocimum basilicum* and determination of its antimicrobial activity. *J. Anim. Plant Sci.* 30 (1): 10.36899.

58 Nethravathi, P.C., Shruthi, G.S., Suresh, D. et al. (2015). *Garcinia xanthochymus* mediated green synthesis of ZnO nanoparticles: photoluminescence, photocatalytic and antioxidant activity studies. *Ceram Int.* 41 (7): 8680–8687.

59 Abdelkhalek, A. and Al-Askar, A.A. (2020). Green synthesized ZnO nanoparticles mediated by *Mentha spicata* extract induce plant systemic resistance against tobacco mosaic virus. *Appl. Sci.* 10 (15): 5054.

60 Khalil, A.T., Ovais, M., Ullah, I. et al. (2017). *Sageretia thea* (Osbeck.) mediated synthesis of zinc oxide nanoparticles and its biological applications. *Nanomedicine* 12 (15): 1767–1789.

61 Jayachandran, A., Aswathy, T.R., and Nair, A.S. (2021). Green synthesis and characterization of zinc oxide nanoparticles using *Cayratia pedata* leaf extract. *Biochem. Biophys. Rep.* 26: 100995.

62 Prasannalakshmi, P., Shanmugam, N., Kumar, A.S., and Kannadasan, N. (2016). Phase-dependent electrochemistry of TiO2 nanocrystals for supercapacitor applications. *J. Electroanal. Chem.* 775: 356–363.

63 Hussain, M., Ceccarelli, R., Marchisio, D.L. et al. (2010). Synthesis, characterization, and photocatalytic application of novel TiO_2 nanoparticles. *Chem. Eng. J.* 157 (1): 45–51.

64 Yadav, L.S.R., Manjunath, K., Kavitha, C., and Nagaraju, G. (2018). An investigation of hydrogen generation and antibacterial activity of TiO_2 nanoparticles synthesized by the ionic liquid aided ionothermal method. *J. Sci.: Adv. Mater. Devices* 3 (2): 181–187.

65 Lan, T., Qiu, H., Xie, F. et al. (2015). Rutile TiO_2 mesocrystals/reduced graphene oxide with high-rate and long-term performance for lithium-ion batteries. *Sci. Rep.* 5 (1): 8498.

66 Naresh Kumar Reddy, P., Shaik, D.P., Ganesh, V. et al. (2019). Structural, optical and electrochemical properties of TiO_2 nanoparticles synthesized using medicinal plant leaf extract. *Ceram. Int.* 45 (13): 16251–16260.

67 Ahmad, W., Jaiswal, K.K., and Soni, S. (2020). Green synthesis of titanium dioxide (TiO_2) nanoparticles by using *Mentha arvensis* leaves extract and its antimicrobial properties. *Inorg. Nano-Met. Chem.* 50 (10): 1032–1038.

68 Nabi, G., Majid, A., Riaz, A. et al. (2021). Green synthesis of spherical TiO_2 nanoparticles using *Citrus limetta* extract: excellent photocatalytic water decontamination agent for RhB dye. *Inorg. Chem. Commun.* 129: 108618.

69 Achudhan, D., Vijayakumar, S., Malaikozhundan, B. et al. (2020). The antibacterial, antibiofilm, antifogging and mosquitocidal activities of titanium dioxide (TiO_2) nanoparticles green-synthesized using multiple plants extracts. *J. Environ. Chem. Eng.* 8 (6): 104521.

70 Silva-Osuna, E.R., Vilchis-Nestor, A.R., Villarreal-Sanchez, R.C. et al. (2022). Study of the optical properties of TiO_2 semiconductor nanoparticles synthesized using *Salvia rosmarinus* and its effect on photocatalytic activity. *Opt. Mater. (Amst).* 124: 112039.

71 Roopan, S.M., Bharathi, A., Prabhakarn, A. et al. (2012). Efficient phyto-synthesis and structural characterization of rutile TiO_2 nanoparticles using *Annona squamosa* peel extract. *Spectrochim. Acta Part A Mol. Biomol. Spectrosc.* 98: 86–90.

72 Ansari, A., Siddiqui, V.U., Rehman, W.U. et al. (2022). Green synthesis of TiO_2 nanoparticles using *Acorus calamus* leaf extract and evaluating its photocatalytic and in vitro antimicrobial activity. *Catalysts* 12 (2): 181.

73 Kaur, H., Kaur, S., Singh, J. et al. (2019). Expanding horizon: green synthesis of TiO_2 nanoparticles using *Carica papaya* leaves for photocatalysis application. *Mater. Res. Express* 6 (9): 095034.

74 Subhapriya, S. and Gomathipriya, P. (2018). Green synthesis of titanium dioxide (TiO_2) nanoparticles by *Trigonella foenum-graecum* extract and its antimicrobial properties. *Microb. Pathogen.* 116: 215–220.

75 Narayanan, M., Devi, P.G., Natarajan, D. et al. (2021). Green synthesis and characterization of titanium dioxide nanoparticles using leaf extract of *Pouteria campechiana* and larvicidal and pupicidal activity on *Aedes aegypti*. *Environ. Res.* 200: 111333.

76 Goutam, S.P., Saxena, G., Singh, V. et al. (2018). Green synthesis of TiO_2 nanoparticles using leaf extract of *Jatropha curcas* L. for photocatalytic degradation of tannery wastewater. *Chem. Eng. J.* 336: 386–396.

77 Truskewycz, A., Shukla, R., and Ball, A.S. (2016). Iron nanoparticles synthesized using green tea extracts for the fenton-like degradation of concentrated dye mixtures at elevated temperatures. *J. Environ. Chem. Eng.* 4 (4): 4409–4417.

78 Kamran, U., Bhatti, H.N., Iqbal, M., and Nazir, A. (2019). Green synthesis of metal nanoparticles and their applications in different fields: a review. *Z. Phys. Chem.* 233 (9): 1325–1349.

79 Bhatti, H.N., Iram, Z., Iqbal, M. et al. (2020). Facile synthesis of zero valent iron and photocatalytic application for the degradation of dyes. *Mater. Res. Express* 7 (1): 015802.

80 Ahmad, W., Kumar Jaiswal, K., and Amjad, M. (2020). *Euphorbia herita* leaf extract as a reducing agent in a facile green synthesis of iron oxide nanoparticles and antimicrobial activity evaluation. *Inorg. Nano-Met. Chem.* 51: 1147–1154.

81 Izadiyan, Z., Shameli, K., Miyake, M. et al. (2020). Cytotoxicity assay of plant-mediated synthesized iron oxide nanoparticles using *Juglans regia* green husk extract. *Arabian J. Chem.* 13 (1): 2011–2023.

82 Groiss, S., Selvaraj, R., Varadavenkatesan, T., and Vinayagam, R. (2017). Structural characterization, antibacterial and catalytic effect of iron oxide nanoparticles synthesised using the leaf extract of *Cynometra ramiflora*. *J. Mol. Struct.* 1128: 572–578.

83 Vasantharaj, S., Sathiyavimal, S., Senthilkumar, P. et al. (2019). Biosynthesis of iron oxide nanoparticles using leaf extract of *Ruellia tuberosa*: antimicrobial properties and their applications in photocatalytic degradation. *J. Photochem. Photobiol., B* 192: 74–82.

84 Makarov, V., Makarova, S.S., Love, A.J. et al. (2014). Biosynthesis of stable iron oxide nanoparticles in aqueous extracts of *Hordeum vulgare* and *Rumex acetosa* plants. *Langmuir* 30 (20): 5982–5988.

85 Rahmani, R., Gharanfoli, M., Gholamin, M. et al. (2020). Plant-mediated synthesis of superparamagnetic iron oxide nanoparticles (SPIONs) using *Aloe vera* and flaxseed extracts and evaluation of their cellular toxicities. *Ceram. Int.* 46 (3): 3051–3058.

86 Miri, A., Najafzadeh, H., Darroudi, M. et al. (2021). Iron oxide nanoparticles: biosynthesis, magnetic behavior, cytotoxic effect. *ChemistryOpen* 10 (3): 327–333.

87 Biswas, A., Vanlalveni, C., Lalfakzuala, R. et al. (2021). *Mikania mikrantha* leaf extract mediated biogenic synthesis of magnetic iron oxide nanoparticles: characterization and its antimicrobial activity study. *Mater. Today Proc.* 42: 1366–1373.

88 Moacă, E.-A., Watz, C.G., Flondor (Ionescu), D. et al. (2022). Biosynthesis of iron oxide nanoparticles: physico-chemical characterization and their in vitro cytotoxicity on healthy and tumorigenic cell lines. *Nanomaterials* 12 (12): 2012.

89 Rajendran, A., Alsawalha, M., and Alomayri, T. (2021). Biogenic synthesis of husked rice-shaped iron oxide nanoparticles using coconut pulp (*Cocos nucifera* L.) extract for photocatalytic degradation of Rhodamine B dye and their in vitro antibacterial and anticancer activity. *J. Saudi Chem. Soc.* 25(9): 101307.

3
Microorganism-Based Synthesis of Nanomaterials and Their Applications

Khairatun Najwa Mohd Amin, Rohana Abu, Sharifah Fathiyah Sy Mohamad, and Afkar Rabbani Hidayatullah Hipeni

Faculty of Chemical and Process Engineering Technology, Universiti Malaysia Pahang Al-Sultan Abdullah. Persiaran Tun Kalil Yaakob, Gambang Kuantan, Pahang, Malaysia

Abstract

The advancement of nanomaterials production has shifted from mechanical and chemical-based synthesis to green synthesis by microorganisms. For the past years, the foundations of microorganism-based synthesis have been laid by scientists to point out the vast potential in terms of reliability and sustainability. This was followed by a demonstration of process enhancement using genetically modified engineering approaches, which contributed to the high yield of desired nanomaterials. Diverse microorganisms, like bacteria and fungi, have been employed and will be discussed in this chapter. This chapter will also highlight the recent development of microorganism-based synthesis of organic materials such as cellulose nanomaterials and inorganic materials like metallic nanomaterials. Different mechanisms for these nanomaterials will also be discussed. Finally, the present work on the potential applications of these microorganism-based synthesis nanomaterials like in medical and advanced manufacturing is also mentioned.

Keywords *microorganism; synthesis; nanomaterials; bacteria; bacteria nanocellulose; gold nanomaterial; silver nanomaterial*

Green Synthesis of Nanomaterials: Biological and Environmental Applications, First Edition.
Edited by Archana Chakravarty, Preeti Singh, Saiqa Ikram, and R.N. Yadava.

3.1 Introduction

The era of globalization has resulted in the rise of nanomaterials utilization in the technology industry. It is, by definition, a nano-sized material (10^{-9} m) covered in various domains from chemistry, biology, and physics to display specific material characterization of nanomaterials with diverse applications [1]. As a result, nanomaterials have been extensively studied in terms of improvement and efficiency to increase their feasibility. This, in turn, leads to the application of microorganisms for synthesizing nanomaterials as it is believed to be an environmentally friendly alternative to conventional synthesis methods for producing nanomaterials [2].

Nanomaterials have been produced traditionally via physical and chemical techniques known for using materials with high toxicity, hazardous solvents, chemicals, and toxic reagents. It is also reported to have a low material conversion and high energy consumption [3]. The standard physical approach in nanomaterial synthesis today includes laser ablation [4], lithography [5], ball milling [6], and high-energy irradiation [7]. Chemical approaches in the synthesis of nanomaterials include chemical reduction [8], electrochemistry [9], and photochemical reduction [10]. Nanomaterials are known to differ in properties compared to their bulk material while macro-sized materials are somewhat like their bulk material. The reduction of material size to nano-sized exhibit enhanced properties such as high surface area to volume ratio compared to large material and the ability to manipulate other intrinsic properties that is size dependent such as strong surface reactivity [11]. Kolahalam et al. stated that the size and shape of the nanomaterial play a vital role in the modification of its properties such as chemical reactivity, emission, excitation, and stability [12].

Nanomaterial properties can be altered or controlled by adjusting their size during synthesis using different methods and techniques. The optimization of process parameters such as temperature [13], time [14], gas ratio [15], substrate concentration [16], and stirring speed [17] was reported in previous studies to understand the synthesis mechanisms. This resulted in detailed proposed designs and conditions to fabricate nanomaterials with the required properties offered. The physical and chemical approach, more commonly known as top-down and bottom-up techniques, is widely utilized for the synthesis of nanomaterials despite the disadvantages especially in terms of environmental hazards. However, microorganisms-based synthesis is one of the main focuses in recent studies in order to reduce the use of toxic and hazardous chemicals while promoting green chemistry principles [18].

It is believed that the current and future studies regarding microorganisms-based synthesis have the potential to enhance biocompatible, high-yield, cheap, and eco-friendly procedures, which will ultimately lead to the replacement of chemical synthesis of nanomaterials with diverse applications specifically in biomedical and pharmaceutical areas [19]. Park et al. have emphasized the ability to conduct synthesis at room temperature for microorganisms-based synthesis is an energy-efficient process while having other essential advantages such as reduction of toxic chemicals usage, high feasibility for scale-up, and capability of coping in extreme conditions [20].

The use of microorganisms such as bacteria [21], fungi [22], yeast [23], and viruses [24] has escalated over the years as an alternative to conventional methods of nanomaterials synthesis. It has been reported worldwide that microorganisms based synthesis is an environmentally friendly method that can offer a nontoxic, cheap and reliable way for nanomaterials synthesis with wide range of physicochemical properties, composition and morphology [25]. Apart from the listed advantages, microorganisms also can act as templates for the synthesis and result in a well-defined structure of nanomaterials [26]. Different microorganisms would require specific mechanisms for the nanomaterial's synthesis. A diverse utilization of microorganism-based synthesis for nanomaterials as a substitute for the conventional top-down and bottom-up approaches for nanomaterials processing will require solutions to the challenges that have occurred. Grasso et al. reported the challenges that need to be overcome include decreasing polydispersity of nanomaterials, full characterization of biocapping layer agents, high efficiency of biocapping layer removal and nanomaterials purifications, the regulation of the procedure for the microbial cell culture for repeatability of the synthesis, production costs and yield [27]. Nanomaterials synthesized from microorganisms offer applications in diverse fields depending on their physicochemical properties. The potential application of nanomaterials is impacted by several factors, including their morphology and physical and chemical properties.

This chapter will focus on the knowledge of microorganism-based synthesis and related studies reported in the last decade. This chapter also details the current development, mechanism, and application of microorganism-based synthesis used to synthesize nanomaterials. The knowledge regarding this topic is briefly reviewed to provide new insights and perspectives to increase the available current studies.

3.2 Microorganism

3.2.1 Bacteria

Bacteria use in microorganism-based synthesis is developing rapidly as studies have been done worldwide in the last decade. This is believed to be due to their unique properties, specifically their variability and capability to adapt to various environmental conditions. Iravani et al. classified the synthesized nanomaterials by the size, shape, and species of bacteria in determining its application based on the properties offered [28].

The shape of the nanomaterials synthesized with bacteria is reported to be in a variety of shapes such as spherical [29], triangles [30], hexagons [31], rods [32], and irregular shaped [33]. Furubayashi reported the use of *Magnetospirillum magneticum*, a magnetotactic bacteria in order to synthesize magnetic nanocrystals that offer attributes such as size uniformity, high magnetic properties, and biocompatibility that can be modified via synthesis reaction parameters [34]. *Pseudomonas genus* is a wide array of bacterial species commonly used in bacteria-based synthesis. It includes *Pseudomonas indica* [35], *Pseudomonas fluorescens* [36], *Pseudomonas aeruginosa* [37], *Pseudomonas rhodesiae* [38], *Pseudomonas monteili* [39], and many others up to more than 100 species. Hasan et al. stated that bacteria that is frequently used to synthesize metallic nanomaterials is *B. subtilis 168, Streptomyces* sp., *E. coli, Lactobacillus* sp., *Beijerinck, K. pneumoniae,* and *Stenotropomonas malophilia* due to the high-value intrinsic properties offered by the synthesized nanomaterials [40].

3.2.2 Yeast

Yeasts are single-cell eukaryote microorganisms that have been employed to synthesize nanomaterials successfully. Yeasts, according to the invention, are classified in phylum Ascomycota, insubphylum Saccharomycotina, in the class Saccharomycetes, in the order Saccharomycetales, in the family Saccharomycetaceae, and in the genus *Yeasts* [41]. A lot of research utilized yeast for preparing the metallic nanomaterials like silver and gold [42]. Yeast like *Saccharomyces cerevisiae* [43] and *Phaffia rhodozyma* [44] has successfully produced silver and gold nanomaterials.

3.2.3 Fungi

In synthesizing metal oxide and metal nanomaterials, fungi are commonly used as a reducing and stabilizing agent. Fungal species like *Penicillium, Aspergillus,* and *Fusarium* are among the most employed fungi in the biosynthesis of gold and silver nanomaterials [2]. Fungi are accountable for synthesizing more nanomaterials than bacteria as the intercellular enzymes in fungi act as biological agents and trigger the metal bioaccumulation, binding, and tolerance capacity [45]. Fungi can also produce extracellular nanomaterials that act as catalysts for downstream processing and handling biomass [46].

3.2.4 Virus

Virus comprises a protein shell (capsid) that encapsulates the acids [47]. The capsid offers an attractive function in synthesizing nanomaterials by providing a highly reactive surface for interaction with like metallic ions [48]. A plant virus is favored in the nanomaterial production due to its special structural integrity, easy management, and lower infectivity to humans [49]. Previous study has used the infamous plant pathogenic virus, *Squash leaf curl China virus* as a biotemplate to fabricate silver and gold nanomaterials [50]. In another research, tobacco mosaic virus was used as an additive material prior to plant extracts for producing metal nanomaterials, which managed to promote reduction and increase the nanomaterials number as compared to those with no viral addition [51]. Thus, the virus can be a suitable group as a biotemplate for material synthesis at the nanoscale.

3.3 Development of Microorganism-Based Synthesis of Nanomaterial

3.3.1 Organic Nanomaterial

3.3.1.1 Bacterial Nanocellulose (BNC)

Brown was the first to report the discovery of bacterial nanocellulose (BNC) in 1886, who observed the formation of a gelatinous white fiber floating at the top of the culture medium during the fermentation of acetic acid [52]. Since then, BNC has attracted great attention among researchers in numerous fields including food industries, biomedical, textiles, cosmetics, and biosensors [53–56]. BNC is free from hemicellulose, lignin, pectin, and wax, that is typically present in plant cellulose. Moreover, plant cellulose is 100 times thicker than BNC, which is formed of ultrafine nanofibers with sizes ranging from 20 to 100 nm [57]. BNC also exhibits other remarkable properties such as high water content (about 99%), high crystallinity (up to 90%), high degree of polymerization (up to 8000), biocompatibility, high surface area, and superior mechanical properties (tensile strength of 200–300 MPa and Young's modulus about 15–35 GPa) [58, 59].

BNC, also commonly known as bacterial cellulose, microbial cellulose, or biocellulose, is a linear, unbranched exopolysaccharide synthesized extracellularly by numerous bacterial genera such as *Rhizobium (R. leguminosarum), Azotobacter, Agrobacterium, Salmonella, Aerobacter, Acetobacter, Achromobacter, Burkholderia, Pseudomonas putida, Sarcina,* and *Escherichia* sp. The ability to synthesize cellulose has also been documented in *Salmonella* and *Enterobacteriaceae* isolated from the human gastrointestinal tract [60]. However, until now, *Acetobacter xylinum*, the acetic acid bacteria of the genus *Komagataeibacter xylinus* (formerly designated Gluconacetobacter), is reported as the most thoroughly studied, as well as the most efficient bacteria for nanocellulose synthesis [61]. In addition, other acetic acid bacteria including *Komagataeibacter medellinensis, Komagataeibacter hansenii, Komagataeibacter oboediens, Komagataeibacter nataicola, Komagataeibacter saccharivorans, Komagataeibacter pomaceti,* and *Komagataeibacter rhaeticus* have also been identified as effective producers of BNC [62, 63]. Recently, there have been increasing reports of bacteria generating BNC in the literature [64–66], implying that cellulose production might be possible in a greater variety of bacteria genera. Sunagawa et al. demonstrated the ability of *Enterobacter* sp. CJF 002 to synthesize a comparable morphological structure of BNC at a higher rate than the *Komagataeibacter* spp. [64]. Khan et al. also discovered that *Lactobacillus hilgardii* can synthesize BNC in a modified Y medium at a rate of 0.45 g/l per day [67].

3.3.2 Inorganic Nanomaterial

Microorganisms, for instance bacteria, fungi, yeast, and viruses, are also known to synthesize inorganic nanomaterials either intracellularly (inside cells) or extracellularly (outside cells). Both syntheses mostly involve oxidoreductase enzyme-mediated mechanisms such as nitrate reductase [68] and sulfide reductase [69] in forming metal/metal oxide nanomaterials and metal sulfide nanomaterials, respectively.

The enzymes catalyze the transfer of electrons from electron donor molecules such as nicotinamide adenine dinucleotide (NADH) (dependent on nitrate reductase) or nicotinamide adenine dinucleotide phosphate (NADPH) (dependent on sulfide reductase) to the electron acceptor such as metal ions to reduce to metal nanomaterials. Microorganisms can also act as capping and stabilizing agents in the metal nanomaterials synthesis. However, the produced nanomaterials' size/shape and dispersity are significantly controlled by microorganism types and biosynthesis conditions such as pH, temperature, and reaction time. Nevertheless, this biosynthesis occurs under mild reaction conditions compared to chemical and physical synthesis. The intracellular synthesis involves metal ion transportation into microbial cells—the reduction of metal ions to metal nanomaterials by the enzymes inside the cell.

After biosynthesis, metal nanomaterials are separated from their cell biomass by chemical processing, centrifugation, and filtration. In contrast, extracellular synthesis involves the reduction of metal ions to metal nanomaterials by the enzymes secreted into the culture medium [70]. Extracellular methods use cell exteriors such as cell biomass, cell-free supernatant, and cell-free extract. Extracellular synthesis is widely used because it requires less downstream processing steps for nanomaterials recovery when compared to intracellular synthesis.

3.3.2.1 Gold Nanomaterials

In the last few years, microorganisms have been used for gold nanomaterials synthesis for many biological applications including biosensing, bioimaging, biodegradation, antimicrobial activity, anticancer activity, and tissue repair [71]. Recently, gold nanomaterials have been synthesized using various types of microorganisms like bacteria *Staphylococcus aureus* [72], bacteria *Vibrio alginolyticus* [73], bacteria *Rhodobacter sphaeroides* [74], bacteria *Burkholderia contaminans ZCC* [75], fungi *Phomopsis* sp. *XP-8* [76], fungi *Phoma* sp. [77], and actinomycete *Nocardiopsis dassonvillei* [78].

3.3.2.2 Silver Nanomaterials

Silver nanomaterials are the most widely used inorganic nanomaterials as antimicrobial agents, especially in medical applications. Nitrate reductase is the enzyme responsible for the bioreduction of silver ions to Ag nanomaterials either present intracellularly or secreted extracellularly. In recent years, Ag nanomaterials can be synthesized by various microorganisms such as bacteria *Pseudomonas aeruginosa* [79], bacteria *Bacillus* sp. [80, 81], fungi *Botryosphaeria rhodina* [82], fungi *Aspergillus oryzae SZ1*, fungi *Rhizoctonia solani* [83], and actinomycete *Streptomyces* sp. [84, 85]. Table 3.1 summarizes the latest research on gold and silver nanoparticle synthesis.

3.3.3 Other Nanomaterials

Metal nanomaterials, such as palladium, selenium, tellurium, and platinum, have gained significant attention in recent nanotechnology. Intracellular or extracellular microorganisms have successfully synthesized these metal nanomaterials for pollutant degradation, catalytic, antibacterial, and anticancer activities.

Table 3.1 Microorganisms-based synthesis for inorganic nanomaterials.

Inorganic nanomaterials	Bacteria	Source of metal/ precursor	Application	Size and shape	Mechanism	Reference
Gold	*Staphylococcus aureus*	Chloride	Functional cardiac patch	5–20 nm, spherical	Cell-free extract	[72]
	Vibrio alginolyticus	Chloroauric acid	Anticancer and antioxidant agent	50–100 nm, irregular	Culture extracellular supernatant	[73]
	Rhodobacter sphaeroides	Tetrachloroauric acid	Nitrophenol degradation	10 nm, spherical	Bacterial biomass	[74]
	Bacteria *Burkholderia contaminans* ZCC	Tetrachloroauric acid	Biorecovery	50 nm	EET, cell suspension or cell-free extract	[75]
	Phomopsis sp. XP-8	Chloroauric acid	Biorecovery	15 nm, spherical	Cell biomass	[76]
	Phoma sp.	Tetrachloroauric acid	Antifungal and antibacterial	10–100 nm, spherical	Cell biomass	[77]
	Actinomycete *Nocardiopsis dassonvillei* NCIM 5124	Tetrachloroauric acid	Functional scaffolds	10–25 nm, spherical	Cell biomass	[78]
Silver	*Bacillus subtilis* KMS2-2	Nitrate	Antibacterial agent	18–100 nm, spherical	Culture extracellular supernatant	[81]
	Bacillus sp. *strain* AW1-2	Nitrate	Antifungal agent	22.33–41.95 nm, spherical	Intracellular	[80]

(*Continued*)

Table 3.1 (Continued)

Inorganic nanomaterials	Bacteria	Source of metal/ precursor	Application	Size and shape	Mechanism	Reference
	Bacillus licheniformis M09	Nitrate	Antibacterial agent, dye degradation	7.27 nm, spherical	Culture extracellular supernatant	[86]
	Pseudomonas aeruginosa	Nitrate	Antibacterial agent	7.27 nm, spherical	Culture extracellular supernatant	[79]
	Aspergillus oryzae SZ1	Nitrate	Antifungal agent	1–50 nm, spherical	Culture extracellular supernatant	[87]
	Trichoderma longibrachiatum DSMZ 16517	Nitrate	Biocidal agent	5–11 nm, spherical, triangular and cuboid	Fungal biomass, extracellular	[88]
	Fungi *Rhizoctonia solani*	Nitrate	Antibacterial agent	10–20 nm, spherical	Cell-free extract	[83]
	Fungi *Botryosphaeria rhodina MAMB-05*	Nitrate	Biocidal agent	1.5–0.8 nm	Fungal biomass, extracellular	[82]
	Streptomyces chiangmaiensis SSUT88A	Nitrate	Antibacterial agent	13.57–30.47 nm, spherical	Intracellular and extracellular supernatant	[85]

3.4 Mechanism of Microorganism-Based Synthesis of Nanomaterial

3.4.1 Organic Nanomaterial

3.4.1.1 Bacterial Nanocellulose (BNC)

BNC is mostly synthesized by a bottom-up process involving oxidative fermentation either in synthetic or natural culture medium at temperatures ranging between 25 °C and 30 °C and pH levels between 3 and 7 [89]. Glucose, sucrose, fructose, mannitol, and glycerol are among the common types of carbon sources utilized by different strains of microorganisms in the fermentation process of BNC [90]. It has been established that BNC formed by different bacterial strains under different growth conditions, such as temperature, pH, dissolved oxygen content, reactor type, cultivation methods, and ratio of medium volume to surface area, play crucial roles in determining the amount, physicochemical characteristics and morphologies of BNC [91–93]. Furthermore, different shapes of BNC can be synthesized depending on the cultivation methods: static or agitated. Static cultures often produce a higher yield of BNC in gelatinous pellicles made of a nanofibrillar network, while the agitated cultures create irregularly shaped sphere-like BNC with a less tightly packed structure [94].

The synthesis of BNC by *A. xylinum* consists of three major steps. Briefly, it starts with the polymerization of glucose into linear β-1,4-glucan chains by the bacteria. The produced chains are then secreted out extracellularly through the microscopic cellular pores in the cell membrane. The cellulose chains are then assembled and crystallized into web-like network strips. As a result, an ultrathin-three-dimensional network of cellulose nanofibers with highly uniaxially oriented nanofibrils was formed. This network is called a pellicle, and its geometry is dictated by internal and external hydrogen bonding, van der Waals, and hydrophobic interactions [59, 93, 95].

Numerous authors have reported the detailed mechanism of carbon metabolism for BNC synthesis by *Acetobacter xylinum* [96, 97]. The synthesis of BNC is a multistep process that requires specific enzymes, catalytic complexes, and regulatory proteins. When glucose is employed as the carbon source involves at least four stages of enzymatic reaction, as shown in Figure 3.1. The first stage is the glucokinase phosphorylation of glucose, followed by the isomerization of glucose-6-phosphate (Glc-6-p) to glucose-1-phosphate (Glc-1-p) by phosphoglucomutase. Third is the conversion of glucose-1-phosphate (Glc-1-p) to uridine diphosphate glucose (UDPGlc) by UDP-glucose pyrophosphorylase (UGPase). This enzyme is triggered by cyclic nucleotide in *Acetobacter xylinum* (c-di-GMP).

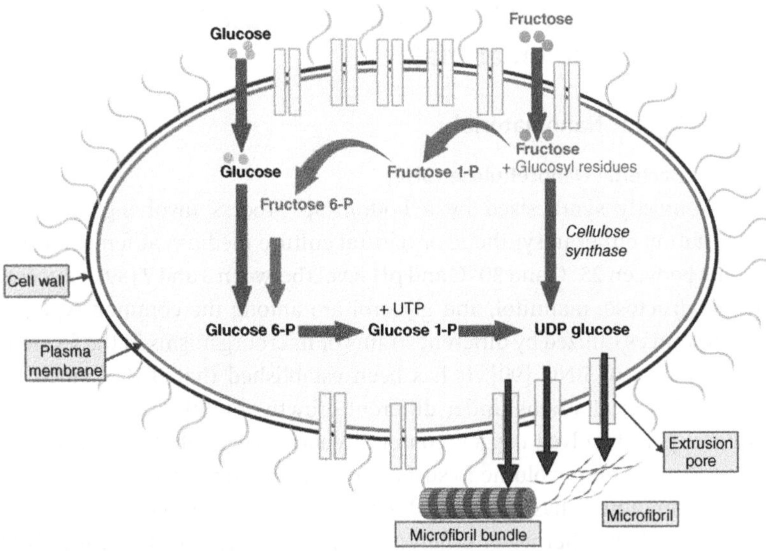

Figure 3.1 Schematic representation of biosynthesis pathways for BNC production.

The enzyme diguanylate cyclize produces the cellulose synthase activator c-di-GMP in *Acetobacter xylinum*, and its concentration is controlled by phosphodiesterases. Finally, the cellulose synthase reaction polymerizes UDPGlc into cellulose [98]. The formation of UDPGlc, prevalent in many organisms, is the direct precursor of BNC synthesis. It is assumed that UGPase, which is about 100 times more active in cellulose producers than that of noncellulose-generating bacteria, plays a key role in the production of BNC. When disaccharides (sucrose and maltose) are utilized by cellulose-producing bacteria as a carbon source, the synthesis of BNC begins with the hydrolysis of disaccharides into monosaccharides (glucose and fructose).

3.4.2 Inorganic Nanomaterial

3.4.2.1 Gold Nanomaterials

The process of Au(III) ions reduction involves Au ions binding into cell wall by functional groups (carboxyl and amine groups) and subsequently reduced to Au^0 by sugars on the biomass. This mechanism was reported in the extracellular synthesis of gold nanomaterials using *Rhodobacter sphaeroides* biomass, and the smaller spherical gold nanomaterials with an average size of 10 nm were formed [74]. The gold nanomaterials exhibited excellent biocatalytic activity in the reduction of nitroaromatic pollutants.

In this work, a low yield of gold nanomaterials was produced intracellularly due to the oxidative stress induced by Au^{3+} ions exposure, even at a low Au concentration of 10 μM. Interestingly, the cells can secrete gold nanomaterials in the extracellular medium to prevent gold toxicity inside the cells.

In another work, spherical gold nanomaterials with a small size of 5–20 nm were well distributed extracellularly on the surface of the *Staphylococcus aureus* cells [72]. The highest formation of gold nanomaterials was achieved under an alkaline pH of 8, while aggregation occurred under acidic conditions. Additionally, the formation of gold nanomaterials decreased with the increase of pH synthesis (pH 10–14), which might be due to AuOH precipitation. However, the irregular shape with a large size of around 50–100 nm gold nanomaterials was synthesized extracellularly using a cell-free supernatant of bacteria *Vibrio alginolyticus* at pH 7 [73]. Furthermore, extracellular synthesis of gold nanomaterials has also been reported using *Burkholderia contaminans* strain ZCC isolated from the soil of a copper-gold mine [75]. The gram-negative bacterium can reduce Au(III) to Au^0 on the cell surface by extracellular electron transfer (EET) through extracellular proteins like riboflavin. This bacterial strain can be used as a tool for recovering precious metal from mining waste, where this subject is the focus of many researchers.

In a similar study, gold can be recovered up to 73% from electronic wastewater using *Phomopsis* sp. XP-8 biomass. The fungal *strains* are capable of synthesizing spherical gold nanomaterials with an average size of 15 nm [76]. They reported that the functional groups (−COO, −C−N, and −N−H) on the mycelia surface were involved in biosorption and reduction of Au(III) ions. In another study, the spherical gold nanomaterials with a size of 10–100 nm were synthesized extracellularly using endophytic fungi, *Phoma species* [77]. The gold nanomaterials were shown to possess strong antifungal and antibacterial properties against *Rhizoctonia solani* and *Xanthomonas oryzae*, respectively. Besides, actinomycete *Nocardiopsis dassonvillei* NCIM 5124 could synthesize spherical gold nanomaterials (10–25 nm) and can be incorporated into scaffolds for tissue repair [78].

3.4.2.2 Silver Nanomaterials

Gram-positive *Bacillus* species are commonly used to synthesize silver nanomaterials either intracellularly or extracellularly. *Bacillus subtilis* KMS2-2 produces spherical silver nanomaterials with a size range of 70–100 nm by using the culture supernatant [81]. Moreover, the spherical silver nanomaterials are also formed at ambient temperature for 36 hours in average size around 18–100 nm using a cell-free supernatant of *Bacillus licheniformis* M09 [86]. The formed silver nanomaterials can remove methyl blue dye via photocatalytic degradation in less than three hours, demonstrating the recent use of microbial nanotechnology in the bioremediation of dye wastewater. In another study, *Bacillus* sp. strain AW1-2 accumulated silver nanomaterials intracellularly with a smaller 22–42 nm size using aerobic incubation at 30 °C [80]. All these silver nanomaterials exhibited a broad-spectrum antimicrobial activity against pathogenic bacteria such as *Pseudomonas fluorescens, Proteus mirabilis, Escherichia coli, Bacillus cereus, Bacillus subtilis, Pseudomonas aeruginosa, Staphylococcus aureus,* and fungi like, *Colletotrichum falcatum*. Similarly, a strong antibacterial agent with spherical silver nanomaterials and a diameter~7.27 nm was formed at room temperature in 24 hours using extracellular gram-negative *Pseudomonas aeruginosa* [79].

In a recent study on fungal biosynthesis, silver nanomaterials were synthesized extracellularly using *Aspergillus oryzae* SZ1 and tested for antifungal activities against clinical isolates of *Candida* species [87]. The spherical and monodispersed silver nanomaterials were obtained with a 1–50 nm size range. Likewise, the smaller spherical shape around 2–5 nm size was achieved under mild conditions by fungal cell-free extract of *Trichoderma reesei* via reduction of silver nitrate [99]. Remarkably, the small spherical silver nanomaterials demonstrated strong antibacterial activity against several bacterial pathogens such as *Staphylococcus aureus, Shigella sonnei, Serratia marcescens, Escherichia coli,* and *Pseudomonas aeruginosa*. Similarly, well-dispersed shaped silver nanomaterials with sizes ranging between 5 and 13 nm were obtained using different *Trichoderma* strains, under optimized conditions (40 °C, pH 12, 24 hours). It has been shown to have high biocidal activity against halotolerant bacteria, making it attractive to prevent biocorrosion, especially in oil and gas industries. In another study, monodisperse spherical silver nanomaterials of 10–20 nm were produced extracellularly by *Rhizoctonia solani* fungi [83]. It was reported that the Ag^+ ions were bound and reduced to Ag^0 by negatively charged functional groups ($-NH_2$, amino group and COO−, carboxylate groups) on the surface of mycelia's cell wall. Functional groups of $-NH_2$ and COO− groups are also accountable for stabilizing the silver nanomaterials. Various functional groups such as amide, carboxyl, carbonyl, methyl, methylene, and phosphate could be involved in the Ag^+ biosorption by *Botryosphaeria rhodina* biomass, and yet less involvement of hydroxyl and amino groups has been reported [82].

Actinomycete *Streptomyces* sp. have been commonly used for the synthesis of silver nanomaterials either intra- or extracellularly. Recently, Mabrouk et al. isolated 34 *Streptomyces* strains from the soil sample of Red Sea Coast, Egypt [84]. However, only *Streptomyces spiralis* and *Streptomyces rochei* could extracellularly synthesize spherical silver nanomaterials around 20–40 nm. In another recent work, intracellular and extracellular *Streptomyces chiangmaiensis* synthesized spherical silver nanomaterials with an average size of 13.57 and 30.47 nm, respectively. Both studies can be utilized as antibacterial agents.

3.4.3 Other Nanomaterials

Palladium-based nanomaterials exhibit excellent catalytic activity in organic and inorganic synthesis. Palladium nanomaterials can be synthesized extracellularly using *Desulfovibrio alaskensis* G20, a sulfate-reducing bacterium [100]. It was also shown to have high catalytic activities in the Suzuki Miyaura reaction of aryl bromides with phenylboronic acid under mild reaction conditions. This bacterial-mediated palladium nanomaterial can catalyze the reaction efficiently (>99%), comparable to palladium catalysts generated using chemical methods. *Shewanella* species (gram-negative bacteria) have been used to synthesize palladium nanomaterials via absorption and reduction of Pd(II) ions by functional groups such as $-NH_2$, $-OH$, $-COO$, and $-PO_3H_2$. The palladium nanomaterials synthesized by *Shewanella oneidensis* MR-1 are capable of reducing inorganic pollutants Cr(IV) to less toxic Cr(III) efficiently at room temperature [101]. However, it was found that the palladium nanomaterials produced by *Shewanella putrefaciens* CN-32 has higher electrocatalytic activity than *Shewanella oneidensis* MR-1 due to finer and uniformly dispersed nanomaterials [102].

Selenium nanomaterial synthesis by microorganisms has been reported recently to possess anticancer properties against colon cancer. These selenium nanomaterials can be synthesized intracellularly from bacteria *Lactobacillus casei* ATCC 393 [103] and *Lactococcus lactis* NZ9000 [104] in the diameter range of 170–550 and 38–152 nm, respectively. This bioreduction of selenite ions to Se^0 was done under anaerobic conditions, and the purified selenium nanomaterials were stabilized by polysaccharides and proteins [104]. Interestingly, the selenium nanomaterials accumulated inside the cells could be secreted into the extracellular space by a special ion transport system [105].

Similarly, tellurium nanomaterials can also be produced under aerobic or anaerobic conditions by microorganisms in the presence of tellurium oxyanions such as TeO_3^{2-} and TeO_4^{2-}. Bacterial *Aeromonas hydrophila* was cultured with the sodium tellurite under anaerobic conditions at pH 6.5 and 30 °C [106]. Remarkably, this novel bacteria strain can tolerate high oxyanion toxicity, producing rod shape of tellurium nanomaterials with average diameter materials of 18 nm. Interestingly, platinum nanomaterials can be synthesized by cell lysate supernatant from various gram-negative and gram-positive bacteria [68]. The smaller material sizes of 3.95, 2.49, and 3.84 nm with higher biogenic activity, such as antioxidant and antibacterial activity, were obtained using *Pseudomonas kunmingensis*, *Psychrobacter faecalis*, and *Vibrio fischeri*, respectively.

3.5 Application of Microorganism-Based Synthesized Nanomaterial

Microorganism-based synthesis nanomaterial is also known as green synthesis. Green synthesis is expected to combat the utilization of hazardous chemicals and may provide solution to existing environmental pollution in producing nanomaterials. As mentioned in the previous subtopic, nanomaterials like gold, silver, and alloy are some common types of organic nanomaterials synthesized via this green process.

These nanomaterials have a high potential to be employed in various fields like agriculture, water treatment, and biomedical industries [107]—major requirements by medical applications mostly suitable with the nanomaterial synthesized via green process. Nanomaterial like gold nanomaterials has been reported to be successfully hybridized/integrated for tumor therapeutic or drug delivery purposes [108]. Meanwhile argentum nanomaterials that have been synthesized using fungi or plant extract may be utilized as coating elements for cardiovascular and orthopedic implants, catheters, or wound dressings.

In water treatment, nanomaterials like nickel oxide, which has been synthesized using fungi and yeast mediators, can work effectively for bioremediation/biosorbent of toxic Ni ions or metals from wastewater [109]. Silver and gold nanomaterials synthesized via the green method have also been reported as having good potential for sensing applications. This sensor is increasing being used in the textile industry [110], environmental pollutants related to soil [111], and medical devices.

Organic or natural nanomaterial like bacterial nanocellulose (BC) has excellent potential for reinforcing polymer composite filler [112]. It is feasible to use recycled paperade of nonwoody cellulose fiber [113]. BC can also be utilized in the biomedical field as dental and vascular implants, wound dressing, in the food industry as edible gels immobilization of probiotics, food packaging, in the electronic industry as capacitors, biosensors, textile industry and for environmental protection as ultrafiltration membrane and sewage treatment [98].

3.6 Conclusion and Perspective

In summary, the development of nanomaterials synthesized by microorganisms was utilized as an alternative to green and environmentally friendly methods. Microorganisms that are commonly used are bacteria, yeast, fungi, and viruses. Most likely, bacteria were the most microorganism used to produce nanomaterial, especially for organic material like nanocellulose. Meanwhile, for inorganic material, microorganisms are randomly found by using yeast, fungi, and viruses as well. A review of inorganic material strongly observed with gold and silver nanomaterials production.

It is suggested for future studies to look at the economic feasibility of producing nanomaterials synthesized by microorganisms. This will have an impact and good impression on the industry by implying this method for high-capacity production. Thus, it will encourage cleaner and greener processes and manage to cope with sustainability issues globally.

Abbreviations

BC	Bacterial cellulose
BNC	Bacterialnanocellulose
EET	Extracellular electron transfer
Glc-1-p	Glucose-1-phosphate
Glc-6-p	Glucose-6-phosphate
NADH	Nicotinamide adenine dinucleotide
NADPH	Nicotinamide adenine dinucleotide phosphate
UDPGlc	Uridine diphosphate glucose
UGPase	UDP-glucose pyrophosphorylase

References

1 Chaudhry, N., Dwivedi, S., Chaudhry, V. et al. (2018). Bio-inspired nanomaterials in agriculture and food: current status, foreseen applications and challenges. *Microb. Pathogen.* 123: 196–200.

2 Shah, M., Fawcett, D., Sharma, S. et al. (2015). Green synthesis of metallic nanoparticles via biological entities. *Materials (Basel)* 8 (11): 7278–7308.

3 Jo, J.H., Singh, P., Kim, Y.J. et al. (2016). *Pseudomonas deceptionensis* DC5-mediated synthesis of extracellular silver nanoparticles. *Artif. Cells Nanomed. Biotechnol.* 44 (6): 1576–1581.

4 Huang, H., Lai, J., Lu, J., and Li, Z. (2019). Pulsed laser ablation of bulk target and particle products in liquid for nanomaterial fabrication. *AIP Adv.* 9 (1): 15307.
5 Schrittwieser, S., Haslinger, M.J., Mitteramskogler, T. et al. (2019). Multifunctional nanostructures and nanopocket particles fabricated by nanoimprint lithography. *Nanomaterials* 9 (12): 1790.
6 Kumar, M., Xiong, X., Wan, Z. et al. (2020). Ball milling as a mechanochemical technology for fabrication of novel biochar nanomaterials. *Bioresour. Technol.* 312: 123613.
7 Fuentes-García, J.A., Santoyo-Salzar, J., Rangel-Cortes, E. et al. (2021). Effect of ultrasonic irradiation power on sonochemical synthesis of gold nanoparticles. *Ultrason. Sonochem.* 70: 105274.
8 Manawi, Y.M., Ihsanullah, S.A., Al-Ansari, T., and Atieh, M.A. (2018). A review of carbon nanomaterials' synthesis via the chemical vapor deposition (CVD) method. *Materials (Basel)* 11 (5): 822.
9 Kamimura, H., Hayashida, M., and Ohgai, T. (2020). CPP-GMR performance of electrochemically synthesized Co/Cu multilayered nanowire arrays with extremely large aspect ratio. *Nanomaterials* 10 (1): 5.
10 Wei, H., Huang, K., Wang, D. et al. (2017). Iced photochemical reduction to synthesize atomically dispersed metals by suppressing nanocrystal growth. *Nat. Commun.* 8 (1): 1–8.
11 Jamkhande, P.G., Ghule, N.W., Bamer, A.H., and Kalaskar, M.G. (2019). Metal nanoparticles synthesis: an overview on methods of preparation, advantages and disadvantages, and applications. *J. Drug Delivery Sci. Technol.* 53: 101174.
12 Kolahalam, L.A., Kasi Viswanath, I.V., Diwakar, B.S. et al. (2019). Review on nanomaterials: synthesis and applications. *Mater. Today Proc.* 18: 2182–2190.
13 Karimi, M., Hesaraki, S., Alizadeh, M., and Kazemzadeh, A. (2016). Synthesis of calcium phosphate nanoparticles in deep-eutectic choline chloride-urea medium: investigating the role of synthesis temperature on phase characteristics and physical properties. *Ceram. Int.* 42 (2): 2780–2788.
14 Gan, Y.X., Jayatissa, A.H., Yu, Z. et al. (2020). Hydrothermal synthesis of nanomaterials. *J. Nanomater.* 2020: 1–3.
15 Aljumaily, M.M., Alsaadi, M.A., Das, R. et al. (2018). Optimization of the synthesis of superhydrophobic carbon nanomaterials by chemical vapor deposition. *Sci. Rep.* 8 (1): 1–12.
16 Nikam, A.V., Prasad, B.L.V., and Kulkarni, A.A. (2018). Wet chemical synthesis of metal oxide nanoparticles: a review. *CrystEngComm* 20 (35): 5091–5107.

17 Othman, N., Masarudin, M.J., Kuen, C.Y. et al. (2018). Synthesis and optimization of chitosan nanoparticles loaded with l-ascorbic acid and thymoquinone. *Nanomaterials* 8 (11): 920.

18 Fawcett, D., Verduin, J.J., Shah, M. et al. (2017). A review of current research into the biogenic synthesis of metal and metal oxide nanoparticles via marine algae and seagrasses. *J. Nanosci.* 2017: 1–15.

19 Barabadi, H., Tajani, B., Moradi, M. et al. (2019). Penicillium family as emerging nanofactory for biosynthesis of green nanomaterials: a journey into the world of microorganisms. *J. Cluster Sci.* 30 (4): 843–856.

20 Park, T.J., Lee, K.G., and Lee, S.Y. (2016). Advances in microbial biosynthesis of metal nanoparticles. *Appl. Microbiol. Biotechnol.* 100 (2): 521–534.

21 Ibrahim, E., Zhang, M., Zhang, Y. et al. (2020). Green-synthesization of silver nanoparticles using endophytic bacteria isolated from garlic and its antifungal activity against wheat fusarium head blight pathogen *Fusarium graminearum*. *Nanomaterials* 10 (2): 219.

22 Rajeshkumar, S., Jeevitha, M., Sheba, D., and Nagalingam, M. (2022). Bacterial and fungal mediated synthesis, characterization and applications of AgNPs. In: *Agri-Waste and Microbes for Production of Sustainable Nanomaterials* (ed. K.A. Abd-Elsalam, R. Periakaruppan, and S. Rajeshkumar), 165–186. Elsevier.

23 Chang, Y., Chen, S., Liu, T. et al. (2021). Yeast cell route: a green and facile strategy for biosynthesis of carbonate nanoparticles. *CrystEngComm* 23 (26): 4674–4679.

24 Zhang, W., Zhang, X.E., and Li, F. (2018). Virus-based nanoparticles of Simian Virus 40 in the field of nanobiotechnology. *Biotechnol. J.* 13 (6): 1700619.

25 Gahlawat, G. and Choudhury, A.R. (2019). A review on the biosynthesis of metal and metal salt nanoparticles by microbes. *RSC Adv.* 9 (23): 12944–12967.

26 Sun, L., Zhang, D., Sun, Y. et al. (2018). Facile fabrication of highly dispersed Pd@Ag core–shell nanoparticles embedded in *Spirulina platensis* by electroless deposition and their catalytic properties. *Adv. Funct. Mater.* 28 (20): 1707231.

27 Grasso, G., Zane, D., and Dragone, R. (2020). Microbial nanotechnology: challenges and prospects for green biocatalytic synthesis of nanoscale materials for sensoristic and biomedical applications. *Nanomaterials* 10 (1): 11.

28 Iravani, S. and Varma, R.S. (2020). Bacteria in heavy metal remediation and nanoparticle biosynthesis. *ACS Sustainable Chem. Eng.* 8 (14): 5395–5409.

29 Ramasubburayan, R., Sumathi, S., Prakash, S. et al. (2017). Synthesis of nano silver by a marine epibiotic bacterium *Bacillus vallismortis* and its potent ecofriendly antifouling properties. *Environ. Nanotechnol. Monit. Manage.* 8: 112–120.

30 Gupta, R. and Padmanabhan, P. (2018). Biogenic synthesis and characterization of gold nanoparticles by a novel marine bacteria *Marinobacter algicola*: progression from nanospheres to various geometrical shapes. *J. Microbiol. Biotechnol. Food Sci.* 8 (1): 732–737.

31 Sidkey, N.M., Arafa, R.A., Moustafa, Y.M. et al. (2017). Biosynthesis of Mg and Mn intracellular nanoparticles via extremo-metallotolerant *Pseudomonas stutzeri*, B4 Mg/W and *Fusarium nygamai*, F4 Mn/S. *J. Microbiol. Biotechnol. Food Sci.* 6 (5): 1181–1187.

32 Presentato, A., Piacenza, E., Anikovskiy, M. et al. (2018). Biosynthesis of selenium-nanoparticles and -nanorods as a product of selenite bioconversion by the aerobic bacterium *Rhodococcus aetherivorans* BCP1. *New Biotechnol.* 41: 1–8.

33 Singh, H., Du, J., Singh, P., and Yi, T.H. (2018). Extracellular synthesis of silver nanoparticles by Pseudomonas sp. THG-LS1.4 and their antimicrobial application. *J. Pharm. Anal.* 8 (4): 258–264.

34 Furubayashi, M., Wallace, A.K., González, L.M. et al. (2021). Genetic tuning of iron oxide nanoparticle size, shape, and surface properties in *Magnetospirillum magneticum*. *Adv. Funct. Mater.* 31 (4): 2004813.

35 Salem, S.S., Ali, O.M., Reyad, A.M. et al. (2022). *Pseudomonas indica*-mediated silver nanoparticles: antifungal and antioxidant biogenic tool for suppressing mucormycosis fungi. *J. Fungi* 8: 126.

36 Gomaa, E.Z., Housseiny, M.M., and Omran, A.A.A.K. (2019). Fungicidal efficiency of silver and copper nanoparticles produced by *Pseudomonas fluorescens* ATCC 17397 against four aspergillus species: a molecular study. *J. Cluster Sci.* 30 (1): 181–196.

37 Arzoo, S., Naqvi, Z., Hussain, M. et al. (2020). Production and antimicrobial activity of silver nanoparticles synthesized from indigenously isolated *Pseudomonas aeruginosa* from rhizosphere. *Pak. J. Pharm. Sci.* 33 (6): 2815–2822.

38 Hossain, A., Hong, X., Ibrahim, E. et al. (2019). Green synthesis of silver nanoparticles with culture supernatant of a bacterium *Pseudomonas rhodesiae* and their antibacterial activity against soft rot pathogen *Dickeya dadantii*. *Molecules* 24: 2303.

39 Panichikkal, J., Thomas, R., John, J.C., and Radhakrishnan, E.K. (2019). Biogenic gold nanoparticle supplementation to plant beneficial *Pseudomonas monteilii* was found to enhance its plant probiotic effect. *Curr. Microbiol.* 76 (4): 503–509.

40 Hasan, M., Ullah, I., Zulfiqar, H. et al. (2018). Biological entities as chemical reactors for synthesis of nanomaterials: progress, challenges and future perspective. *Mater. Today Chem.* 8: 13–28.

41 Lachance, M.A. (2016). Paraphyly and (yeast) classification. *Int. J. Syst. Evol. Microbiol.* 66 (12): 4924–4929.

42 Ahmed, S.F., Mofijur, M., Rafa, N. et al. (2022). Green approaches in synthesising nanomaterials for environmental nanobioremediation: technological advancements, applications, benefits and challenges. *Environ. Res.* 204: 111967. https://www.sciencedirect.com/science/article/pii/S0013935121012627.

43 Kasemets, K., Käosaar, S., Vija, H. et al. (2019). Toxicity of differently sized and charged silver nanoparticles to yeast *Saccharomyces cerevisiae* BY4741: a nano-biointeraction perspective. *Nanotoxicology* 13 (8): 1041–1059.

44 Rónavári, A., Igaz, N., Gopisetty, M.K. et al. (2018). Biosynthesized silver and gold nanoparticles are potent antimycotics against opportunistic pathogenic yeasts and dermatophytes. *Int. J. Nanomed.* 13: 695.

45 Kalpana, V.N. and Devi Rajeswari, V. (2018). A review on green synthesis, biomedical applications, and toxicity studies of ZnO NPs. *Bioinorg. Chem. Appl.* 2018: 3569758.

46 Saxena, J., Sharma, M.M., Gupta, S., and Singh, A. (2014). Emerging role of fungi in nanoparticle synthesis and their applications. *World J. Pharm. Sci.* 3 (9): 1586–1613.

47 Chaudhary, A. and Yadav, R.D. (2019). A review on virus protein self-assembly. *J. Nanopart. Res.* 21 (11): 1–13.

48 Zhang, D., Ma, X.L., Gu, Y. et al. (2020). Green synthesis of metallic nanoparticles and their potential applications to treat cancer. *Front. Chem.* 8: 799.

49 Douglas, T. and Young, M. (1998). Host-guest encapsulation of materials by assembled virus protein cages. *Nature* 393 (6681): 152–155.

50 Thangavelu, R.M., Ganapathy, R., Ramasamy, P., and Krishnan, K. (2020). Fabrication of virus metal hybrid nanomaterials: an ideal reference for bio semiconductor. *Arabian J. Chem.* 13 (1): 2750–2765.

51 Love, A.J., Makarov, V., Yaminsky, I. et al. (2014). The use of tobacco mosaic virus and cowpea mosaic virus for the production of novel metal nanomaterials. *Virology* 449: 133–139.

52 Brown, A.J. (1886). Brown on acetic ferment. *J. chem. Soc. Trans.* 49 (432): 432–439.

53 Chen, C., Ding, W., Zhang, H. et al. (2022). Bacterial cellulose-based biomaterials: from fabrication to application. *Carbohydr. Polym.* 278: 118995.

54 Pandit, A. and Kumar, R. (2021). A review on production, characterization and application of bacterial cellulose and its biocomposites. *J. Polym. Environ.* 29 (9): 2738–2755.

55 Azeredo, H.M.C., Barud, H., Farinas, C.S. et al. (2019). Bacterial cellulose as a raw material for food and food packaging applications. *Front. Sustainable Food Syst.* 3: 7.

56 Azeredo, H.M.C., Rosa, M.F., and Mattoso, L.H.C. (2017). Nanocellulose in bio-based food packaging applications. *Ind. Crops Prod.* 97: 664–671.

57 Abral, H., Lawrensius, V., Handayani, D., and Sugiarti, E. (2018). Preparation of nano-sized particles from bacterial cellulose using ultrasonication and their characterization. *Carbohydr. Polym.* 191: 161–167.

58 Mishra, S., Singh, P.K., Pattnaik, R. et al. (2022). Biochemistry, synthesis, and applications of bacterial cellulose: a review. *Front. Bioeng. Biotechnol.* 10: 780409.

59 Klemm, D., Cranston, E.D., Fischer, D. et al. (2018). Nanocellulose as a natural source for groundbreaking applications in materials science: today's state. *Mater. Today* 21 (7): 720–748.

60 Zogaj, X., Bokranz, W., Nimtz, M., and Römling, U. (2003). Production of cellulose and curli fimbriae by members of the family Enterobacteriaceae isolated from the human gastrointestinal tract. *Infect. Immun.* 71 (7): 4151–4158.

61 Ryngajłło, M., Kubiak, K., Jędrzejczak-Krzepkowska, M. et al. (2019). Comparative genomics of the Komagataeibacter strains – efficient bionanocellulose producers. *Microbiologyopen* 8 (5): e00731.

62 Castro, C., Zuluaga, R., Álvarez, C. et al. (2012). Bacterial cellulose produced by a new acid-resistant strain of Gluconacetobacter genus. *Carbohydr. Polym.* 89 (4): 1033–1037.

63 Castro, C., Cleenwerck, I., Trček, J. et al. (2013). *Gluconacetobacter medellinensis* sp. nov., cellulose- and non-cellulose-producing acetic acid bacteria isolated from vinegar. *Int. J. Syst. Evol. Microbiol.* 63 (PART3): 1119–1125.

64 Sunagawa, N., Tajima, K., Hosoda, M. et al. (2012). Cellulose production by Enterobacter sp. CJF-002 and identification of genes for cellulose biosynthesis. *Cellulose* 19 (6): 1989–2001.

65 Jahn, C.E., Selimi, D.A., Barak, J.D., and Charkowski, A.O. (2011). The *Dickeya dadantii* biofilm matrix consists of cellulose nanofibres, and is an emergent property dependent upon the type III secretion system and the cellulose synthesis operon. *Microbiology* 157 (10): 2733–2744.

66 Kumagai, A., Mizuno, M., Kato, N. et al. (2011). Ultrafine cellulose fibers produced by *Asaia bogorensis*, an acetic acid bacterium. *Biomacromolecules* 12 (7): 2815–2821.

67 Khan, H., Kadam, A., and Dutt, D. (2020). Studies on bacterial cellulose produced by a novel strain of Lactobacillus genus. *Carbohydr. Polym.* 229: 115513.

68 Eramabadi, P., Masoudi, M., Makhdoumi, A., and Mashreghi, M. (2020). Microbial cell lysate supernatant (CLS) alteration impact on platinum nanoparticles fabrication, characterization, antioxidant and antibacterial activity. *Mater. Sci. Eng., C* 117: 111292.

69 Su, Z., Li, X., Xi, Y. et al. (2022). Microbe-mediated transformation of metal sulfides: mechanisms and environmental significance. *Sci. Total Environ.* 825: 153767.

70 Hulkoti, N.I. and Taranath, T.C. (2014). Biosynthesis of nanoparticles using microbes – a review. *Colloids Surf., B* 121: 474–483.

71 Patil, M.P. and Do, K.G. (2018). Marine microorganisms for synthesis of metallic nanoparticles and their biomedical applications. *Colloids Surf., B* 172: 487–495.

72 Qiu, R., Xiong, W., Hua, W. et al. (2021). A biosynthesized gold nanoparticle from *Staphylococcus aureus* – as a functional factor in muscle tissue engineering. *Appl. Mater. Today* 22: 100905.

73 Shunmugam, R., Renukadevi Balusamy, S., Kumar, V. et al. (2021). Biosynthesis of gold nanoparticles using marine microbe (*Vibrio alginolyticus*) and its anticancer and antioxidant analysis. *J. King Saud Univ. Sci.* 33 (1): 101260.

74 Italiano, F., Agostiano, A., Belviso, B.D. et al. (2018). Interaction between the photosynthetic anoxygenic microorganism *Rhodobacter sphaeroides* and soluble gold compounds. From toxicity to gold nanoparticle synthesis. *Colloids Surf., B* 172 (June): 362–371.

75 Wang, Y., You, L.X., Zhong, H.L. et al. (2022). Au(III)-induced extracellular electron transfer by *Burkholderia contaminans* ZCC for the bio-recovery of gold nanoparticles. *Environ. Res.* 210 (January): 112910.

76 Xu, X., Yang, Y., Zhao, X. et al. (2019). Recovery of gold from electronic wastewater by Phomopsis sp. XP-8 and its potential application in the degradation of toxic dyes. *Bioresour. Technol.* 288 (April): 121610.

77 Soltani Nejad, M., Samandari Najafabadi, N., Aghighi, S. et al. (2022). Evaluation of Phoma sp. biomass as an endophytic fungus for synthesis of extracellular gold nanoparticles with antibacterial and antifungal properties. *Molecules* 27 (4): 1181.

78 Bennur, T., Javdekar, V., Tomar, G.B., and Zinjarde, S. (2020). Gold nanoparticles biosynthesized by *Nocardiopsis dassonvillei* NCIM 5124 enhance osteogenesis in gingival mesenchymal stem cells. *Appl. Microbiol. Biotechnol.* 104 (9): 4081–4092.

79 Jimoh, A.A., Akpeji, B., Azeez, S.O. et al. (2022). Biosynthesis of Ag and TiO_2 nanoparticles and the evaluation of their antibacterial activities. *Inorg. Chem. Commun.* 141: 109503.

80 Ajaz, S., Ahmed, T., Shahid, M. et al. (2021). Bioinspired green synthesis of silver nanoparticles by using a native Bacillus sp. strain AW1-2: characterization and antifungal activity against *Colletotrichum falcatum* Went. *Enzyme Microb. Technol.* 144 (July 2020): 109745.

81 Mathivanan, K., Selva, R., Chandirika, J.U. et al. (2019). Biologically synthesized silver nanoparticles against pathogenic bacteria: synthesis, calcination and characterization. *Biocatal. Agric. Biotechnol.* 22 (October): 101373.

82 Muñoz, A.J., Espínola, F., Ruiz, E. et al. (2021). Biosorption mechanisms of Ag(I) and the synthesis of nanoparticles by the biomass from *Botryosphaeria rhodina* MAMB-05. *J. Hazard. Mater.* 420: 126598.

83 Potbhare, A.K., Chouke, P.B., Mondal, A. et al. (2019). *Rhizoctonia solani* assisted biosynthesis of silver nanoparticles for antibacterial assay. *Mater. Today Proc.* 29: 939–945.

84 Mabrouk, M., Elkhooly, T.A., and Amer, S.K. (2021). Actinomycete strain type determines the monodispersity and antibacterial properties of biogenically synthesized silver nanoparticles. *J. Genet. Eng. Biotechnol.* 19 (57): 1–12.

85 Rosyidah, A., Weeranantanapan, O., Chudapongse, N. et al. (2022). *Streptomyces chiangmaiensis* SSUT88A mediated green synthesis of silver nanoparticles: characterization and evaluation of antibacterial action against clinical drug-resistant strains. *RSC Adv.* 12 (7): 4336–4345.

86 Momin, B., Rahman, S., Jha, N., and Annapure, U.S. (2019). Valorization of mutant *Bacillus licheniformis* M09 supernatant for green synthesis of silver nanoparticles: photocatalytic dye degradation, antibacterial activity, and cytotoxicity. *Bioprocess. Biosyst. Eng.* 42 (4): 541–553.

87 Zainab, S., Hamid, S., Sahar, S., and Ali, N. (2021). Fluconazole and biogenic silver nanoparticles-based nano-fungicidal system for highly efficient elimination of multi-drug resistant Candida biofilms. *Mater. Chem. Phys.* 2022 (276): 125451.

88 Omran, B.A., Nassar, H.N., Younis, S.A. et al. (2019). Physiochemical properties of *Trichoderma longibrachiatum* DSMZ 16517-synthesized silver nanoparticles for the mitigation of halotolerant sulphate-reducing bacteria. *J. Appl. Microbiol.* 126 (1): 138–154.

89 Mohammadkazemi, F., Azin, M., and Ashori, A. (2015). Production of bacterial cellulose using different carbon sources and culture media. *Carbohydr. Polym.* 117: 518–523.

90 Huang, Y., Zhu, C., Yang, J. et al. (2014). Recent advances in bacterial cellulose. *Cellulose* 21 (1): 1–30.

91 García-Sánchez, M.E., Robledo-Ortíz, J.R., Jiménez-Palomar, I. et al. (2020). Production of bacterial cellulose by *Komagataeibacter xylinus* using mango waste as alternative culture medium. *Rev. Mex. Ing. Chim.* 19 (2): 851–865.

92 Feng, X., Ullah, N., Wang, X. et al. (2015). Characterization of bacterial cellulose by *Gluconacetobacter hansenii* CGMCC 3917. *J. Food Sci.* 80 (10): E2217–E2227.

93 Jozala, A.F., Pértile, R.A.N., dos Santos, C.A. et al. (2015). Bacterial cellulose production by *Gluconacetobacter xylinus* by employing alternative culture media. *Appl. Microbiol. Biotechnol.* 99 (3): 1181–1190.

94 Singhsa, P., Narain, R., and Manuspiya, H. (2018). Bacterial cellulose nanocrystals (BCNC) preparation and characterization from three bacterial cellulose sources and development of functionalized BCNCs as nucleic acid delivery systems. *ACS Appl. Nano Mater.* 1 (1): 209–221.

95 Jozala, A.F., de Lencastre-Novaes, L.C., Lopes, A.M. et al. (2016). Bacterial nanocellulose production and application: a 10-year overview. *Appl. Microbiol. Biotechnol.* 100 (5): 2063–2072.

96 Römling, U. (2002). Molecular biology of cellulose production in bacteria. *Res. Microbiol.* 153 (4): 205–212.

97 Brown, R.M. (1999). Cellulose structure and biosynthesis. *Pure Appl. Chem.* 71 (5): 767–775.

98 Reshmy, R., Philip, E., Thomas, D. et al. (2021). Bacterial nanocellulose: engineering, production, and applications. *Bioengineered* 12 (2): 11463.

99 Kumari, M., Pandey, S., Giri, V.P. et al. (2017). Tailoring shape and size of biogenic silver nanoparticles to enhance antimicrobial efficacy against MDR bacteria. *Microb. Pathogen.* 105: 346–355.

100 Era, Y., Dennis, J.A., Wallace, S., and Horsfall, L.E. (2021). Micellar catalysis of the Suzuki Miyaura reaction using biogenic Pd nanoparticles from *Desulfovibrio alaskensis*. *Green Chem.* 23 (22): 8886–8890.

101 Zhang, Y., Zhao, Q., and Chen, B. (2022). Reduction and removal of Cr(VI) in water using biosynthesized palladium nanoparticles loaded *Shewanella oneidensis* MR-1. *Sci. Total Environ.* 805: 150336.

102 Li, Q., Zhang, S., Xuan, W. et al. (2021). Microbial synthesis of highly dispersed nano-Pd electrocatalyst for oxygen reduction reaction. *Int. J. Hydrogen Energy* 46 (53): 26886–26896.

103 Spyridopoulou, K., Tryfonopoulou, E., Aindelis, G. et al. (2021). Biogenic selenium nanoparticles produced by: *Lactobacillus casei* ATCC 393 inhibit colon cancer cell growth in vitro and in vivo. *Nanoscale Adv.* 3 (9): 2516–2528.

104 Xu, C., Qiao, L., Ma, L. et al. (2019). Biosynthesis of polysaccharides-capped selenium nanoparticles using *Lactococcus lactis* NZ9000 and their antioxidant and anti-inflammatory activities. *Front. Microbiol.* 10: 1632.

105 Wang, Y., Shu, X., Zhou, Q. et al. (2018). Selenite reduction and the biogenesis of selenium nanoparticles by *Alcaligenes faecalis* se03 isolated from the gut of *Monochamus alternatus* (Coleoptera: Cerambycidae). *Int. J. Mol. Sci.* 19 (9): 2799.

106 Castro, L., Li, J., González, F. et al. (2020). Green synthesis of tellurium nanoparticles by tellurate and tellurite reduction using *Aeromonas hydrophila* under different aeration conditions. *Hydrometallurgy* 196: 105415.

107 Bhatnagar, N. (2022). A review on biosynthesis of nanoparticles and their applications. In: *Intelligent Computing Techniques for Smart Energy Systems* (ed. A. Tripathi, A. Soni, A. Shrivastava, et al.), 339–346. Springer.

108 Huo, M., Wang, L., Chen, Y., and Shi, J. (2020). Nanomaterials/microorganism-integrated microbiotic nanomedicine. *Nano Today* 32: 100854.

109 Ahmad, W., Bhatt, S.C., Verma, M. et al. (2022). A review on current trends in the green synthesis of nickel oxide nanoparticles, characterizations, and their applications. *Environ. Nanotechnol. Monit. Manage.* 18: 100674.

110 Kapoor, A., Shankar, P., and Ali, W. (2020). Green synthesis of metal nanoparticles for electronic textiles. In: *Green Nanomaterials*, Advanced Structured Materials (ed. S. Ahmed and W. Ali), 81–97. Springer.

111 Ahmad, R. and Kumar, V. (2020). Microorganism based biosensors to detect soil pollutants. *Plant Arch.* 20 (2): 2509–2516.

112 Omran, A.A.B., Mohammed, A.A.B.A., Sapuan, S.M. et al. (2021). Micro-and nanocellulose in polymer composite materials: a review. *Polymers (Basel)* 13 (2): 231.

113 Skočaj, M. (2019). Bacterial nanocellulose in papermaking. *Cellulose* 26 (11): 6477–6488.

4

Biopolymer-Based Nanomaterials and Their Applications

Baranya Murugan[1], Is Fatimah[2], MA Motalib Hossain[3], and Suresh Sagadevan[1,2]

[1] *Nanotechnology & Catalysis Research Centre, University of Malaya, Kuala Lumpur, Malaysia*
[2] *Department of Chemistry, Faculty of Mathematics and Natural Sciences, Universitas Islam Indonesia, Yogyakarta, Indonesia*
[3] *Institute of sustainable energy, Universiti Tenaga Nasional, Kajang, Selangor, Malaysia*

Abstract

Biopolymers and their nano-structured analogs play a vital role in hindering an issue of feedstock lack and in generating an eco-friendly environment. A fast-evolving multidisciplinary science known as nanotechnology enables the manufacturing of polymers at the nanoscale scale for various medical applications. As compared to synthetic ones, biopolymers used in biomedical applications ensure biocompatibility, biodegradability, and minimal immunogenicity. A biopolymer such as silk fibroins, collagen, gelatin, albumin, starch, cellulose, and chitosan can be readily produced into suspensions which act as carriers for dual-large and small medicinal molecules. The techniques, commonly used to make the biopolymer, include spray-drying, electrospraying, desolvation, supercritical fluid extraction, layer-by-layer self-assembly, freeze drying, and microemulsion for different applications in medicine such as nanocarrier for drug delivery systems, tissue engineering, regenerative medicine, water treatment, energy sector, and food industry. We also discussed the various characterization techniques and importance of biopolymers in nanotechnology-based biomedical applications.

Keywords *antimicrobial activity; biomedical applications; characterization methods; nanomaterials; natural and synthetic polymers; biocompatibility; nanoscience*

Green Synthesis of Nanomaterials: Biological and Environmental Applications, First Edition.
Edited by Archana Chakravarty, Preeti Singh, Saiqa Ikram, and R.N. Yadava.

4.1 Introduction

Biopolymers are covalently required monomeric components that are assembled into molecules to generate polymeric biomolecules [1]. Biopolymers are biodegradable materials produced by living organisms, as designated by the word "bio" [2]. Biopolymer is a term used to describe a broad variety of substances that are frequently derived from biological sources such as microbes, plants, or trees. Biopolymers are also materials originating from synthetic chemistry from biological sources mainly vegetable oils, sugars, lipids, resins, proteins, amino acids, and so on [3]. In addition to superior qualities, biopolymers' molecular makeup is identical to those of macromolecules found in the extracellular matrix (ECM), which makes them compatible, functional with, and useful for the host [4]. Natural polymers are those extracted from natural sources of polymers, including plants, animals, and microorganisms. Due to their homology to the extracellular matrix, mechanical stability, biocompatibility, and high-water holding capacity. natural polymers are broadly used in a variety of biomedical applications, significantly for the regeneration of skin tissue [5]. Natural polymers are used as encapsulation, microparticles, matrix systems, scaffolds, inhalations, and viscous liquid formulations in drug delivery systems [6]. This polymeric material also acts as a drug release carrier, binding agent, emulsifying agent, and bioadhesive.

Synthetic polymers have a very simple and irregular structure as compared to biopolymers, are three-dimensional structures, and have complex molecular assemblies. These properties of biopolymers make them very active molecules *in vivo* [7]. The precise, defined shape and structure are known to be a vital function of biopolymers. Biopolymers are mostly synthesized from starch, sugar, natural fibers, and many other organic biodegradable components with different compositions [8]. These biopolymers deteriorate when coming into contact with microorganisms found in compost, soil, or sea debris [9]. The oxygen and nitrogen atoms involved in the structural backbone of biopolymers make them easily biodegradable and allow them to be produced from renewable resources [10]. Upon biodegradation, they are transformed into CO_2, water, biomass, humid matter, and other organic compounds. Thus, biological processes spontaneously recycle biopolymers [11]. Synthetic biopolymers comprised of poly (L-lactic acid), poly (e-caprolactone), poly (vinyl alcohol), and poly (butylene succinate), have attracted a lot of interest in the biological fields [12]. Biopolymers produced over microbial fermentation consist of microbial polyesters, like poly (3-hydroxybutyrate-co-3-hydroxy valerate), poly (hydroxy alkanoates) (PHAs), and poly (β hydroxybutyrate) (PHB), and other microbial polyesters comprising curdlan and pullulan are also widely used for biomedical applications [13].

A current trend and advancement in nanoscience and technology is the aid of single molecules for cell regeneration. A vital part of developing smart biopolymer scaffolds for biomedical applications is enhancing smart biomaterial research [14]. The most commonly preferred biopolymers are biodegradable polymeric materials because they have definite physical, chemical, biological, biomechanical, and degradation properties. This chapter concentrates on the history and current developments of the application of biopolymers for the breakthrough in the biomedical field in the following sections: drug transport, tissue engineering, wound healing, and finally the use of both natural and manmade materials in various biomedical applications round out the list. The use of biopolymers in drug delivery, tissue engineering, infections, and wound healing have received particular attention because they show how diverse biopolymers have advanced and been used in current biomedical applications [15].

4.2 Classification of Biopolymers

Biopolymers are divided into three wide groups depending on the characteristics of repeating monomeric units such as polysaccharides, polynucleotides, and polypeptides. A glycosidic bond is formed when monosaccharide units are bonded together, and a biopolymer produced is called a polysaccharide. These polysaccharides include starch, cellulose, glycogen, chitin/chitosan, pectin, and alginate [16]. Polysaccharide biopolymers are further categorized into four groups.

- Sugar-based biopolymer
- Starch-based biopolymer
- Cellulose-based biopolymer
- Lignin-based biopolymer

4.2.1 Sugar-Based Biopolymer

Sugar-based biopolymers such as sucrose or starch are employed as raw materials in the fabrication of polyhydroxybutyrate. They can be fabricated through extrusion, blowing, injection, and vacuum forming. Lactose, taken out from potatoes, wheat, and sugar beet, is converted into lactic acid polymers known as poly lactides [17]. Sugar-based biopolymers also paid lots of attention to developing macromolecule- and nanoparticle-based cancer imaging and therapy [18].

4.2.2 Starch-Based Biopolymer

Starch functions as a natural polymer, such as in wheat, tapioca, and potatoes, which is comprised of glucose by melting starch. Starch-based polymers are also seen in vegetables including tapioca, corn, wheat, and potatoes. Dextran is a group of low molecular weight carbohydrates produced by enzymatic immobilization of *Enterococcus faecalis* dextransucrase onto biopolymer carriers [19, 20]. Starches also gain a lot of attention due to their economical, biodegradable, and ability to be refined with standard plastic processing machinery [21].

4.2.3 Cellulose-Based Biopolymers

Cellulose-based biopolymer is made of natural polymers like glucose and other natural resources such as cotton, timber, wheat, and corn [22]. Cellulose-based biopolymers are attracted much in biomedical applications because of their properties such as strength and stiffness, biodegradability, and renewability [23]. The downside of cellulose fibers includes moisture absorption, quality variations, low thermal stability, and poor compatibility with the hydrophobic polymer matrix [24].

4.2.4 Lignin-Based Biopolymers

The other significant biopolymer that is not derived from polysaccharides is lignin. A heterogeneous natural polymer with an aromatic structure, lignin accounts for 15–35% of the biomass in lignocellulosic forestry. DNA (deoxyribonucleic acid), RNA (ribonucleic acid), and polynucleotides are some examples of biopolymers made up of nucleotide units. A few lignin-based biopolymers include silk, collagen, and keratin [25].

4.2.5 Biopolymers Based on Synthetic Materials

Biopolymers are produced from synthetic compounds utilized for making biodegradable polymers namely aliphatic-aromatic copolyesters procured from petroleum. These are completely biodegradable and compostable [26]. There are non-biodegradable-based biopolymers and biodegradable-based biopolymers. This is known as "bioplastic." The classification of biopolymers is shown in Figure 4.1.

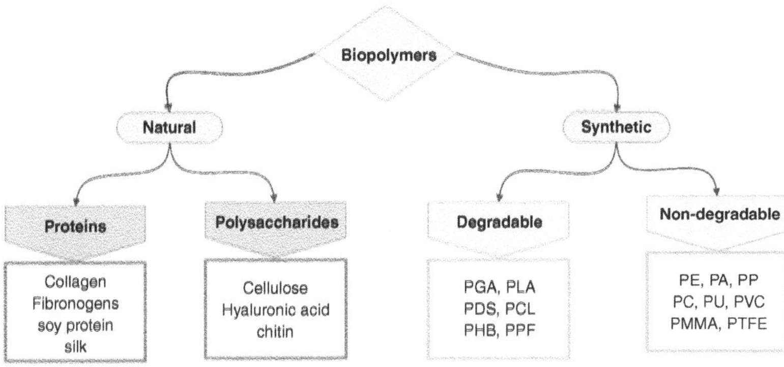

Figure 4.1 Classification of biopolymers.

4.3 Synthesis Methods of Biopolymers

There are different synthesis methods for a variety of biopolymers. Table 4.1 elaborates on the biopolymers which have been prepared by different synthesis route and their biomedical applications.

4.4 Characterization Methods of Biopolymers

The most commonly used characterization techniques for biopolymers include scanning electron microscopy (SEM), Fourier transforms infrared spectroscopy (FTIR), X-ray diffraction (XRD), and thermogravimetric analysis (TGA). The natural and synthetic polymers were synthesized and characterized using various characterization techniques thoroughly depending on the properties elaborated above promising for use in biomedical applications [7]. Following the characterization of the biopolymers, the properties of the material were assessed based on the American Society for Testing and Materials norms on account of their biodegradability, and environment-friendly to touch with quality standards [38]. Characterization techniques used for biopolymers are shown in Figure 4.2.

Thakur and his group demonstrated the development of green composites; natural polymers were graft-copolymerized under the impact of free radicals. During the time of graft-copolymer synthesis process, various factors were affected and standardized. FTIR was used to assess the functional groups of the material and also to find the physical, chemical, thermal, swelling, and functional modifications in the developed natural grafted copolymers, SEM, and TGA. Natural composites synthesized from raw

Table 4.1 Describes the common biopolymer synthesis method and application.

Biopolymers	Preparation method	Applications	Reference
Alginate	Emulsification and gelation method	Edible film Gelling agent	[27]
Gelatin	Gelation and extraction method	Food packaging	[28]
Chitosan	Solution intercalation (aqueous) method	Food packaging	[29]
Polycaprolactone (PCL)	Polycondensation, polymerization, and electrospinning method	Food packaging, filtration	[30]
Polylactic acid/poly ethylene glycol/chitosan	Extrusion method	Mainly applied in bone and dental implants	[31]
Potato starch/wheat gluten	Compression molding	It is used in the production of bio-based plastics	[32]
Polyvinyl alcohol/chitosan	Electrospinning	It is used for drug delivery and food packaging applications	[33]
Nanocellulose/carbon nanotubes	Cast molding, good electrical conductivity	It acts the part of sensor applications	[34]
Polyamide-11/halloysite nanotubes/lysozyme	Electrospinning	Used for the assembly of clay bio-active compounds	[35]
Gum/cloister/ cinnamaldehyde	Solvent casting	Used for the fabrication of bio-composite materials	[36]
Corn zein/cloisite	Solution intercalation	Used for yielding of bio-composite materials	[37]

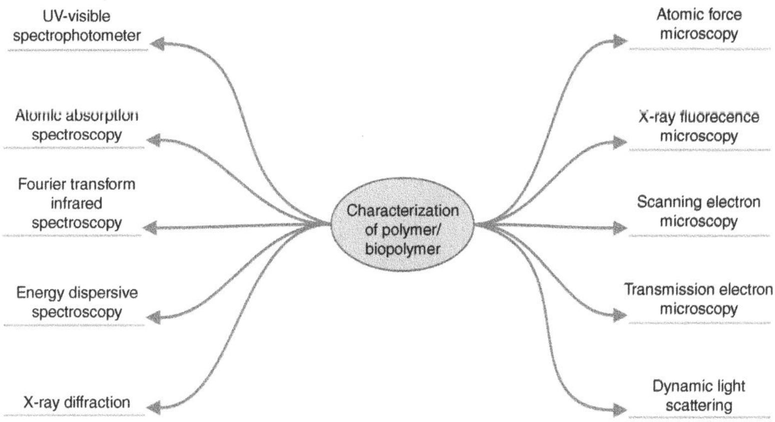

Figure 4.2 Characterization methods used for biopolymers.

polymers were found to have better tensile and mechanistic properties when compared to the parent natural polymers [39]. Fukushima and his group demonstrated a process of melt blending for the preparation of polylactic acid (PLA) and polycaprolactone (PCL) with the addition of fumed silica. Characterization of PLA and PCL was assessed using SEM, and the thermal stability of the polymer was estimated using thermos gravimetric analysis. Dynamic mechanical analysis was used to reveal the thermomechanical degradation of the addition of fumed silica on both the polymers, PLA and PCL [40]. Swain and his group demonstrated the synthesis and characterization of biopolymers and their composites to estimate the thermal stability, morphological and mechanical properties using TGA, SEM, FTIR, XRD, and other techniques. Depending on the characterization techniques analysis aids in figuring out the properties and progress of the biopolymers [41].

4.5 Nanotechnology-Based Applications of Biopolymers

The recent developments in biopolymers have gained a lot of attention from industries and simplicity in commercialization. Collagen, keratin, chitosan, silk, and elastin are commonly available biopolymers combined with synthetic polymers for an enhanced effect of these biopolymers as composites. These biopolymers such as dialysis, wound dressing materials, prosthetic three-dimensional scaffolds, implantable medical devices in the

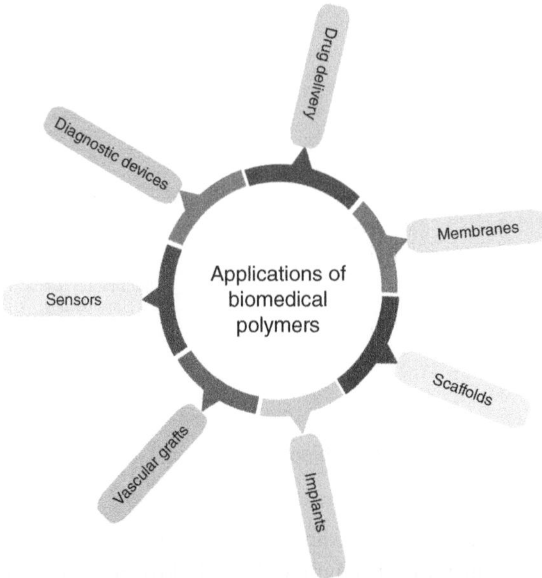

Figure 4.3 Different biomedical applications of biopolymers.

fields of cell-based transplantation, bone tissue engineering, and gene therapy are broadly utilized in the pharmaceutical industries [42]. Various biomedical applications of biopolymers are shown in Figure 4.3.

4.5.1 Drug Delivery Systems

Biopolymers were developed for drug delivery systems because of their structural diversity and unique physiological and therapeutic applications. These biopolymers have been widely used in the pharmaceutical industry with numerous advantages such as the formulation of drug delivery vehicles [11]. The most familiar biopolymers such as collagen, starch, fibroin, chitosan and cellulose, and gelatin can be prepared in suspensions to transfer to divergent-sized molecules. There are different synthesis methods used for the preparation of biopolymers, namely electrospinning, electrospraying, microemulsion, freeze drying, and supercritical fluid extraction to deliver the drug molecules to various drug delivery systems, especially ocular and nasal systems [43]. Biopolymers would turn out to be a better platform for drug delivery systems due to their decreased toxicity and biodegradability [44]. These biopolymers have also been used to develop microcapsules, microspheres, or nanocapsules and have been shown to be stable *in vitro* and *in vivo* conditions. They also showed enhanced

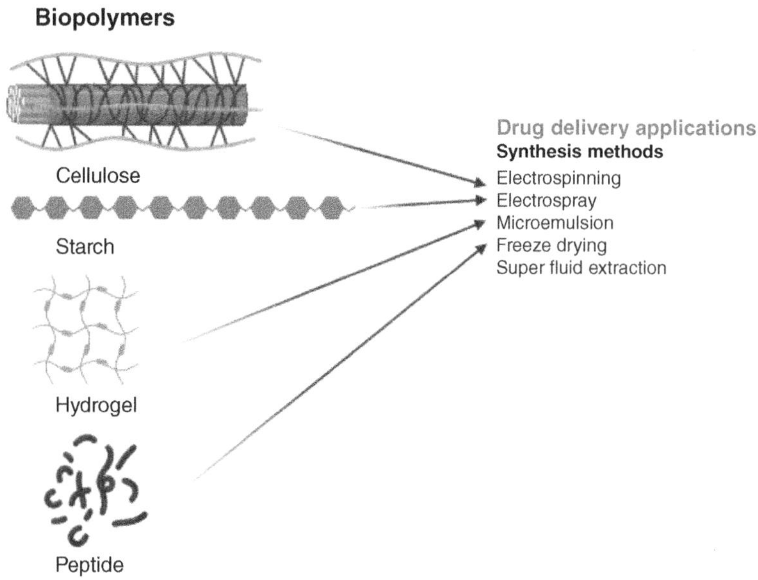

Figure 4.4 Biopolymers in drug delivery applications. *Source:* Created with BioRender.com.

biocompatibility, biodegradability, and simplicity of use when tested *in vivo*, minimized adverse effects, and improved the health condition [45]. The various biopolymers used for drug delivery applications are shown in Figure 4.4.

4.5.2 Medical Implants

For implantable medical devices, various biopolymers have been used such as chitosan and PLA because of their compatibility and degradability. For example, chitosan nanoparticles are used for medical implants, especially in ophthalmology, cardiac valves, and tissue regeneration [46]. Another biopolymer known as collagen is very commonly used for bone scaffolds and cardiovascular implants. There are numerous applications for polyhydroxy alkanoates such as cartilage regeneration and esophageal implants. In otolaryngology, hyaluronic acid implants are used to advance the tissues in vocal folds and cartilages. Biopolymers derived from polysaccharides, proteins, and biopolymers from microbial polymers were developed using preparation methods such as casting, electrospinning, freeze drying, 3D bioprinting, and are keenly used as medical implants [47]. Biopolymers in medical implants are shown in Figure 4.5.

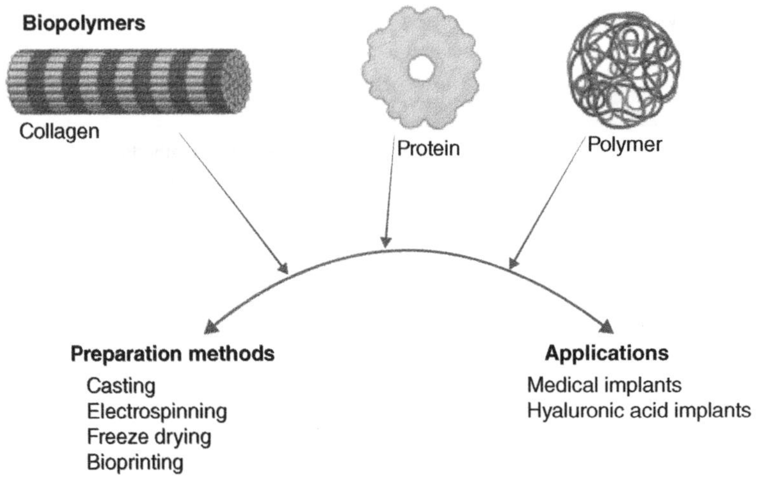

Figure 4.5 Biopolymers in medical implants. *Source:* Created with BioRender.com.

4.5.3 Antimicrobial Activity of Biopolymers

Nthunya et al. and his group synthesized the biopolymer-metal nanoparticles. The nanoparticles obtained were estimated for wastewater disinfectants. Biopolymer-metal nanoparticles were subjected to an antimicrobial test by the microdilution method. These nanoparticles showed a superior biocidal effect against the bacterial strains used [48]. Leudjo Taka et al. and his group used metal nanoparticles and polymer-metal nanoparticle composites subjected to antimicrobial activity using the traditional disc diffusion method. The outcomes are unreliable for evaluating nanoparticle antimicrobial activity. This is because the low diffusivity and poor solubility of the nanomaterials in dimethyl sulphoxide avert the penetration via the pore in the culture media [49].

4.5.4 Wound Healing

This type of injury usually involves cutting, tearing, or puncturing the skin. In the past, honey, animal fat, and grease were used to cure wounds. Cleaning the wound site, removing necrotic tissue, dressing the wound, taking antibiotics to avoid infection, and compression are some of the typical treatments that can be used to treat the wound [50]. Nowadays, dressing and wound healing have undergone significant improvements. Depending on the physical, chemical, and biological characteristics of the substance used in clinical trials, there are several types of wound dressings [51]. With recent advancements, there has been an establishment toward creating the appropriate dressing model for all types of wounds. An acute type of dressings like gauze, bandages, cotton, and tulle are the most commonly used items for this purpose [52]. For treating both acute and chronic wounds, polymeric dressings such as films, foams, hydrogels, scaffolds, alginates, hydrocolloids, and fibers have been paid a lot of attention. Based on recent developments in this field, polymer-based biomaterials have become a good platform for resolving clinical issues in the real world [53].

4.5.5 Tissue Engineering Applications

Biopolymer as biomaterials plays a greater role in cell seeding, proliferation, control of structure, morphology, and chemistry as rational substitutes for mimicking extracellular matrix. These biopolymers were found suitable and more effective on human skin fibroblast cells for tissue regeneration. Collagen, hyaluronic acid, and chitosan are the most popular biopolymers utilized in tissue engineering applications such as cartilage regeneration, vascular engineering, and skin regeneration [54]. A few of the frequently used scaffolds for organs and tissues are made up of PLA, PGA, and PCL. These polymers as biomaterials produce scaffolds that imitate extracellular matrix processes such as the development of organs and tissues [55]. Biopolymer nanocomposites are used to develop tissues and organs, which can be printed in three dimensions. 3D bioprinting can fabricate 3D structures, and interior structures hinge on the given geometries have paid great attention in tissue engineering applications, especially patient-specific tissues, and coupled with nanotubes, nanostructures, and nanofibers [56].

4.5.6 Food Packaging Material

Biopolymers have played a greater role in food industries, especially in food packaging material. For food packaging material applications, there is a variety of plant polysaccharides and protein-derived biopolymers, including cellulose, chitin, and starch [57]. Biopolymers as biomaterials namely PLA, PHA, PCL, PHB, PHAs, PVAs, PVOH, and EVOH were also assessed. Biopolymers such as starch, cellulose, pectin, and soy, allowed to coat in the form of a matrix tend to show more efficacious properties than other polymeric materials [58]. Biopolymers in food packaging applications are shown in Figure 4.6.

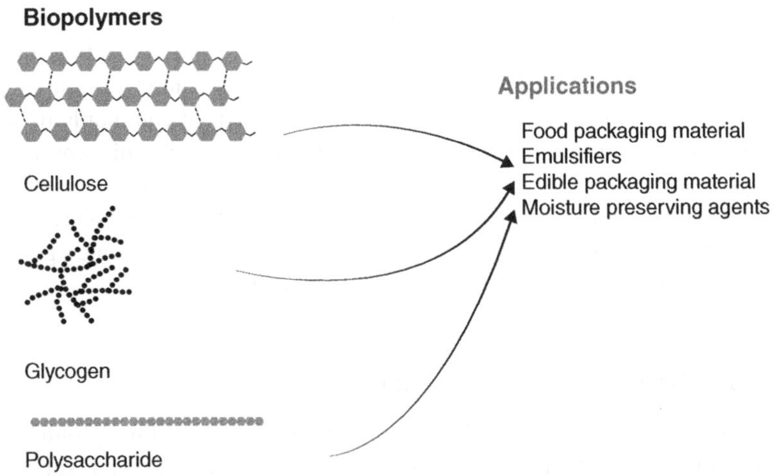

Figure 4.6 Biopolymers in food packaging applications. *Source:* Created with BioRender.com.

4.6 Conclusions

Biopolymeric materials are synthesized and characterized using microscopic, spectroscopic, and diffraction studies. Recent advancements in biopolymeric materials upon sustainability and the environment have emerged in fabricating green materials. Once the biopolymers are characterized, the usefulness of the biopolymer and based composites has been used for various applications such as drug delivery systems, organs, tissue scaffolds, tissue regeneration, cartilage regeneration, wound dressing materials, implants, and the food industry as packaging material. The physical characteristics of biopolymers, such as flexibility, viscosity, density, elevated melting and boiling points, thermal stability, conductivity, degree of crystallinity, mechanical strength, elasticity, and optical properties, must be evaluated which are necessary features to be analyzed for their use in the pharmaceutical industry. Researchers have also taken enormous efforts to produce a biopolymer with minimal cost. The foremost challenge is to enhance the properties of biopolymers comparable to other types of synthetic biopolymers. The modification of biopolymers or a combination of biopolymers with other components could be a promising strategy for eco-friendly and achievable commercialization. In conclusion, biopolymers as biomaterials have played a better role in their various applications perspectives and an utter suppression of synthetic polymers.

Acknowledgments

The authors are grateful to the University of Malaya for funding this work under grant number ST010-2023.

Conflict of Interest

The authors declare no conflict of interest.

References

1 Carlini, A.S., Adamiak, L., and Gianneschi, N.C. (2016). Biosynthetic polymers as functional materials. *Macromolecules* 49 (12): 4379–4394. https://doi.org/10.1021/acs.macromol.6b00439, PMID 27375299.

2 Baranwal, J., Barse, B., Fais, A. et al. (2022). Biopolymer: a sustainable material for food and medical applications. *Polymers* 14 (5): 983. https://doi.org/10.3390/polym14050983, PMID 35267803.

3 Hosseini, S.M.H., Ghiasi, F., and Jahromi, M. (2017). Nanocapsule formation by complexation of biopolymers. In: *Nanoencapsulation Technologies for the Food and Nutraceutical Industries* (ed. S.M. Jafari), 447–492. Elsevier.

4 Freeman, R., Boekhoven, J., Dickerson, M.B. et al. (2015). Biopolymers and supramolecular polymers as biomaterials for biomedical applications. *MRS Bull.* 40 (12): 1089–1101. https://doi.org/10.1557/mrs.2015.270, PMID 26989295.

5 Srivastava, P. and Kalam, S.A. (2019). Natural polymers as potential antiaging constituents. In: *Pharmacognosy – Medicinal Plants* (ed. S. Perveen and A. Al-Taweel). Intech Open. http://dx.doi.org/10.5772/intechopen.80808.

6 Carrêlo, H., Soares, P.I.P., Borges, J.P., and Cidade, M.T. (2021). Injectable composite systems based on microparticles in hydrogels for bioactive cargo-controlled delivery. *Gels* 7 (3): 147. https://doi.org/10.3390/gels7030147, PMID 34563033.

7 Reddy, M.S.B., Ponnamma, D., Choudhary, R., and Sadasivuni, K.K. (2021). A comparative review of natural and synthetic biopolymer composite scaffolds. *Polymers (Basel)* 13 (7): 1105. https://doi.org/10.3390/polym13071105, PMID 33808492.

8 Ling, S., Chen, W., Fan, Y. et al. (2018). Biopolymer nanofibrils: structure, modeling, preparation, and applications. *Prog. Polym. Sci.* 85: 1–56. https://doi.org/10.1016/j.progpolymsci.2018.06.004, PMID 31915410.

9 Siracusa, V. (2019). Microbial degradation of synthetic biopolymers waste. *Polymers (Basel)* 11 (6): 1066. https://doi.org/10.3390/polym11061066, PMID 31226767.

10 Yadav, P., Yadav, H., Shah, V.G. et al. (2015). Biomedical biopolymers, their origin and evolution in biomedical sciences: a systematic review. *J. Clin. Diagn. Res.* 9 (9): ZE21–ZE25. https://doi.org/10.7860/JCDR/2015/13907.6565, PMID 26501034.

11 Gheorghita, R., Anchidin-Norocel, L., Filip, R. et al. (2021). Applications of biopolymers for drugs and probiotics delivery. *Polymers* 13 (16): 2729. https://doi.org/10.3390/polym13162729, PMID 34451268.

12 Baek, S.W., Song, D.H., Lee, H.I. et al. (2021). Poly (l-lactic acid) composite with surface-modified magnesium hydroxide nanoparticles by biodegradable oligomer for augmented mechanical and biological properties. *Materials (Basel)* 14 (19): 5869. https://doi.org/10.3390/ma14195869, PMID 34640265.

13 McAdam, B., Brennan Fournet, M., McDonald, P., and Mojicevic, M. (2020). Production of polyhydroxybutyrate (PHB) and factors impacting its chemical and mechanical characteristics. *Polymers (Basel)* 12 (12): 2908. https://doi.org/10.3390/polym12122908, PMID 33291620.

14 Nikolova, M.P. and Chavali, M.S. (2019). Recent advances in biomaterials for 3D scaffolds: a review. *Bioact. Mater.* 4 (4): 271–292. https://doi.org/10.1016/j.bioactmat.2019.10.005, PMID 31709311.

15 Song, R., Murphy, M., Li, C. et al. (2018). Current development of biodegradable polymeric materials for biomedical applications. *Drug Des. Dev. Ther.* 12: 3117–3145. https://doi.org/10.2147/DDDT.S165440, PMID 30288019.

16 Dassanayake, R.S., Acharya, S., and Abidi, N. (2018). Biopolymer-based materials from polysaccharides: properties, processing, characterization and sorption applications. In: *Advanced Sorption Process Applications* (ed. S. Edebali), 1–24. IntechOpen.

17 Lisitsyn, A., Semenova, A., Nasonova, V. et al. (2021). Approaches in animal proteins and natural polysaccharides application for food packaging: edible film production and quality estimation. *Polymers (Basel)* 13 (10): 1592. https://doi.org/10.3390/polym13101592, PMID 34063360.

18 Eroglu, M.S., Oner, E.T., Mutlu, E.C., and Bostan, M.S. (2017). Sugar-based biopolymers in nanomedicine; new emerging era for cancer imaging and therapy. *Curr. Top. Med. Chem.* 17 (13): 1507–1520. https://doi.org/10.2174/1568026616666161222101703, PMID 28017156.

19 Jiang, T., Duan, Q., Zhu, J. et al. (2020). Starch-based biodegradable materials: challenges and opportunities. *Adv. Ind. Eng. Polym. Res.* 3 (1): 8–18. https://doi.org/10.1016/j.aiepr.2019.11.003.

20 Khalikova, E., Susi, P., and Korpela, T. (2005). Microbial dextran-hydrolyzing enzymes: fundamentals and applications. *Microbiol. Mol. Biol. Rev.* 69 (2): 306–325. https://doi.org/10.1128/MMBR.69.2.306-325.2005, PMID 15944458.

21 Sikora, J.W., Majewski, Ł., and Puszka, A. (2021). Modern biodegradable plastics-processing and properties Part II. *Materials (Basel)* 14 (10): 2523. https://doi.org/10.3390/ma14102523, PMID 34066278.

22 Amulya, K., Katakojwala, R., Ramakrishna, S., and Venkata Mohan, S. (2021). Low carbon biodegradable polymer matrices for sustainable future. *Composites, Part C: Open Access* 4: 100111.

23 Bhaladhare, S. and Das, D. (2022). Cellulose: a fascinating biopolymer for hydrogel synthesis. *J. Mater. Chem. B* 10 (12): 1923–1945. https://doi.org/10.1039/d1tb02848k, PMID 35226030.

24 Zimniewska, M., Wladyka-Przybylak, M., and Mankowski, J. (2011). Cellulosic bast fibers, their structure and properties suitable for composite applications. In: *Cellulose Fibers: Bio- and Nano-Polymer Composites* (ed. S. Kalia, B. Kaith, and I. Kaur), 97–119. Berlin, Heidelberg: Springer.

25 Vásquez-Garay, F., Carrillo-Varela, I., Vidal, C. et al. (2021). A review on the lignin biopolymer and its integration in the elaboration of sustainable materials. *Sustainability* 13 (5): 2697. https://doi.org/10.3390/su13052697.

26 Vroman, I. and Tighzert, L. (2009). Biodegradable polymers. *Materials* 2 (2): 307–344. https://doi.org/10.3390/ma2020307.

27 Senturk Parreidt, T.S., Müller, K., and Schmid, M. (2018). Alginate-based edible films and coatings for food packaging applications. *Foods* 7 (10): 170. https://doi.org/10.3390/foods7100170, PMID 30336642.

28 Lu, Y., Luo, Q., Chu, Y. et al. (2022). Application of gelatin in food packaging: a review. *Polymers* 14 (3): 436. https://doi.org/10.3390/polym14030436, PMID 35160426.

29 Cazón, P. and Vázquez, M. (2019). Applications of chitosan as food packaging materials. In: *Sustainable Agriculture Reviews 36*, vol. 36 (ed.G. Crini and E. Lichtfouse), 81–123. Springer.

30 Shengchang, Z., Campagne, C., and Salaün, F. (2019). Influence of solvent selection in the electrospraying process of polycaprolactone. *Appl. Sci.* 9 (3): 402. https://doi.org/10.3390/app9030402.

31 Sheikh, Z., Najeeb, S., Khurshid, Z. et al. (2015). Biodegradable materials for bone repair and tissue engineering applications. *Materials (Basel)* 8 (9): 5744–5794. https://doi.org/10.3390/ma8095273, PMID 28793533.

32 Jiménez-Rosado, M., Zarate-Ramírez, L.S., Romero, A. et al. (2019). Bioplastics based on wheat gluten processed by extrusion. *J. Cleaner Prod.* 239: 117994. https://doi.org/10.1016/j.jclepro.2019.117994.

33 Fathollahipour, S., Abouei Mehrizi, A., Ghaee, A., and Koosha, M. (2015). Electrospinning of PVA/chitosan nanocomposite nanofibers containing gelatin nanoparticles as a dual drug delivery system. *J. Biomed. Mater. Res. A* 103 (12): 3852–3862. https://doi.org/10.1002/jbm.a.35529, PMID 26112829.

34 Khalid, M.Y., Al Rashid, A., Arif, Z.U. et al. (2021). Recent advances in nanocellulose-based different biomaterials: types, properties, and emerging applications. *J. Mater. Res. Technol.* 14: 2601–2623. https://doi.org/10.1016/j.jmrt.2021.07.128.

35 Pinto, L., Bonifacio, M.A., De Giglio, E. et al. (2021). Biopolymer hybrid materials: development, characterization, and food packaging applications. *Food Packag. Shelf Life* 28: 100676. https://doi.org/10.1016/j.fpsl.2021.100676.

36 Kaur, R., Sharma, A., Puri, V., and Singh, I. (2019). Preparation and characterization of biocomposite films of carrageenan/locust bean gum/montmorrillonite for transdermal delivery of curcumin. *BioImpacts* 9 (1): 37–43. https://doi.org/10.15171/bi.2019.05, PMID 30788258.

37 Alcântara, A.C., Darder, M., Aranda, P., and Ruiz-Hitzky, E. (2016). Effective intercalation of zein into Na-montmorillonite: role of the protein components and use of the developed biointerfaces. *Beilstein J. Nanotechnol.* 7: 1772–1782. https://doi.org/10.3762/bjnano.7.170, PMID 28144527.

38 Pires, J.R.A., Souza, V.G.L., Fuciños, P. et al. (2022). Methodologies to assess the biodegradability of bio-based polymers—current knowledge and existing gaps. *Polymers* 14 (7): 1359. https://doi.org/10.3390/polym14071359, PMID35406232.

39 Thakur, V.K., Thakur, M.K., and Gupta Raju, K. (2014). Graft copolymers of natural fibers for green composites. *Carbohydr. Polym.* 104: 87–93. https://doi.org/10.1016/j.carbpol.2014.01.016, PMID 24607164.

40 Fukushima, K., Tabuani, D., Abbate, C. et al. (2011). Preparation, characterization and biodegradation of biopolymer nanocomposites based on fumed silica. *Eur. Polym. J.* 47 (2): 139–152. https://doi.org/10.1016/j.eurpolymj.2010.10.027.

41 Swain, S.K., Pattanayak, A.J., and Sahoo, A.P. (2018). Functional biopolymer composites. In: *Functional Biopolymers*, Springer Series on Polymer and Composite Materials (ed. V. Thakur and M. Thakur), 159–182. Cham: Springer doi:10.1007/978-3-319-66417-0_6.

42 Aslam Khan, M.U., Abd Razak, S.I., Al Arjan, W.S. et al. (2021). Recent advances in biopolymeric composite materials for tissue engineering and regenerative medicines: a review. *Molecules* 26 (3): 619. https://doi.org/10.3390/molecules26030619, PMID 33504080.

43 Mikušová, V. and Mikuš, P. (2021). Advances in chitosan-based nanoparticles for drug delivery. *Int. J. Mol. Sci.* 22 (17): 9652. https://doi.org/10.3390/ijms22179652, PMID 34502560.

44 Mansoor, S., Kondiah, P.P.D., Choonara, Y.E., and Pillay, V. (2019). Polymer-based nanoparticle strategies for insulin delivery. *Polymers* 11 (9): 1380. https://doi.org/10.3390/polym11091380, PMID 31443473.

45 Vikulina, A.S. and Campbell, J. (2021). Biopolymer-based multilayer capsules and beads made via templating: advantages, hurdles and perspectives. *Nanomaterials (Basel)* 11 (10): 2502. https://doi.org/10.3390/nano11102502, PMID 34684943.

46 Notario-Pérez, F., Martín-Illana, A., Cazorla-Luna, R. et al. (2022). Applications of chitosan in surgical and post-surgical materials. *Mar. Drugs* 20 (6): 396. https://doi.org/10.3390/md20060396, PMID 35736199.

47 Ambekar, R.S. and Kandasubramanian, B. (2019). Progress in the advancement of porous biopolymer scaffold: tissue engineering application. *Ind. Eng. Chem. Res.* 58 (16): 6163–6194. https://doi.org/10.1021/acs.iecr.8b05334.

48 Petrovova, E., Tomco, M., Holovska, K. et al. (2021). PHB/CHIT scaffold as a promising biopolymer in the treatment of osteochondral defects – an experimental animal study. *Polymers (Basel)* 13 (8): 1232.

49 Nthunya, L.N., Masheane, M.L., Malinga, S.P. et al. (2017). A greener approach to prepare electrospun antibacterial β-cyclodextrin/cellulose acetate nanofibers for removal of bacteria from water. *ACS Sustainable Chem. Eng.* 5 (1): 153–160. https://doi.org/10.1021/acssuschemeng.6b01089.

50 Leudjo Taka, A., Doyle, B.P., Carleschi, E. et al. (2020). Spectroscopic characterization and antimicrobial activity of nanoparticle doped cyclodextrin polyurethane bionanosponge. *Mater. Sci. Eng. C Mater. Biol. Appl.* 115: 111092. https://doi.org/10.1016/j.msec.2020.111092, PMID 32600696.

51 Sood, A., Granick, M.S., and Tomaselli, N.L. (2014). Wound dressings and comparative effectiveness data. *Adv. Wound Care (New Rochelle)* 3 (8): 511–529. https://doi.org/10.1089/wound.2012.0401, PMID 25126472.

52 Dhivya, S., Padma, V.V., and Santhini, E. (2015). Wound dressings – a review. *Biomedicine (Taipei)* 5 (4): 22. https://doi.org/10.7603/s40681-015-0022-9, PMID 26615539.

53 Mir, M., Ali, M.N., Barakullah, A. et al. (2018). Synthetic polymeric biomaterials for wound healing: a review. *Prog. Biomater.* 7 (1): 1–21. https://doi.org/10.1007/s40204-018-0083-4, PMID 29446015.

54 Chen, F.M. and Liu, X. (2016). Advancing biomaterials of human origin for tissue engineering. *Prog. Polym. Sci.* 53: 86–168. https://doi.org/10.1016/j.progpolymsci.2015.02.004, PMID 27022202.

55 Dhandayuthapani, B., Yoshida, Y., Maekawa, T., and Sakthi Kumar, D.S. (2011). Polymeric scaffolds in tissue engineering application: a review. *Int. J. Polym. Sci.* 2011: 290602. https://doi.org/10.1155/2011/290602.

56 Marew, T. and Birhanu, G. (2021). Three-dimensional printed nanostructure biomaterials for bone tissue engineering. *Regen. Ther.* 18: 102–111. https://doi.org/10.1016/j.reth.2021.05.001, PMID 34141834.

57 Shaikh, S., Yaqoob, M., and Aggarwal, P. (2021). An overview of biodegradable packaging in food industry. *Curr. Res. Food Sci.* 4: 503–520.

58 Iñiguez-Moreno, M., Ragazzo-Sánchez, J.A., and Calderón-Santoyo, M. (2021). An extensive review of natural polymers used as coatings for postharvest shelf-life extension: trends and challenges. *Polymers (Basel)* 13 (19): 3271. https://doi.org/10.3390/polym13193271, PMID 34641086.

5

Photoinduced Synthesis of Nanoparticles

Nowsheenah Farooq, Abu Taha, and Athar Adil Hashmi

Bioinorganic Lab, Department of Chemistry, Jamia Millia Islamia, New Delhi, DL, India

Abstract

The unforeseen modification in surface properties of materials to develop an interesting and emerging class of nanosized material has attained much attention from the scientific community, due to their unique and promising role in various fields such as medicinal chemistry, environmental chemistry, ayurvedic sciences, energy, nutrition fields, and biomedical and electronic fields. Among the various developed methods for the synthesis of nanomaterials, the photoinduced method is considered a versatile and convenient process with distinctive advantages. Here, in this book chapter, we have started by giving a brief historical background of nanomaterials and the importance of photochemical synthesis of nanomaterials in the introductory part. In the middle portion, we have elaborated on various methods for synthesizing nanomaterials, including direct photoreduction, photosensitization, photocatalytic fabrication of metal nanomaterials, and so on. In the last portion, we have penned down the application of nanoparticles in various fields and conclusion of the book chapter.

Keywords *photoreduction; photosensitization; photocatalytic fabrication; nanomaterials; photoinduced synthesis*

Green Synthesis of Nanomaterials: Biological and Environmental Applications, First Edition.
Edited by Archana Chakravarty, Preeti Singh, Saiqa Ikram, and R.N. Yadava.

5.1 Introduction

Nanomaterials are substances of the size of 100 nm. This is the least size in one direction; these materials are strongly fervent on thermo-dynamics and kinetic parameters for formation. The process of formation and size of nanomaterials are done by taking controlled measures in the synthesis. Nanomaterials are unique with respect to their shape, size, and chemical composition stability.

- Capping agents
- Precursor concentration
- Rate of dose
- Solvent properties
- Intensity and concentration (photochemical synthesis)
- Temperature

Some basic types of nanomaterials are

- Nano silver
- Carbon nanotubes
- 3D printer
- Quantum dots
- Nano pesticides

In the process of photochemical synthesis, the size of the nanomaterial is reduced, resulting in a different and varied shape of the nanomaterial. This process of varied styles of nanomaterial is known as confinement; this process generally takes place in semiconductors. Confinement in a single direction forms thin films, while confinement in two directions forms quantum rods, and confinement in all directions forms quantum dots.

5.1.1 Role of Nanomaterials

- Nanomaterials have played a very important role in the everyday life of human beings in particular and living beings in general.
- Nanomaterials are used in designing pharmaceuticals that can target specific organs or cells in the body. This in turn enhances the impact of treatment.
- Nanomaterials are very useful when added to cement; it makes cement very strong and lighter.
- Nanomaterial can be added to cloth for strength.
- The size of nanomaterials makes them very useful in electronic devices.
- Nanomaterials are used for environmental remediation; they help in the cleaning-up of toxins.

- Nanomaterials have been very effective in cleaning up drinking water at a very low cost.
- In most cases, nanomaterials are eco-friendly.
- The problem of mosquitoes was solved by blocking them using a nanomaterial called graphene films.
- Nanomaterials have become hugely useful in day-to-day human life, but the effects of nanomaterials on the ecosystem are still not known.
- The researchers are investigating the interactions of nanomaterials with biological processes, which is very important to be known for the betterment of biodiversity.
- It is very different as of now to understand the impact of nanomaterials or the disadvantages of using nanomaterials since their chemical properties are very different and rare, their reactivity is high, and they do not dissolve in water or any other liquid medium.

Nanosized particles are also known to be versatile in nature like lotus and lily plant leaves staying afloat on the water surface. The top of the leaf doesn't get damp, and even when a water droplet falls on the leaf surface it retains its shape and doesn't fan out. The techniques like scanning tunneling microscope (STM) and transmission electron microscope (TEM) bring to light that the upper leaf area of the lotus is made of arrays of nano-shaped structures of leaf pigments [1, 2].

The analysts strived to prepare various nanoparticles having different morphology and sizes by various modes of synthesis. Among all these methods, the photochemical and radiation chemical methods of preparing nanomaterials are regarded as environment friendly, as these methods do not require harmful chemicals, elevated temperature, and a nonreactive atmosphere which if used create havoc to the environment.

Radiation-mediated preparation of metal nanoparticles such as gold and silver has been explored by several research groups [3–7]. Radiation-induced preparation of other common metal nanoparticles such as nickel and copper has also been reported in literatures [8–12]. Also, in semiconducting nanomaterials like CdS, CdSe, and ZnS, preparation by radiation-induced methods has been extensively studied by several researchers throughout the globe [13–21]. Metal oxide nanoparticle preparation by radiation-induced method has also been reported by several groups [22–27]. Photochemical synthesis of gold and silver, as well as other transition metal nanoparticles, has been reported by different research groups [28–33]. Similarly, photochemical preparation of semiconducting nanomaterials has also been studied by several groups [34, 35]. The photochemical preparation of metal oxide nanomaterials has been investigated by several research groups [36, 37]. Ichimura and coworkers have shown photochemical settlement of semiconductors and metal oxide thin films over a suitable base dipped in the reaction medium by photo-irradiation with ultraviolet radiation [38, 39]. These two methods are gaining great attention from researchers across the globe for nanomaterial synthesis and their use in different fields.

5.2 Methods of Synthesis

There are several methods for the synthesis of nanomaterials. These methods can be broadly divided into two classes.

- Physical methods
- Chemical methods

5.2.1 Physical Synthesis of Nanomaterials

It is a top-down approach for nanomaterial synthesis in which large substances are broken to the atomic level and afterward put together for making nanomaterials as films, quantum dots, and quantum wells. The process is hectic, energy-consuming, and costly. Contradictory to this, chemical methods support the bottom-up approach, where nanomaterials are prepared by assembling atoms and molecules that are present in the solution.

5.2.2 Chemical Synthesis of Nanomaterials

- Chemical Vapor Depositions (CVD)
- Chemical Bath Deposition (CBD)
- Hydrothermal Synthesis
- Sono Chemical Synthesis
- Organometallic Routes
- Photochemical Methods
- Catalytic Methods
- Radiation Chemical Routes

Photochemical and radiation chemical methods are considered good for the environment. The photochemical and radiation chemical reactions resulting in the formation of nanomaterials use different kinds of radiations such as ultraviolet, visible, X-ray, and gamma ray. Ultraviolet and visible lights having energy in the range of 1–4 eV are used for photochemical reactions while X-ray and gamma rays are used for radiation chemical methods. These rays are a part of the electromagnetic spectrum. Photochemical reactions use high-energy photons like gamma and X-rays. Radiation chemical reactions may also be done by high-energy electron beam and photon beam (KeV to MeV).

Ionizing photon beams are produced by accelerators like LINAC (linear collider), generally used to produce high-energy rays, when radiation of both high and low energy meet precursor solution, many events take place here and finally result in the synthesis of nanomaterials in the range of quantum dots, nanoribbons, core sheet nanoparticles, nanoflowers, and nanorods in colloidal solution.

In both methods, there is a presence of a transient intermediate substance that reacts with the precursors to form nanomaterials that are desired. In photochemical processes, the light photon to get excited to the higher energy state is absorbed by a specific molecule. This light photon on getting absorbed by the specific molecule gets excited to a higher state like a singlet or triplet where photochemical reactions take place, resulting in free radicals and stable nanomaterials. In the process of chemical radiation, high-energy radiation transfers energy into the solvent, and then super excited ions and radicals of solvent molecules are formed immediately which then convert into free radicals [40]. Water as a solvent forms free radicals like $OH^{\cdot}$ (hydroxyl radicals), H (hydrogen atom), and $e_{(aq)}^{-}$ (hydrated electrons) including the molecular species like H_2, H_3O^+, and H_2O_2 in microseconds. Water radiolysis is shown in Figure 5.1.

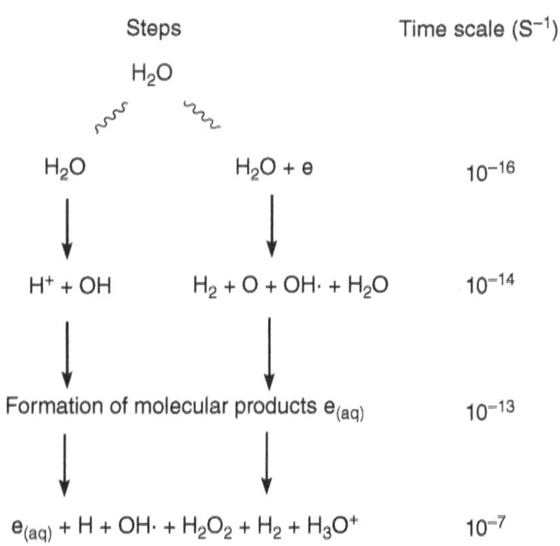

Figure 5.1 Radiolysis of water.

In this instance of the relation, it can be affirmed that in both the reactions (i.e. photochemical reactions and radiation chemical reactions) free radicals are formed, which are then used for the effective synthesis of assorted styles of nanoparticles.

5.3 Photochemical Synthesis of Nanomaterials

In this method, ultraviolet rays are absorbed by a certain species; these species generate some free radicals having reducing properties. In the case of ultraviolet absorption, acetone is generally used with propane-2-ol in which acetone takes light and gets photo-excited. During the photo-excitation, acetone is of oxidizing nature, hence it borrows one hydrogen atom from propan-2-ol molecule nearby, which results in the formation of 2-hydroxy-2-propyl radicals $(CH_3)2C^*OH$ in solution as shown in Figure 5.2.

In photo radiation with sunlight, acetophenone in place of acetone (as a scintillator) is used and gets photo-excited where it abstracts a proton from propan-2-ol and makes two types of radicals which are reducing in nature. In uranyl, the uranyl ion gets photo-excited by ultraviolet or visible rays, and it is also in oxidizing nature when in a photo-excited state where it abstracts a hydrogen atom from any nearby molecule, even water, to form free radicals.

In the photochemical preparation of CdS, CuS, and ZnS, the precursor for sulfur being put into use is $Na_2S_2O_3$, and during the reaction $S_2O_3^{2-}$ ions take the ultraviolet light with the release of sulfur atom and electrons.

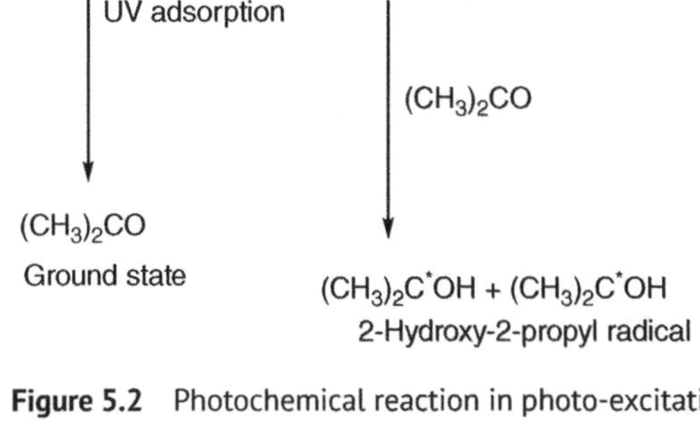

Figure 5.2 Photochemical reaction in photo-excitation of acetone.

Additionally, they combine with metal ions forming metal sulfides in the form of thin films or nanoparticles [38, 39]. Due to its spatial and temporal control techniques, the photochemical process has got a huge attention for the production of metal nanoparticles [28, 41]. In this process, the solution containing metal precursors is exposed to ultraviolet rays or visible rays. Hence the photochemical reactions or routes are very efficient and less hazardous in nature, because:

- Harmful or toxic compounds are not used in it.
- Not dependable upon expensive instruments
- Not dependable upon highly skilled personnel
- Can be done out at ambient conditions (room temperature, air pressure) [42–44]

Formation of metal nanoparticles in solution uses many reagents like metal precursor, salt or complex, reducing agents, and stabilizing agents [45]. It begins with the reductions of metal from positive oxidation (M^{n+}) to zero (M^{O}) oxidation state. This is done using the photo-catalyzed reducer [41, 46]. When the metallic precursor is reduced, the M^{O} nuclei then grow or aggregate to nanoparticle size [47]. Stabilizers and capping agents are very important as they control the preparation of homogeneous nanoparticles of required morphology and size [48, 49]. This prevents their agglomeration and improves their colloidal stability [50–53]. For capping, polymers are very useful in trapping and saving the nanoparticles from coalescence and oxidation [54, 55]. Various chemical properties of polymers enable them for specific interactions with metal surface, which triggers a significant change in the shape and the size of the metallic NPs obtained [54]. Many studies revealed the effects of assorted polymers on the properties of silver and gold nanoparticles.

1) Poly(vinyl pyrrolidone) (PVP) [56–58].
2) Poly methacrylate (PMA) [59].
3) Polysaccharide, monosaccharides, and proteins (chitosan, glucose, dextrose, gelatin) [28, 57, 60, 61].
4) Sodium dodecyl sulfate (SDS) [62].
5) Natural rubber latex [54].

Nanoparticles morphology and features are determined by the parameters of the experiment using

- The stabilizer and its nature
- Precursor concentration
- Exposure time
- Light and wavelength [63, 64].

5.3.1 Synthesis of Gold Nanoparticles Using Ultraviolet Light

The ultraviolet radiation whose wavelength is in the range of 100–400 nm be generally categorized into three major groups: Ultraviolet-A (315–400 nm), Ultraviolet-B (290–315 nm), Ultraviolet-C (100–290 nm) [56]. Since wavelength and energy are inversely proportional, Ultraviolet-C radiation generally has the best energy which can get us irreversible cell damage. Hence photosensitizer is used to enhance the light absorption by the reaction medium. The photosensitizer in the appropriate solvents that are directed to it generates radicals which are then involved in the Mn^{+} process of reduction and formation of subsequent metallic NPs. The most frequently used photosensitizer is a ketone family with metallic NP synthesis [41].

Ketones form ketyl radicals under ultraviolet radiation which then behave as a reducing agent. For example, benzoin is a ketone that forms ketyl radicals in photolysis. Within the benzoin Norrish type I occurs, generating triplet ketyl which is short-lived; it concomitantly triggers the cationic metal precursor reduction due to radical depletion. This helps in efficiently converting metal ions to metal nanoparticles [41, 65]. The sources of ultraviolet radiation are mercury lamps and ultraviolet light-emitting diodes (LED) that are placed in photo reactors [66]. The size and morphology of synthesized metal nanoparticles are influenced by experimental parameters like pH, illumination time, and reagent's concentration. Gabriel et al. performed an experiment on silver which was a spherical nanoparticle on chitosan/clay nanocomposite. They used varying amounts of chitosan, due to a stabilizing agent [67], and observed that nanoparticle having uniform size (2.7–6.3 nm) is formed at a low clay content, and nanoparticles of the larger size distribution (2.20 nm) are formed at higher rate.

The ultraviolet photochemical preparation of gold (Au) NPs is done by:

- Using sunshine-absorbing solutes.
- Without a reducing agent, the reduction of metal ions can take place.
- A continuing and predictable reduction rate is given typically by well-defined photon-mediated generation of reducing equivalents.
- No sophisticated instruments are required, and it is a low-cost process [55].

5.3.1.1 Influence of pH

NPs of silver are isotropic (nanosphere) and anisotropic (nanorods, nanowires, nanoshells, etc.). The nanosphere of silver is characterized by the SPR band. The maxima of the SPR band are found at approx. 520 nm, and the nanorods usually exhibit SPR two absorption bands. The mode of transversal is approx. 520 nm, the position of it depends on the ratio of length to width [31, 68, 69].

The morphology of gold nanoparticle is mainly regulated by adjusting different parameters of reaction conditions and exposure time of radiation. To check the influence of pH on gold nanoparticles formation, Rodríguez et al. have done an experiment using ultraviolet lamps (256 nm) at different pH and lots of reagents using tetrachloro auric acid (precursor) $HAuCl_4$, (Nitrates) $AgNO_3$, hexadecyltrimethylammonium bromide (cationic surfactant CTAB), ascorbic acid and hydrochloric acid, ammonium hydroxide, cyclohexane, and acetone [70], and it was found that gold nanospheres dominate in pH of 9, and the ultraviolet-visible spectrum shows one SPR band (515 nm) while nanorods of gold display two SPR bands in acidic solutions (pH 3 or 5). Cheng et al. [71] made observations on gold and found that nanorods are obtained at low pH values, while the nanoparticles become spherical because the pH does increase. Unal et al. followed the same methodology and screened nanoparticles of gold and its effects. They used red cabbage extract renewable as both the stabilizer and chemical agent. A pH of 2.5 yields aggregated nanoparticles of 5–70 nm in size, while basic pH produced uniform nanoparticles of 18–30 nm in size [72]. This can be mainly due to the occurrence of nucleation at alkaline pH yielding isotropic growth of small nanomaterials. Acidic pH causes the nuclei to grow anisotropically [70]. Hence nanoparticle's size and morphology can be managed by controlling the pH of the reaction medium.

5.3.1.2 Influence of Precursor Concentration

The most widely used precursor for gold is tetrachloroauric acid ($HAuCl_4$) [28]. Gold nanoparticles are affected by the precursor amount. Anisotropic nanoparticles formation is favored by the presence of silver nitrate as silver ions are associated with the presence of a longitudinal SPR band. The effect of the concentration of tetrachloroauric acid and silver nitrate on the morphology of silver nanoparticles was studied by Sanabria Cala and coworkers at 256 nm wavelength [73]. The decrease in the expansion of micelles of cationic surfactant CTAB upon increasing HAuCl concentration was noticed, and spherical nanoparticles were formed. The formation of anisotropic gold nanorods by an increase in the amount of silver nitrate was observed. Reducing and stabilizing agents are required by many photochemical methods [74, 75]. Shiraishi et al. observed the effect of acid amount on the dimension of gold nanoparticles which are formed under irradiation at 254 nm [74]. As depicted by the SPR intensity at 530 nm approx., with an increase in the amount of citric acid, there is a sharp increase in the yield of gold nanoparticles indicating a quick nucleation process. In consequence, small NPs of gold are formed. Ultraviolet rays can also be used without any reductant, and this was studied by Texeira and teammates using 254 nm wavelength and polyethyleneimine as a stabilizing agent [76]. Stabilizing agents play a very important role in the nucleation process of nanomaterial formation. The surfactants are also known to serve as stabilizing agents and play a key role in gold nanoparticle formation [77, 78]. Surfactants can be cationic like dodecyl trimethyl ammonium bromide, alkyl triethylammonium bromide, and anionic (e.g. sodium dodecylbenzene sulfonated) also, non-ionic analogs (tween x-80, hydrogenate purgative (HCO), triton x1100, etc.) [79]. The nature and structure of surfactants alter the size of nanoparticles. Shang and coworker synthesized gold nanoparticles using $HAuCl_4$ and some other surfactants [80]. Results indicated that nanoparticle size can be controlled by various surfactants. In positively charged surfactants, the positive side attracts $AuCl4^-$ ion, therefore leading to the formation of large gold nanoparticles, while in anionic surfactants, the negatively charged side repels $AuCl4^-$ ion and results in the formation of small nanoparticles.

5.3.2 Synthesis of Silver Nanoparticles Using Ultraviolet Light

Photochemical synthesis of nanoparticles using an ultraviolet light source is an authentic method. This method is known to be nature friendly as the solvent used is water and the capping agents used are not toxic [80]. These demonstrations are done at normal temperature and normal pressure where silver nitrate is used as a precursor [81–83], and the stabilizers used are green materials [84].

5.3.2.1 Influence of pH

To control the dimensions of Ag NPs, the pH of the reaction medium plays a very important role [85, 86]. Studies have shown that the dimensions of silver nanoparticles decrease in basic media due to the fast nucleation process, and large silver nanoparticles are formed in an acidic media because the reaction is slower [87, 88]. Absorption of light is affected by pH, hence the localized SPR (LSPR) band position [89]. Babusca et al. prepared silver nanoparticles by a two-step process: in the first step aqueous mixture of silver nitrate was heated with trisodium citrate. In the second step, it is irradiated by a UV lamp for 45 minutes [86]. In the very beginning, silver nanoparticles are reduced, pH increases from 5 to 6, and SPR band maxima shift from 425 to 418 nm. These observations indicate small-sized silver nanoparticles formation [89, 90].

5.3.2.2 Influence of Reducing Agents

Rheima and the team irradiated a solution of glucose ($C_6H_{12}O_6$), which is used in reducing as well as stabilizing the silver nitrate under a mercury ultraviolet lamp (365 nm, 125 W) for the time period of 30 minutes to form crystalline hexagonal and spherical silver nanoparticles, having a mean size of 20 nm [91]. Valandra et al. reported utilizing other materials, for example 10-oxo-10-H-di benzene thio-pyran-3-4-dicarboximide chitosan and triethanolamine which again is stabilizing as well as a reducing agent [28]. When Ultraviolet-LED of a wavelength of about 365 nm radiated for four long hours on silver nitrate solution which contains thio-pyran-3-4-dicarboximide chitosan, spherical, self-assembled silver nanoparticles of 2–24 nm size were found. On another side, a 15-minute irradiation on a solution made of thio-3-4-dicarboximide chitosan and triethanolamine is enough to get stable spherical silver nanoparticles having a diameter of 2.5 nm. Plant extracts and polymers are universally used reducing agents to prepare silver nanoparticles under UV irradiation in addition to the green reducing agents. The presence of PVP (Polyvinylpyrrolidone) is known as a decent protective agent and inhibitor of nanoparticle aggregation.

Rado'n and Lukoweic irradiated silver nitrate by using UV light of wavelength 365 nm for about 10 minutes [28]. The nanoparticles formed were cubic, rod, and spherical in shape when PVP was used as a reducing agent, with a mean size of 50.3 nm. While using T chloramine, unstable and irregular Ag NPs are formed and have a mean size of 11.7 ± 7.2 nm. With increasing amount of T chloramine, large and irregular silver nanoparticles are formed. The studies proved that reducing agent is not always necessary for nanoparticle synthesis. Huang and Yang used silver nitrate as a precursor and aqueous laponite as a stabilizer [92]. Darroudi et al. utilized gelatin as a stabilizing agent with silver nitrate under ultraviolet radiation for 48 hours [28].

5.3.3 Synthesis of Gold Nanoparticles Under Visible Light

Visible light can be used in the synthesis of nanoparticles being pollution-free and nontoxic. Visible light while used in the chemical reaction does not trace in chemical reactions [93].

Like ultraviolet-assisted photochemical preparation of nanoparticles, silver and gold nanoparticles can be synthesized using visible light [94]. Annadhasan et al. synthesized silver and gold nanoparticles by visible light irradiation using N-cholyl-I-valine as a stabilizer as well as a reducing agent at basic pH [95]. Formation of nanoparticles was achieved in 20 minutes, and the SPR band was obtained at 524 nm. Pienpinijtham et al. used starch (reducing and stabilizing agent) and nanoparticles of various shapes (triangular, hexagonal) [96].

5.3.4 Synthesis of Silver Nanoparticles Under Visible Light

Tang and his coworkers described the preparation of silver nanomaterial using visible light [97]. The aqueous solution of silver nitrate was radiated with visible light for approximately a time of one hour at room temperature. SDS was used as a surfactant [94, 98, 99].

5.4 Photochemical Synthesis of UO_2 Nanoparticles in Aqueous Solutions

The preparation of UO_2 nanoparticles by ultraviolet rays has been extensively studied. Nenoff et al. and Roth et al. studied the radiolytic synthesis of UO_2 nanoparticles [24]. The steps taking place in preparation of UO_2 nanoparticles are shown in Figure 5.3 [100–103]. Rath and his teammates explored a single-step preparation of UO_2 nanoparticles using an aqueous solution of uranyl nitrate under ultraviolet radiation [36]. Irradiation by ultraviolet light on (CH_3C^*OH) radicals in acetone in the presence of propan-2-ol is an excellent reducing agent. The reaction proceeds as shown in Figure 5.4.

$$(CH_3)_2CO \xrightarrow{h\nu} {}^*(CH_3)_2CO$$

$$(CH_3)_2CO + (CH_3)_2CHOH \longrightarrow 2(CH_3)_2C^{\cdot}OH$$

$$2(CH_3)_2C^*OH + UO_2^{2+} \longrightarrow UO_2^{2+} + (CH_3)_2CO + H^+$$

$$UO_2^+ + UO_2^+ \longrightarrow UO_2^+ + UO_2^{2+}$$

$$(CH_3)_2C^{\cdot}OH + UO_2^+ \longrightarrow UO_2 + (CH_3)_2CO + H^+$$

$$UO_2^{2+} \xrightarrow{h\nu} {}^*UO_2^{2+}$$

$${}^*UO_2^{2+} + (CH_3)_2CHOH \longrightarrow UO_2^+ + (CH_3)_2C^{\cdot}OH + H^+$$

$$UO_2^+ + (CH_3)_2C^{\cdot}OH \longrightarrow UO_2 + (CH_3)_2CO + H^+$$

$$UO_2 + nUO_2 \longrightarrow \underset{\text{Nanoparticles}}{(UO_2)}$$

Figure 5.3 Formation of UO_2 nanoparticles.

$$(CH_3)_2CHOH + O_2 \longrightarrow (CH_3)_2CO + HO_2^{\bullet}$$

$$HO_2^{\bullet} + HO_2^{\bullet} \longrightarrow H_2O_2 + O_2$$

$$(CH_3)_2C^{\bullet}OH + H_2O_2 \longrightarrow (CH_3)_2CO + OH^{\bullet} + H_2O$$

$$OH^{\bullet} + (CH_3)_2CHOH \longrightarrow (CH_3)_2C^{\bullet}OH + H_2O$$

Figure 5.4 Reaction in presence of oxygen.

During the formation of UO_2 nanoparticles, a decrease in the induction time on increase in propan-2-ol amount was found. Also, upon the long exposure of uranyl nitrate without propan-2-ol or acetone, no UO_2 nanoparticles were formed. This suggests the key role of 2-propanol. It was found that the synthesis of UO_2 nanoparticles occurs in both aerated and non-aerated solutions. Thus, in the synthesis of UO_2 oxygen does not have any negative role as shown in Figure 5.4.

5.5 Photochemical Synthesis of ZnO Nanoparticles

Zinc oxide is an inorganic material having unique features, that is, band gap of 3.3 eV, wide radiation absorption range, high catalytic property, and semiconducting property [104]. In addition, the smaller size of ZnO nanoparticles (ZnO NPs) makes them very reactive, and for the same reason, it has got much attention in recent years. ZnO NPs have a wide range of applications including biomedical, electronics, optics, cosmetics industry, food industry, and agriculture [105, 106]. Muhammad Zia and the team synthesized ZnO NPs using different monochromatic light (blue, green, yellow, and red) and daylight. Synthesized ZnO NPs were analyzed using different analytic tools to determine their crystal structure, morphology, and size. They evaluated the biological and antioxidant properties of synthesized nanoparticles such as antimicrobial activity, enzyme inhibition, and free-radical scavenging [107]. In other work, P. Sadhukhan et al. used microwave light for ZnO NPs formation and evaluated its biomedical application for drug carriers, anticancer drugs, and antimicrobial activity. ZnO NPs was loaded with hydrophobic drug quercetin which exhibits pH-dependent drug delivery. ZnO NPs show anticancer activity on human breast cancer cell lines, while antibacterial activity was tested on gram (−) bacteria *Escherichia coli* [108]. Various applications of photoinduced nanoparticles are given in Table 5.1.

Table 5.1 Applications of photoinduced nanoparticles.

S No.	Nanoparticles	Method of synthesis	Applications	Characterizations	References
1	ZnS NPs	Gamma ray	Optical property	UV, TEM	[14]
2	Ag NPs	UV irradiation	Photoluminescence	UV, TEM	[49]
3	Ag NPs	UV irradiation	Bacteria inhibitor	UV, TEM, SEM, FTIR, XRD	[42]
4	Au NPs	IR radiation	Photothermal agent for hyperthermia	UV, TEM, XRD	[50]
5	Ag NPs	Solar radiation	Antibacterial activity	UV, IR, TEM, XRD	[46]
6	Ag NPs	Mercury lamp	Antibacterial activity	UV, TEM, XRD, XPS	[62]
7	Au NPs	UV irradiation	Catalytic activity	UV, SEM, DLS	[72]
8	Au NPs	Sunlight irradiation	Colorimetric detection of Ni^{2+} and Co^{2+}	TEM, DLS, EDX, XPS, FT-IR, TGA	[95]
9	Ag NPs	Sunlight irradiation	Antioxidant and antibacterial films in food packaging	UV, TEM, SEM, XRD	[98]
10	ZnO NPs	Different light regimes (daylight, blue, yellow, and red LEDs)	Antioxidant, antibacterial, nanozyme, anti-amylase, antiurease, and antilipase	SEM, EDX, XRD, FTIR	[107]

5.6 Conclusion

In this chapter, we discussed fabrication of nanoparticles using photo radiation, characterization, and their applications. Through characterization techniques like UV-visible, TEM, SEM, and DLS, the formation of these nanoparticles was revealed. The shape and size of nanoparticles make them fit for various applications. The photoinduced synthesis of nanoparticles has successfully produced stable nanoparticles with a significant role, due to diverse applications in the healthcare sector and environment. The colorimetric sensing of heavy metals by gold nanoparticles offers a rapid and pocket-effective procedure for water sample analysis. In the formation of silver nanoparticles by sunlight irradiation, the citrate solution affected the nanoparticles' shapes, and the amount produced is affected by silver precursor concentration, number of accelerators, and the reaction time. Silver nanoparticles are well known for being antibacterial, antimicrobial, and so on. ZnO is a multifunctional material, and the ZnO nanoparticles have profound anticancer activities. Further studies of free-radical photochemical-based synthesis of nanoparticles are being studied widely nowadays.

Abbreviations

LED	light-emitting diode
PMA	poly methacrylate
PVP	polyvinyl pyrrolidone
SDS	sodium dodecyl sulfate

References

1 Guo, Z., Zhou, F., Hao, J., and Liu, W. (2005). Stable biomimetic super-hydrophobic engineering materials. *J. Am. Chem. Soc.* 127 (45): 15670–15671. https://pubs.acs.org/doi/10.1021/ja0547836.

2 Ensikat, H.J., Ditsche-Kuru, P., Neinhuis, C., and Barthlott, W. (2011). Superhydrophobicity in perfection: the outstanding properties of the lotus leaf. *Beilstein J. Nanotechnol.* 2: 152–161. https://www.beilstein-journals.org/bjnano/articles/2/19.

3 Abedini, A., Daud, A.R., Abdul Hamid, M.A. et al. (2013). A review on radiation-induced nucleation and growth of colloidal metallic nanoparticles. *Nanoscale Res. Lett.* 8 (1): 474. https://nanoscalereslett.springeropen.com/articles/10.1186/1556-276X-8-474.

4 Hayes, D., Micic, O.I., Nenadovic, M.T. et al. (1989). Radiolytic production and properties of ultrasmall cadmium sulfide particles. *J. Phys. Chem.* 93 (11): 4603–4608. https://pubs.acs.org/doi/10.1021/j100348a040.

5 Swayambunathan, V., Hayes, D., Schmidt, K.H. et al. (1990). Thiol surface complexation on growing cadmium sulfide clusters. *J. Am. Chem. Soc.* 112 (10): 3831–3837. https://pubs.acs.org/doi/abs/10.1021/ja00166a017.

6 Marignier, J.L., Belloni, J., Delcourt, M.O., and Chevalier, J.P. (1985). Microaggregates of non-noble metals and bimetallic alloys prepared by radiation-induced reduction. *Nature* 317 (6035): 344–345. http://www.nature.com/articles/317344a0.

7 Belloni, J., Marignier, J.-L., and Mostafavi, M. (2020). Mechanisms of metal nanoparticles nucleation and growth studied by radiolysis. *Radiat. Phys. Chem.* 169: 107952. https://linkinghub.elsevier.com/retrieve/pii/S0969806X18306509.

8 Joshi, S.S., Patil, S.F., Iyer, V., and Mahumuni, S. (1998). Radiation induced synthesis and characterization of copper nanoparticles. *Nanostructured Mater.* 10 (7): 1135–1144. https://linkinghub.elsevier.com/retrieve/pii/S0965977398001536.

9 Belloni, J. (2006). Nucleation, growth and properties of nanoclusters studied by radiation chemistry. *Catal. Today* 113 (3–4): 141–156. https://linkinghub.elsevier.com/retrieve/pii/S0920586105008801.

10 Joshi, R. and Mukherjee, T. (2007). The reduction process of phytic acid–silver ion system: a pulse radiolysis study. *Radiat. Phys. Chem.* 76 (5): 811–817. https://linkinghub.elsevier.com/retrieve/pii/S0969806X06003306.

11 Kapoor, S., Joshi, R., and Mukherjee, T. (2003). Synthesis and stabilization of cadmium and thallium metal nanoparticles in a polymer matrix. *J. Colloid Interface Sci.* 267 (1): 74–77. https://linkinghub.elsevier.com/retrieve/pii/S0021979703007598.

12 Kapoor, S., Joshi, R., Mukherjee, T., and Mittal, J.P. (2001). Pulse radiolytic studies of metallic clusters in a quaternary microemulsion system. *Res. Chem. Intermed.* 27 (7–8): 747–754. http://link.springer.com/10.1163/15685670152622022.

13 Mostafavi, M., Liu, Y., Pernot, P., and Belloni, J. (2000). Dose rate effect on size of CdS clusters induced by irradiation. *Radiat. Phys. Chem.* 59 (1): 49–59. https://linkinghub.elsevier.com/retrieve/pii/S0969806X99005216.

14 Souici, A.H., Keghouche, N., Delaire, J.A. et al. (2006). Radiolytic synthesis and optical properties of ultra-small stabilized ZnS nanoparticles. *Chem. Phys. Lett.* 422 (1–3): 25–29. https://linkinghub.elsevier.com/retrieve/pii/S0009261406001928.

15 Atinault, E., De Waele, V., Belloni, J. et al. (2010). Radiolytic yield of U^{IV} oxidation into U^{VI}: a new mechanism for U^{V} reactivity in acidic solution. *J. Phys. Chem. A* 114 (5): 2080–2085. https://pubs.acs.org/doi/10.1021/jp9051177.

16 Guleria, A., Singh, S., Rath, M.C. et al. (2012). Tuning of photoluminescence in cadmium selenide nanoparticles grown in CTAB based quaternary water-in-oil microemulsions. *J. Lumin.* 132 (3): 652–658. https://linkinghub.elsevier.com/retrieve/pii/S0022231311005606.

17 Singh, S., Rath, M.C., Singh, A.K. et al. (2011). CdSe nanoparticles grown via radiolytic methods in aqueous solutions. *Radiat. Phys. Chem.* 80 (6): 736–741. https://linkinghub.elsevier.com/retrieve/pii/S0969806X11000399.

18 Singh, S., Rath, M.C., Singh, A.K. et al. (2010). Reversible formation of CdSe nanoparticles in aqueous solutions by high-energy electron beam irradiation. *Mater. Chem. Phys.* 124 (1): 6–9. https://linkinghub.elsevier.com/retrieve/pii/S0254058410004517.

19 Singh, S., Guleria, A., Rath, M.C. et al. (2013). Sea urchin like shaped CdSe nanoparticles grown in aqueous solutions via electron beam irradiation. *J. Nanosci. Nanotechnol.* 13 (8): 5365–5373. http://openurl.ingenta.com/content/xref?genre=article&issn=1533-4880&volume=13&issue=8&spage=5365.

20 Singh, S., Guleria, A., Singh, A.K. et al. (2013). Radiolytic synthesis and spectroscopic investigations of cadmium selenide quantum dots grown in cationic surfactant based quaternary water-in-oil microemulsions. *J. Colloid Interface Sci.* 398: 112–119. https://linkinghub.elsevier.com/retrieve/pii/S0021979713001355.

21 Singh, S., Rath, M.C., and Sarkar, S.K. (2011). Investigation of dynamics of radiolytic formation of CdSe nanoparticles in aqueous solutions. *J. Phys. Chem. A* 115 (46): 13251–13258. https://pubs.acs.org/doi/10.1021/jp206028x.

22 Rath, M.C., Sunitha, Y., Ghosh, H.N. et al. (2009). Investigation of the dynamics of radiolytic formation of ZnO nanostructured materials by pulse radiolysis. *Radiat. Phys. Chem.* 78 (2): 77–80. https://linkinghub.elsevier.com/retrieve/pii/S0969806X08002478.

23 Roth, O., Hasselberg, H., and Jonsson, M. (2009). Radiation chemical synthesis and characterization of UO_2 nanoparticles. *J. Nucl. Mater.* 383 (3): 231–236. https://linkinghub.elsevier.com/retrieve/pii/S0022311508005357.

24 Nenoff, T.M., Jacobs, B.W., Robinson, D.B. et al. (2011). Synthesis and low temperature in situ sintering of uranium oxide nanoparticles. *Chem. Mater.* 23 (23): 5185–5190. https://pubs.acs.org/doi/10.1021/cm2020669.

25 Rath, M.C., Naik, D.B., and Sarkar, S.K. (2013). Reversible growth of UO_2 nanoparticles in aqueous solutions through 7MeV electron beam irradiation. *J. Nucl. Mater.* 438 (1–3): 26–31. https://linkinghub.elsevier.com/retrieve/pii/S002231151300411X.

26 Rath, M.C. and Naik, D.B. (2014). Post-irradiation induction time in the radiolytic synthesis of UO_2 nanoparticles in aqueous solutions. *J. Nucl. Mater.* 454 (1–3): 54–59. https://linkinghub.elsevier.com/retrieve/pii/S0022311514003869.

27 Rath, M.C., Keny, S.J., and Naik, D.B. (2016). Electron beam induced synthesis of uranium dioxide nanoparticles: effect of solvent composition. *Radiat. Phys. Chem.* 126: 85–89. https://linkinghub.elsevier.com/retrieve/pii/S0969806X16301542.

28 Scaiano, J.C., Billone, P., Gonzalez, C.M. et al. (2009). Photochemical routes to silver and gold nanoparticles. *Pure Appl. Chem.* 81 (4): 635–647. https://www.degruyter.com/document/doi/10.1351/PAC-CON-08-09-11/html.

29 Marin, M.L., McGilvray, K.L., and Scaiano, J.C. (2008). Photochemical strategies for the synthesis of gold nanoparticles from Au(III) and Au(I) using photoinduced free radical generation. *J. Am. Chem. Soc.* 130 (49): 16572–16584. https://pubs.acs.org/doi/10.1021/ja803490n.

30 Gu, H., Yang, Y., Tian, J., and Shi, G. (2013). Photochemical synthesis of noble metal (Ag, Pd, Au, Pt) on graphene/ZnO multihybrid nanoarchitectures as electrocatalysis for H_2O_2 reduction. *ACS Appl. Mater. Interfaces* 5 (14): 6762–6768. https://pubs.acs.org/doi/10.1021/am401738k.

31 Dong, S.-A. and Zhou, S.-P. (2007). Photochemical synthesis of colloidal gold nanoparticles. *Mater. Sci. Eng. B* 140 (3): 153–159. https://linkinghub.elsevier.com/retrieve/pii/S0921510707001420.

32 Kapoor, S. (1998). Preparation, characterization, and surface modification of silver particles. *Langmuir* 14 (5): 1021–1025. https://pubs.acs.org/doi/10.1021/la9705827.

33 Kapoor, S., Palit, D.K., and Mukherjee, T. (2002). Preparation, characterization and surface modification of Cu metal nanoparticles. *Chem. Phys. Lett.* 355 (3–4): 383–387. https://linkinghub.elsevier.com/retrieve/pii/S0009261402002932.

34 Kumar, V.N., Suriakarthick, R., Shyju, T.S., and Gopalakrishnan, R. (2013). Deposition of CdS thin films by chemical bath and photochemical deposition methods and its characterization. *AIP Conf. Proc.* 347–348. http://aip.scitation.org/doi/abs/10.1063/1.4810243.

35 Singh, A., Kunwar, A., and Rath, M.C. (2018). L-cysteine capped CdSe quantum dots synthesized by photochemical route. *J. Nanosci. Nanotechnol.* 18 (5): 3419–3426. http://www.ingentaconnect.com/content/10.1166/jnn.2018.14687.

36 Rath, M.C., Keny, S., and Naik, D.B. (2016). Direct synthesis of UO_2 nanoparticles in aqueous solutions through photochemical method. *J. Nanosci. Nanotechnol.* 16 (9): 9575–9582. http://openurl.ingenta.com/content/xref?genre=article&issn=1533-4880&volume=16&issue=9&spage=9575.

37 Su, C.-Y., Lan, W.-J., Chu, C.-Y. et al. (2016). Photochemical green synthesis of nanostructured cobalt oxides as hydrogen peroxide redox for bifunctional sensing application. *Electrochim. Acta* 190: 588–595. https://linkinghub.elsevier.com/retrieve/pii/S0013468615310173.

38 Ichimura, M., Kobayashi, R., and Miyawaki, T. (2004). Photochemical deposition of ZnS thin films by intermittent illumination. *Jpn. J. Appl.*

Phys. 43 (9A): L1196. https://iopscience.iop.org/article/10.1143/JJAP.43.L1196.

39 Podder, J., Kobayashi, R., and Ichimura, M. (2005). Photochemical deposition of CuxS thin films from aqueous solutions. *Thin Solid Films* 472 (1–2): 71–75. https://linkinghub.elsevier.com/retrieve/pii/S0040609004008934.

40 Dispenza, C., Grimaldi, N., Sabatino, M.A. et al. (2015). Radiation-engineered functional nanoparticles in aqueous systems. *J. Nanosci. Nanotechnol.* 15 (5): 3445–3467. http://openurl.ingenta.com/content/xref?genre=article&issn=1533-4880&volume=15&issue=5&spage=3445.

41 Sakamoto, M., Fujistuka, M., and Majima, T. (2009). Light as a construction tool of metal nanoparticles: synthesis and mechanism. *J. Photochem. Photobiol. C: Photochem. Rev.* 10 (1): 33–56. https://linkinghub.elsevier.com/retrieve/pii/S1389556708000865.

42 Elsupikhe, R.F., Ahmad, M.B., Shameli, K. et al. (2016). Photochemical reduction as a green method for the synthesis and size control of silver nanoparticles in κ-carrageenan. *IEEE Trans. Nanotechnol.* 15 (2): 209–213. http://ieeexplore.ieee.org/document/7372480/.

43 Reddy, G.B., Madhusudhan, A., Ramakrishna, D. et al. (2015). Green chemistry approach for the synthesis of gold nanoparticles with gum kondagogu: characterization, catalytic and antibacterial activity. *J. Nanostructure. Chem.* 5 (2): 185–193. http://link.springer.com/10.1007/s40097-015-0149-y.

44 dos Santos, M.A., Paterno, L.G., Moreira, S.G.C., and Sales, M.J.A. (2019). Original photochemical synthesis of Ag nanoparticles mediated by potato starch. *SN Appl. Sci.* 1 (6): 554. http://link.springer.com/10.1007/s42452-019-0586-1.

45 El-Sheikh, M.A. (2014). A novel photosynthesis of carboxymethyl starch-stabilized silver nanoparticles. *Sci. World J.* 2014: 1–11. http://www.hindawi.com/journals/tswj/2014/514563/.

46 Mathew, S., Prakash, A., and Radhakrishnan, E.K. (2018). Sunlight mediated rapid synthesis of small size range silver nanoparticles using *Zingiber officinale* rhizome extract and its antibacterial activity analysis. *Inorg. Nano-Met. Chem.* 48 (2): 139–145. https://www.tandfonline.com/doi/full/10.1080/24701556.2017.1373295.

47 Mallick, K., Wang, Z.L., and Pal, T. (2001). Seed-mediated successive growth of gold particles accomplished by UV irradiation: a photochemical approach for size-controlled synthesis. *J. Photochem. Photobiol. A: Chem.* 140 (1): 75–80. https://linkinghub.elsevier.com/retrieve/pii/S1010603001003896.

48 Darroudi, M., Ahmad, M.B., Zak, A.K. et al. (2011). Fabrication and characterization of gelatin stabilized silver nanoparticles under UV-light.

Int. J. Mol. Sci. 12 (9): 6346–6356. http://www.mdpi.com/1422-0067/12/9/6346.

49 Kempa, T., Farrer, R.A., Giersig, M., and Fourkas, J.T. (2006). Photochemical synthesis and multiphoton luminescence of monodisperse silver nanocrystals. *Plasmonics* 1 (1): 45–51. http://link.springer.com/10.1007/s11468-006-9008-5.

50 Huang, W.-C. and Chen, Y.-C. (2008). Photochemical synthesis of polygonal gold nanoparticles. *J. Nanoparticle Res.* 10 (4): 697–702. http://link.springer.com/10.1007/s11051-007-9293-8.

51 Sengani, M., Grumezescu, A.M., and Rajeswari, V.D. (2017). Recent trends and methodologies in gold nanoparticle synthesis – a prospective review on drug delivery aspect. *OpenNano* 2: 37–46. https://linkinghub.elsevier.com/retrieve/pii/S2352952017300038.

52 Slepička, P., Slepičková Kasálková, N., Siegel, J. et al. (2019). Methods of gold and silver nanoparticles preparation. *Materials (Basel)* 13 (1): 1. https://www.mdpi.com/1996-1944/13/1/1.

53 Ribeiro, K.C., Fernandes, L., Bechlin, M.A. et al. (2023). Effect of the excitation wavelength on the photochemical synthesis of citrate-capped Au nanoparticles. *Braz. J. Phys.* 53 (2): 41. https://link.springer.com/10.1007/s13538-022-01251-z.

54 Abu Bakar, N.H.H., Ismail, J., and Abu Bakar, M. (2007). Synthesis and characterization of silver nanoparticles in natural rubber. *Mater. Chem. Phys.* 104 (2–3): 276–283. https://linkinghub.elsevier.com/retrieve/pii/S0254058407001757.

55 Courrol, L.C., de Oliveira Silva, F.R., and Gomes, L. (2007). A simple method to synthesize silver nanoparticles by photo-reduction. *Colloids Surf. A Physicochem. Eng. Asp.* 305 (1–3): 54–57. https://linkinghub.elsevier.com/retrieve/pii/S0927775707003585.

56 Lin, S.K. and Cheng, W.T. (2020). Fabrication and characterization of colloidal silver nanoparticle via photochemical synthesis. *Mater. Lett.* 261: 127077. https://linkinghub.elsevier.com/retrieve/pii/S0167577X19317094.

57 Łukowiec, D. and Radoń, A. (2020). Self-organization of silver nanoparticles during synthesis of Ag–Au nanoalloy by UV irradiation method. *J. Mater. Sci.* 55 (7): 2796–2801. http://link.springer.com/10.1007/s10853-019-04136-w.

58 Wang, L., Wei, G., Guo, C. et al. (2008). Photochemical synthesis and self-assembly of gold nanoparticles. *Colloids Surf. A Physicochem. Eng. Asp.* 312 (2–3): 148–153. https://linkinghub.elsevier.com/retrieve/pii/S0927775707006656.

59 Dubas, S.T. and Pimpan, V. (2008). Green synthesis of silver nanoparticles for ammonia sensing. *Talanta* 76 (1): 29–33. https://linkinghub.elsevier.com/retrieve/pii/S0039914008000982.

60 Gabriel, J.S., Gonzaga, V.A.M., Poli, A.L., and Schmitt, C.C. (2017). Photochemical synthesis of silver nanoparticles on chitosans/ montmorillonite nanocomposite films and antibacterial activity. *Carbohydr. Polym.* 171: 202–210. https://linkinghub.elsevier.com/retrieve/pii/S014486171730526X.

61 Valandro, S., Poli, A., Neumann, M., and Schmitt, C. (2019). Photochemical synthesis of Ag and Au nanoparticles using a thioxanthone substituted chitosan as simultaneous photoinitiator and stabilizer. *J. Braz. Chem. Soc.* 30: 2642–2648. http://jbcs.sbq.org.br/audiencia_pdf.asp?aid2=5687&nomeArquivo=2019-0065AR_BrazWoman.pdf.

62 Kora, A.J., Manjusha, R., and Arunachalam, J. (2009). Superior bactericidal activity of SDS capped silver nanoparticles: synthesis and characterization. *Mater. Sci. Eng. C* 29 (7): 2104–2109. https://linkinghub.elsevier.com/retrieve/pii/S0928493109001283.

63 Zhang Z, Yu J, Zhang J, Lian Y, Shi Z, Cheng Z, et al. pH-controlled growth of triangular silver nanoprisms on a large scale. *Nanoscale Adv.* 2019; 1(12): 4904–8. http://xlink.rsc.org/?DOI=C9NA00635D

64 Lee, K.-H., Huang, K.-M., Tseng, W.-L. et al. (2007). Manipulation of the growth of gold and silver nanomaterials on glass by seeding approach. *Langmuir* 23 (3): 1435–1442. https://pubs.acs.org/doi/10.1021/la061880j.

65 Eftekhari-Kenzerki, Z., Fardid, R., and Behzad-Behbahani, A. (2019). Impact of silver nanoparticles on the ultraviolet radiation direct and bystander effects on TK6 cell line. *J. Med. Phys.* 44 (2): 118–125. http://www.ncbi.nlm.nih.gov/pubmed/31359930.

66 Bárta, J., Procházková, L., Vaněček, V. et al. (2019). Photochemical synthesis of nano- and micro-crystalline particles in aqueous solutions. *Appl. Surf. Sci.* 479: 506–511. https://linkinghub.elsevier.com/retrieve/pii/S0169433219304295.

67 Huang, L., Zhai, M.L., Long, D.W. et al. (2008). UV-induced synthesis, characterization and formation mechanism of silver nanoparticles in alkalic carboxymethylated chitosan solution. *J. Nanoparticle Res.* 10 (7): 1193–1202. http://link.springer.com/10.1007/s11051-007-9353-0.

68 Abdelrasoul, G.N., Cingolani, R., Diaspro, A. et al. (2014). Photochemical synthesis: effect of UV irradiation on gold nanorods morphology. *J. Photochem. Photobiol. A: Chem.* 275: 7–11. https://linkinghub.elsevier.com/retrieve/pii/S1010603013004590.

69 Kundu, S., Panigrahi, S., Praharaj, S. et al. (2007). Anisotropic growth of gold clusters to gold nanocubes under UV irradiation. *Nanotechnology* 18 (7): 075712. https://iopscience.iop.org/article/10.1088/0957-4484/18/7/075712.

70 Rodríguez, G.R.C., Gauthier, G.H., Ladeira, L.O. et al. (2017). Effect of pH and chloroauric acid concentration on the geometry of gold nanoparticles obtained by photochemical synthesis. *J. Phys. Conf. Ser.* 935: 012027. https://iopscience.iop.org/article/10.1088/1742-6596/935/1/012027.

71 Cheng, J., Ge, L., Xiong, B., and He, Y. (2011). Investigation of pH effect on gold nanorod synthesis. *J. Chin. Chem. Soc.* 58 (6): 822–827. https://onlinelibrary.wiley.com/doi/10.1002/jccs.201190128.

72 Unal, I.S., Demirbas, A., Onal, I. et al. (2020). One step preparation of stable gold nanoparticle using red cabbage extracts under UV light and its catalytic activity. *J. Photochem. Photobiol. B, Biol.* 204: 111800. https://linkinghub.elsevier.com/retrieve/pii/S1011134419310036.

73 Jara, N., Milán, N.S., Rahman, A. et al. (2021). Photochemical synthesis of gold and silver nanoparticles – a review. *Molecules* 26 (15): 4585. https://www.mdpi.com/1420-3049/26/15/4585.

74 Shiraishi, Y., Tanaka, H., Sakamoto, H. et al. (2017). Photoreductive synthesis of monodispersed Au nanoparticles with citric acid as reductant and surface stabilizing reagent. *RSC Adv.* 7 (11): 6187–6192. http://xlink.rsc.org/?DOI=C6RA27771C.

75 Almatroudi, A. (2020). Silver nanoparticles: synthesis, characterisation and biomedical applications. *Open Life Sci.* 15 (1): 819–839. https://www.degruyter.com/document/doi/10.1515/biol-2020-0094/html.

76 Teixeira, P.R., Santos, M.S.C., Silva, A.L.G. et al. (2016). Photochemically-assisted synthesis of non-toxic and biocompatible gold nanoparticles. *Colloids Surf. B: Biointerfaces* 148: 317–323. https://linkinghub.elsevier.com/retrieve/pii/S0927776516306440.

77 Huang, Y. and Kim, D.-H. (2012). Light-controlled synthesis of gold nanoparticles using a rigid, photoresponsive surfactant. *Nanoscale* 4 (20): 6312. http://xlink.rsc.org/?DOI=c2nr31717f.

78 Pal, A. (2004). Photochemical synthesis of gold nanoparticles via controlled nucleation using a bioactive molecule. *Mater. Lett.* 58 (3–4): 529–534. https://linkinghub.elsevier.com/retrieve/pii/S0167577X03005408.

79 Liao, L., Lv, G., Cai, D., and Wu, L. (2016). The sequential intercalation of three types of surfactants into sodium montmorillonite. *Appl. Clay Sci.* 119: 82–86. https://linkinghub.elsevier.com/retrieve/pii/S0169131715300697.

80 Shang, Y., Min, C., Hu, J. et al. (2013). Synthesis of gold nanoparticles by reduction of HAuCl4 under UV irradiation. *Solid State Sci.* 15: 17–23. https://linkinghub.elsevier.com/retrieve/pii/S1293255812003044.

81 Mittelman, A.M., Fortner, J.D., and Pennell, K.D. (2015). Effects of ultraviolet light on silver nanoparticle mobility and dissolution. *Environ. Sci. Nano* 2 (6): 683–691. http://xlink.rsc.org/?DOI=C5EN00145E.

82 Spadaro, D., Barletta, E., Barreca, F. et al. (2010). Synthesis of PMA stabilized silver nanoparticles by chemical reduction process under a two-step UV irradiation. *Appl. Surf. Sci.* 256 (12): 3812–3816. https://linkinghub.elsevier.com/retrieve/pii/S0169433210000589.

83 Berti, L., Alessandrini, A., and Facci, P. (2005). DNA-templated photoinduced silver deposition. *J. Am. Chem. Soc.* 127 (32): 11216–11217. https://pubs.acs.org/doi/10.1021/ja052461w.

84 Filip, G.A., Moldovan, B., Baldea, I. et al. (2019). UV-light mediated green synthesis of silver and gold nanoparticles using Cornelian cherry fruit extract and their comparative effects in experimental inflammation. *J. Photochem. Photobiol. B, Biol.* 191: 26–37. https://linkinghub.elsevier.com/retrieve/pii/S1011134418311436.

85 Marciniak, L., Nowak, M., Trojanowska, A. et al. (2020). The effect of pH on the size of silver nanoparticles obtained in the reduction reaction with citric and malic acids. *Materials (Basel)* 13 (23): 5444. https://www.mdpi.com/1996-1944/13/23/5444.

86 Babusca, D., Popescu, L., Sacarescu, L. et al. (2020). Two phase photochemical synthesis of silver nanoparticles and their impact on the chlorophylls. *Mol. Cryst. Liq. Cryst.* 698 (1): 56–64. https://www.tandfonline.com/doi/full/10.1080/15421406.2020.1731087.

87 Chutrakulwong, F., Thamaphat, K., and Limsuwan, P. (2020). Photo-irradiation induced green synthesis of highly stable silver nanoparticles using durian rind biomass: effects of light intensity, exposure time and pH on silver nanoparticles formation. *J. Phys. Commun.* 4 (9): 095015. https://iopscience.iop.org/article/10.1088/2399-6528/abb4b5.

88 Zhou, W., Liu, Y.-L., Stallworth, A.M. et al. (2016). Effects of pH, electrolyte, humic acid, and light exposure on the long-term fate of silver nanoparticles. *Environ. Sci. Technol.* 50 (22): 12214–12224. https://pubs.acs.org/doi/10.1021/acs.est.6b03237.

89 Oprica, L., Andries, M., Sacarescu, L. et al. (2020). Citrate-silver nanoparticles and their impact on some environmental beneficial fungi. *Saudi J. Biol. Sci.* 27 (12): 3365–3375. https://linkinghub.elsevier.com/retrieve/pii/S1319562X20304009.

90 Alqadi, M.K., Abo Noqtah, O.A., Alzoubi, F.Y. et al. (2014). pH effect on the aggregation of silver nanoparticles synthesized by chemical reduction. *Mater. Sci.* 32 (1): 107–111. http://link.springer.com/10.2478/s13536-013-0166-9.

91 Ali, A.A., Al-Hassani, R.M., Hussain, D.H. et al. (2019). Fabrication of solar cells using novel micro- and nano-complexes of triazole schiff base derivatives. *J. Southwest Jiaotong Univ.* 54 (6): 1–19. http://jsju.org/index.php/journal/article/view/399.

92 Huang, H. and Yang, Y. (2008). Preparation of silver nanoparticles in inorganic clay suspensions. *Compos. Sci. Technol.* 68 (14): 2948–2953. https://linkinghub.elsevier.com/retrieve/pii/S0266353807003946.

93 Luo, Y. (2007). Size-controlled preparation of polyelectrolyte-protected gold nanoparticles by natural sunlight radiation. *Mater. Lett.* 61 (11–12): 2164–2166. https://linkinghub.elsevier.com/retrieve/pii/S0167577X06010548.

94 Bhaduri, G.A., Little, R., Khomane, R.B. et al. (2013). Green synthesis of silver nanoparticles using sunlight. *J. Photochem. Photobiol. A: Chem.* 258: 1–9. https://linkinghub.elsevier.com/retrieve/pii/S1010603013000579.

95 Annadhasan, M., Kasthuri, J., and Rajendiran, N. (2015). Green synthesis of gold nanoparticles under sunlight irradiation and their colorimetric detection of Ni^{2+} and Co^{2+} ions. *RSC Adv.* 5 (15): 11458–11468. http://xlink.rsc.org/?DOI=C4RA14034F.

96 Pienpinijtham, P., Han, X.X., Suzuki, T. et al. (2012). Micrometer-sized gold nanoplates: starch-mediated photochemical reduction synthesis and possibility of application to tip-enhanced Raman scattering (TERS). *Phys. Chem. Chem. Phys.* 14 (27): 9636. http://xlink.rsc.org/?DOI=c2cp40330g.

97 Tang, B., Sun, L., Li, J. et al. (2015). Sunlight-driven synthesis of anisotropic silver nanoparticles. *Chem. Eng. J.* 260: 99–106. https://linkinghub.elsevier.com/retrieve/pii/S1385894714010985.

98 Nguyen, V.T. (2020). Sunlight-driven synthesis of silver nanoparticles using pomelo peel extract and antibacterial testing. *J. Chem.* 2020: 1–9. https://www.hindawi.com/journals/jchem/2020/6407081/.

99 Susilowati, E. and Maryani, A. (2018). Sunlight-assisted synthesis of colloidal silver nanoparticles using chitosan as reducing agent. *IOP Conf. Ser.: Mater. Sci. Eng.* 349: 012019. https://iopscience.iop.org/article/10.1088/1757-899X/349/1/012019.

100 Ballhausen, C.J. (1965). Spectroscopy and photochemistry of uranyl compounds. *Nucl. Phys.* 65 (3): 527–528. https://linkinghub.elsevier.com/retrieve/pii/0029558265900295.

101 Tsushima, S. (2009). Photochemical reduction of UO_2^{2+} in the presence of alcohol studied by density functional theory calculations. *Inorg. Chem.* 48 (11): 4856–4862. https://pubs.acs.org/doi/10.1021/ic900066z.

102 Park, Y.-Y., Harada, M., Tomiyasu, H., and Ikeda, Y. (1991). Photochemical reactions of uranium(VI) in aqueous phosphoric acid solutions. *J. Nucl. Sci. Technol.* 28 (5): 418–425. http://www.tandfonline.com/doi/abs/10.1080/18811248.1991.9731377.

103 Salomone, V.N., Meichtry, J.M., Schinelli, G. et al. (2014). Photochemical reduction of U(VI) in aqueous solution in the presence of 2-propanol. *J. Photochem. Photobiol. A: Chem.* 277: 19–26. https://linkinghub.elsevier.com/retrieve/pii/S1010603013005030.

104 Wu, C., Qiao, X., Chen, J. et al. (2006). A novel chemical route to prepare ZnO nanoparticles. *Mater. Lett.* 60 (15): 1828–1832. https://linkinghub.elsevier.com/retrieve/pii/S0167577X05012747.

105 Kołodziejczak-Radzimska, A. and Jesionowski, T. (2014). Zinc oxide – from synthesis to application: a review. *Materials (Basel)* 7 (4): 2833–2881. http://www.mdpi.com/1996-1944/7/4/2833.

106 Newman, M.D., Stotland, M., and Ellis, J.I. (2009). The safety of nanosized particles in titanium dioxide– and zinc oxide–based sunscreens. *J. Am. Acad. Dermatol.* 61 (4): 685–692. https://linkinghub.elsevier.com/retrieve/pii/S0190962209005398.

107 Sajjad, A., Bhatti, S.H., Ali, Z. et al. (2021). Photoinduced fabrication of zinc oxide nanoparticles: transformation of morphological and biological response on light irradiance. *ACS Omega* 6 (17): 11783–11793. https://pubs.acs.org/doi/10.1021/acsomega.1c01512.

108 Sadhukhan, P., Kundu, M., Rana, S. et al. (2019). Microwave induced synthesis of ZnO nanorods and their efficacy as a drug carrier with profound anticancer and antibacterial properties. *Toxicol. Rep.* 6: 176–185. https://linkinghub.elsevier.com/retrieve/pii/S2214750018307844.

6

Green Nanomaterials in Textile Industry

Indu Kumari[1], Sarabjeet Kaur[2], and Ratnesh Das[3]

[1] *CT Group of Institutions, Jalandhar, India*
[2] *Surface Chemistry and Catalysis: Characterisation and Application Team (COK-KAT), Leuven (Arenberg), Leuven, Belgium*
[3] *Department of Chemistry, Dr. Harisingh Gour Central University, Sagar, MP, India*

Abstract

The textile industry is growing worldwide at a fast pace and revealing its major influence on the environment. However, conventional methods such as chemical and processing methods used in the textile industry cause environmental pollution and toxicity. Therefore, to overcome these drawbacks, green chemistry plays a major role in protecting human health, textiles, and environmental risks in an economically better way. For this, green nanomaterials play a key role in technological evolution owing to their surface-related properties, especially high tensile strength, soft hand, durability, water repellency, antibacterial applications, flame-retardant properties, ultraviolet-visible protection, etc. Therefore, the current chapter intends to review the synthesis of nanomaterials using a green synthetic approach and their role in the textile industry. The various applications of green nanomaterials in the textile industry will be discussed. This chapter aims to provide new insights into developing green nanomaterials for managing, preventing, and controlling human health and environmental risks.

Keywords *textiles; green nanomaterials; green chemistry; green synthesis; sustainability*

Green Synthesis of Nanomaterials: Biological and Environmental Applications, First Edition.
Edited by Archana Chakravarty, Preeti Singh, Saiqa Ikram, and R.N. Yadava.

6.1 Introduction

In recent years, "nano" has gained popularity all around the globe. In 1959, Richard Feynman initially offered the thought of nanotechnology during his speech "There's Plenty of Room at the Bottom," conveyed at an American Physical Society conference at the California Institute of Technology [1]. The ideas Feynman introduced were unnoticed until 1974, when Norio Taniguchi coined the "nanotechnology" term [2]. Nanomaterials exhibit various inherent characteristics. Due to the huge surface-to-volume ratio of nanomaterials compared to bulk materials, nanomaterials occupy extensive applications in various areas, including electronics, biomedical applications, textiles, information technology, and wastewater treatment. Particles ranging from 1–100 nm have been generally regarded as nanomaterials, and these materials' augmented surface area has been chiefly utilized. These nanomaterials can be produced through many methods that can control and produce nanostructures exhibiting favored size, morphology, and crystalline structure. The two typical fabrication methods used are "top-down" and "bottom-up" [3]. Generally, the "bottom-to-top approach" is used to synthesize the nanoparticles through the accumulation of atoms or ions, and continuous size augmentation of the ensuing nuclei in the nano range. The widely used approach is chemical methods used for the preparation of range of nanoparticles and their related applications. However, the "top-to-bottom approach" is also utilized to generate nanomaterials via mechanical procedures such as the high-energy ball-milling method. Nowadays, nanotechnology contributes to developing improved materials with advanced characteristics for utilization in various application fields. The atoms present in the nanomaterials are absolutely ordered. Thus, while the dimensions of materials vary from macro to nano-size, wide-ranging variations take place in the properties of the materials [4].

Nanomaterials utilized in textile functionalization are mainly produced through the chemical methods of the "bottom-to-top" method. The textile industry affects the financial system and the nearby environment of any region to a huge level. With the world's population increase and developed clothing needs, the textile industry is well-hassled for fabricating high-quality textiles. Nanoparticles or nanomaterials are reported broadly in the literature for their exploitation in antimicrobial properties and UV-protective clothing in the field of textile industry [5]. At present, environmental security has been measured by clients concurrently with the ease and excellence of their outfits. Nanotechnology is a moderately novel technology, and a few obstacles have been noticed; thus, it has a wide future for advanced research regarding human and environmental protection measures. Owing to the reality that the nanomaterials may not be identifiable after their release into the environment, these materials could lead to a range of conservation troubles only if the remediation system is insecure. The research related to emission and environmental measures of nanomaterials of the nanotextiles is tremendously significant. Investigations are critical for technically relating the structure–function relation of nanomaterials in terms of the basic chemistry in terms of working, illustration, and toxicity. Complete threat solutions must be done on nanomaterials, exhibiting a genuine experience risk throughout its production or practice. Therefore, green nanomaterials are one of the best solutions to reduce the plausible environmental hazards and individual medical problems through the fabrication and usage of nanomaterials and to expand the replacement of widespread things with advanced ecofriendly nanomaterials [6].

Considering this, the current chapter discusses the various applications of green nanomaterials in the field of textiles. This kind of research can be helpful for the suitable development of new green nanomaterials with advanced applications. However, to the best of our knowledge, not so much research is available on the role of green nanomaterials in textile industries. The chapter mainly highlights the sustainable use of green nanomaterials in the field of textiles, discharge from textiles, and the various methods for investigating nanomaterial toxicity. The dangerous effects of nanomaterials on human health and the environment are also discussed.

6.2 Nanomaterials Consistent with Textiles

The preparation and utility of various inorganic nanomaterials have been reported in the literature. However, transition metals and metal oxide (i.e. 3d-transition metals covalently bonded to oxygen atoms forming different structures) nanomaterials show better compatibility with textiles for textile functionalization and effluent remediation [5]. Among transition metal nanomaterials, silver and gold nanoparticles show remarkable properties in textile industry. For instance, the exclusive physical and chemical characteristics of silver nanoparticles are responsible for their efficacy in different fields including food, medical, and textiles. Silver and gold nanoparticles are mainly considered for the functioning of textiles including UV protection, antimicrobial, etc., with concurrent coloration. Recent research on the green synthesis of silver nanoparticles (AgNPs) for advanced coloration applications in textiles is available in the literature [7].

In case of metal oxide nanoparticles such as ZnO, TiO_2 have been extensively investigated for effluent treatment and dye degradation of textiles in wastewater. Photocatalytic mechanisms have been observed as the best suited mechanism for dye degradation in wastewater treatment. However, nanomaterials such as iron oxides have been utilized for heavy metal ion adsorption due to their high efficiency of adsorption.

6.3 Techniques Related to Textile Functionalization

6.3.1 Pad Dry Cure Method

In literature, the pad dry cure method is considered the most convenient textile finishing method. This technique can be used to deposit a range of coatings to textiles (Figure 6.1). Using this technique, the deposition of nanomaterials on textile fibers can be performed without destroying the substrates. In Figure 6.1, the fibers are sunken in the nanomaterial solution (coating solution), and after that, the unwanted is squeezed out in the rollers that are dried and restored. TiO_2 nanosol has been utilized and directly applied on the textile surface by using the pad dry cure technique that aids in overcoming obstacles of available techniques to control the industrial scalability of the procedure.

Various reports have been documented for textile finishing using the pad dry cure method in the literature. In one such research, fiber samples were cleaned by washing in an ultrasound bath for half an hour, and ready samples were then coated in titania nano sol (3 wt.%), allowed to infuse for three minutes, which is then traveled via a two-roller laboratory padder followed by oven drying at 100 °C and cured at 130 °C. Finally, final washing in water has been given to remove any nanoparticles not physicochemically adsorbed [8]. Another study synthesized an antimicrobial finish for cotton fabric from silver chloride dispersed in a reactive organic–inorganic binder. The pad dry cure and exhaustion techniques have been utilized in the cotton textile for the smooth application of sols. The consequences showed that the antimicrobial activities of the coated layers were found to be more effective than for fungi [9].

Figure 6.1 Textile treatment through Pad dry cure method.

6.3.2 *In Situ* Preparation

In previous studies, the preparation and dispersion of nanomaterials on the surface of textiles have been performed concurrently using the *in situ* technique. Numerous reports have been available on the synthesis and applications of nanoparticles using different reducing agents [10]. Alkali-treated fibers are coated with the solution of silver nitrate and go into chemical reduction for *in situ* generation of AgNPs on the fiber surface. The alkali pretreatment effect on loading and stabilizing AgNPs was established and observed to be competent for higher silver agents. Fabrics, after treatment, own high washing durability and strong antibacterial properties against *E. coli* and *S. aureus* [11]. The dispersion of metal nanoparticles including gold, platinum, or palladium onto the natural cellulose fibers via different chemical techniques has been studied. The electrostatic association of citrate-stabilized metal nanoparticles onto the cationic surfaces of cellulose and then the negative metal complex ions adsorption over the cationic cellulose has been performed [12]. Cotton fabrics with polymer-silver nanocomposite have been synthesized by oxidative polymerization using pyrrole and silver nitrate. AgNPs dispersed on the polypyrrole or cotton matrix layer, and the binding with silver was performed by adsorption or electrostatic forces of attraction. Conductivity and antimicrobial performance against gram-positive *S. aureus* and gram-negative *E. coli* bacteria of the fiber have been augmented, exhibiting various applications in nanomaterials [13].

6.3.3 Green Nanotechnology

Generally, the most common methods followed for nanoparticle synthesis include gel–sol, solvothermal, hydrothermal, and chemical vapor precipitation, sonochemical methods, radiation chemical reduction, and microwave-assisted preparation [6]. These methods are lengthy, costly, and utilize hazardous substances that lead to harmful reactions [6]. However, techniques involving green synthesis utilize natural items such as leaves, roots, seeds, latex, and stems or microorganisms to decrease/remove the utilization and manufacture of poisonous substances produced by nanoparticle preparation [6]. Using this new approach of green synthesis, excellent results can be obtained via substances that exist in the environment; however, equivalent to the methods available for the chemical preparation of nanoparticles. Since the plant extracts are not pathogenic and biomolecules formed in the biosynthesis process undergo an extremely controlled assembly, they have received attention in green synthesis [6]. For the synthesis of nanoparticles using plant leaf extract, the extract is added to metal substrate solutions with varying reaction conditions. The rate of nanoparticle formation and their yield can be controlled by determining the parameters related to the situation of plant leaf extract. The phytochemicals present in plant leaf extracts possess strange potential to remove metal ions in much less time than fungi and bacteria [6]. Thus, plant leaf extracts are an excellent source for preparing different nanoparticles. For instance, Luque et al. [14] synthesized SnO_2 nanoparticles via green synthesis using an extract of *Camellia sinensis* leaves in different ratios. A small amount of tin chloride ($SnCl_2$-$2H_2O$) was added to the solution of prepared extracts. To obtain a homogeneous dispersion, the solutions were adequately mixed, and then excess water was removed using a thermal bath that formed a material with a plastic or pasty texture. The resulting material was distributed over a porcelain capsule and calcined. Lastly, the material in the capsule was converted into a very fine powder using an agate mortar. The nanoparticles of SnO_2 were characterized using different spectroscopic techniques, and their photocatalytic activity was also evaluated in the degradation of organic dyes.

Additionally, the green preparation of AgNPs on the cotton fiber has been performed using Tollen's reagent and concurrently adsorbed. Silver nitrate ($AgNO_3$) was converted to Ag_2O, and then silver nanoparticles were prepared on the cotton fiber [13]. Hasan et al. [7] reported the method for the coloration of polyester fabric by using green synthesized silver nanoparticles (G-AgNPs@PET) using chitosan. In this method, chitosan-mediated green silver nanoparticles (G-AgNPs) were generated by adding $AgNO_3$ to chitosan, and the final solution was magnetically stirred for half an hour. The color change of the solution from colorless to yellow confirmed the synthesis of G-AgNPs, termed yellow-colored G-AgNPs solution. The further addition of different ratios of $AgNO_3$ in ascorbic acid leads to the two colored AgNPs solution, i.e. red colored G-AgNPs and blue-colored G-AgNPs solutions. The green coloration of the PET fabric was started by adding the PET to the prepared AgNP solution. The coloration was done in a dying machine bath. After coloration, the fabric was rinsed and dried using an oven dryer. Scheme 6.1 represents the preparation and method of coloration. The characteristics of fabric such as coloration and fastness were considerably enhanced.

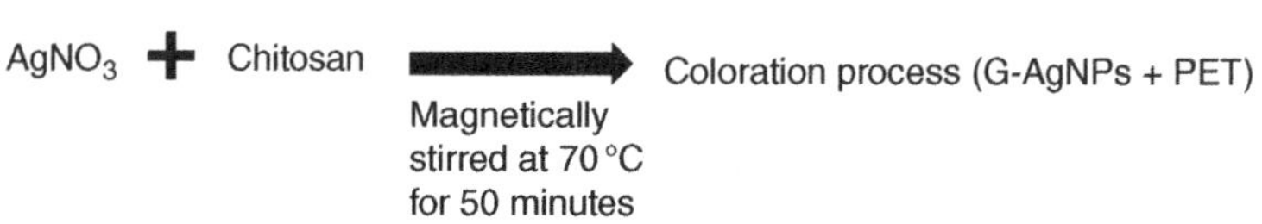

Scheme 6.1 Synthesis and coloration of G-AgNPs onto PET fabric.

Moreover, the major bacterial reduction after many washing cycles indicates the outstanding antibacterial characteristics of fabrics. This leads to low-cost production of sustainable textiles. Hiremath et al. [15] synthesized magnetite nanoparticles via green synthesis using ultrasonication method that exhibits effective microbe protection.

6.4 Utilization of Nanotechnology in Textile Industry

The main four areas by which green nanoparticles find utilization in the textile industry.

6.4.1 Nanofinishing

Applying colloidal solution or ultrafine spreading of nanomaterials on textile material is the basic idea of nanofinishing that helps improve textile functionality [16–20]. It exhibits some advantages as compared to conventional finishing. For instance, nanofinishing needs a slight amount of nanomaterials as compared to bulk materials needed in conventional nanofinishing to get the same result. The aesthetic feel of the textile materials and the durability of textile materials are not influenced by using this technique. Some functionalities that are difficult to obtain using conventional finishes could probably be carried out by nanofinishing [21].

6.4.2 Nanofibers

Nanofibers can be synthesized using different techniques such as electrospinning, melt blowing, force spinning, etc. [21]. Among these, electrospinning is the most suitable because of its low cost, better porosity, and productivity. Qi et al. [22] designed and synthesized a stretchable piezoresistive carbon nanotube containing nanofiber sensing yarn for the first time. Also, the nanofibers show a high tendency as effective layers of face masks to tackle harmful diseases such as COVID-19 [23].

6.4.3 Nanocoating

In a nanocoating method, thin layers are added to the precursor to enhance some characteristics for exhibiting superior functionality. Nanocoatings can solve the problems faced by traditional coatings including durability, abrasion resistance, strength, flexibility, and adhesion among the

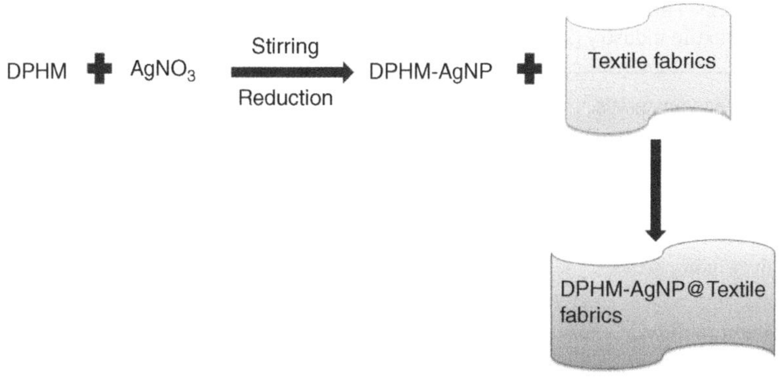

Scheme 6.2 Preparation of diphosphate malonate-silver nanoparticle (DPHM-AgNP) nanocomposites and their effect on textile fabrics.

substrate and coating layer [24]. The coating of nanomaterial on fabrics does not affect their breathability [25].

6.4.4 Nanocomposite

The solid material with different phases is termed nanocomposite only if at least one dimension of the reinforcing phase lies at the nano-level [26]. Additionally, polymer-based nanocomposites are basically nanomaterials that are deposited onto polymer matrices. The polymer nanocomposite-based coatings and fibers have huge prospects in producing advanced textiles with great functionality. Significant nanocomposite materials have been produced using facile one-pot technology. Attia et al. [27] developed nanocomposites based on AgNPs and diphosphate malonate (DPHM) as organic phosphates (Scheme 6.2). Different textile fabrics have been mixed with the nanocomposites that enhance their significant characteristics including fire retardancy (Table 6.1).

6.5 Nanomaterials with Different Functional Textiles

Nanomaterials occupy huge place in the textile industry for a variety of applications. Different nanomaterial includes different functional textiles, such as UV-protective textiles, flame-retardant textiles, antimicrobial textiles, antiodor textiles, repellent textiles, wrinkle resistance textiles, etc. Vigneshwaran et al. [34] established the basic procedures followed in using nanomaterials to express the efficient characteristics of cotton textiles.

Table 6.1 Few recently synthesized green nanomaterials and their applications in the textile industry [28–33].

Green nanomaterials	Applications
Zinc oxide nanoparticles	Photocatalytic degradation of dyes
Copper oxide nanoparticles	Nano-bioremediation of textile industry wastewater
Silver nanoparticles	Sustainable and energy-efficient photocatalytic degradation of textile dye
Titanium dioxide nanocatalysts	Photocatalytic performance in removing of methylene blue dye from effluent
Graphene oxide nanoparticles	Photodegradation of organic dyes
Chitosan-based green nanomaterials	Treatment of textile industry dyes

6.5.1 UV-Protective Textiles

Since ultraviolet radiation enters the earth's surface and causes the depletion of the ozone layer, a notable harmful effect on both human skin and clothes, more exposure to UV radiation increases the chances of various dangerous diseases, including cancer-related problems. Therefore, defense against UV radiation is essential in the textile industry [35]. Few metal oxide nanomaterials such as titanium dioxide nanoparticles, magnetite nanoparticles, zinc oxide nanoparticles, and nano-ceria obstruct UV radiations promise a sustainable as well as improved performance than organic UV absorbers [36–40]. Owing to the special properties including safety and stability in UV radiation, the aforementioned nano-inorganic-UV additives are generally favored compared to the organic additives. Other characteristics such as composition, particle size, crystallinity, and crystal geometry also affect the UV-blocking characteristics of nano-sized UV additives [41].

Noorian et al. [42] recently synthesized zinc oxide nanoparticles dispersed on the advanced cotton fabric to develop fabrics with many functions. This zinc oxide-4-aminobenzoic acid ligand oxidized cotton fabric confirmed better UV protection and considerable antibacterial effectiveness that can be utilized in novel textiles. In another study, UV-blocking and fire-resistant cotton fabric was designed by coating polyurethane-based MnO_2-$FeTiO_3$ nanocomposites [35]. The MnO_2-$FeTiO_3$ coated cotton fabrics represented a strong UV-blocking capacity and enhanced fire-resistant characteristics. Moreover, the coated cotton fabric possesses its characteristics after many cycles, hence contributing to sustainable, hard-wearing fabric in the textile industry.

6.5.2 Flame-Retardant Textile

Safety needs are a main concern in every part of life and thus also for textiles. Textiles are made up of polymeric structures and thus fairly active toward fire. Traditional flame-retardant additives are being neglected owing to their supposed unfavorable impact on nature, and thus, novel, effectual, and nature-friendly flame-retardant systems are required. Flame-retardant systems are planned to restrain the polymer combustion process by inquisitive with different processes such as pyrolysis and propagation of thermal degradation [43, 44]. Scientists are making efforts to discover ecofriendly substitutes such as the use of nanotechnology to trade traditional flame-retardant systems. Norouzi et al. [45] investigated the effect of nanoparticles on the flame retardation of various textile polymers. The results suggested that most of the nanoparticles could improve the textile polymers' flame-retardant characteristics and thermal constancy. However, the improvement will depend on various factors such as composition, morphology, migration speed, dispersion of the nanoparticles, and suitability between polymer and nanoparticle. Yazhini et al. [46] synthesized crosslinked cotton coated with nanocomposites for UV- protection and flame retardation. The polypyrrole-zinc oxide-carbon nanotube composite-coated cotton displayed enhanced characteristics than uncoated cotton. In another study by Fanglong et al. [47], flame-retardant combinations of traditional intumescent flame-retardant and nanosilica has been developed and were intro duced onto cotton fabric to investigate the synergistic effect of nanosilica on the fire resistance as well as the thermal stability of the intumescent flame-retardant system. The results indicate that the appropriate addition of nanosilica into the conventional intumescent flame-retardant system, to a definite extent, also decreases the thermal stability of the system.

6.5.3 Repellent Textiles

In the textile industry, repellency toward water and oil has been observed as the main condition for whole clothes, generating main concerns for textile producers and researchers for years [48]. Advanced nanocoatings or nanofinishings are meeting similar market demands with water, oil repellents, or super hydrophobic textiles. In literature, the advanced concept i.e. "self-cleaning textiles" has been discussed which shows that the textiles has the tendency to be cleaned without any laundry treatment [49]. Two different methods are available for the development of self-cleaning textiles, i.e. photocatalytic action and the lotus effect. Hydrophobicity is measured in terms of the contact angle of the water drops to the surface of textiles

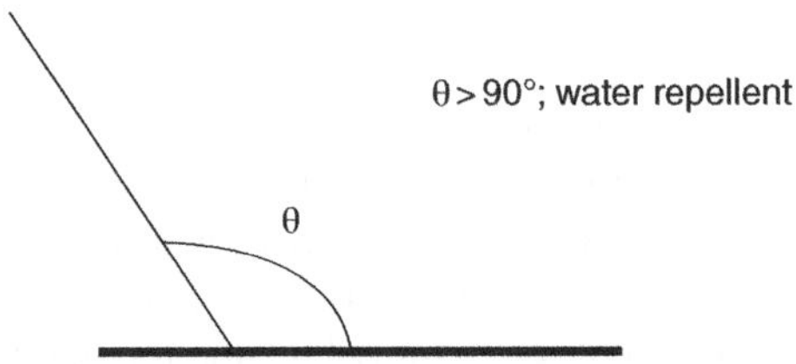

Figure 6.2 Mechanism of hydrophobicity by nanoparticles on the surface of textiles.

(Figure 6.2). A contact angle greater than 90° is termed as hydrophobic, and more than 150° superhydrophobic. This hydrophobic nature is mimicked from the lotus leaf and is also termed as the "lotus effect." The lotus effect is generated by the surface variation of the textile fabric by nanocoating or nanofinishing, generally by utilizing surface-modified carbon nanotubes, nano-zirconia, and zinc oxide nanorods [50]. In photocatalytic method, zinc oxide, and titanium dioxide nanoparticle-based coatings are used to prepare self-cleaning textiles. For instance, Wang et al. [51] reported that gold/titanium dioxide/silicon dioxide nanosol is a better photocatalyst than titanium dioxide nanosol, responsible for the superior self-cleaning property. Fluorine-free superhydrophobic cotton fabrics with the self-cleaning photocatalytic tendency have been fabricated by combining superhydrophobic SiO_2 and photoactive titanium dioxide[52].

6.5.4 Antibacterial and Antimicrobial Textiles

Textile fabrics consisting of cellulose fibers such as linen, cotton, lyocell, and viscose tend to be attacked by microorganisms such as algae, fungi, viruses, protozoa, and bacteria. Recently, various metal oxides and metal nanoparticles have gained great attention from scientist's community as prospective antimicrobial agents. Ullah et al. [53] studied the ecofriendly properties of cotton fibers doped with AgNPs synthesized from natural Chinese Holly plant extracts. These cotton fibers' antimicrobial properties, including silver nanoparticles, were assessed against gram-negative *E. coli* bacteria. The cotton fibers showed fine antibacterial effectiveness, making it suitable for medical purposes.

6.5.5 Wrinkle-Free Textiles

Wrinkle-free functionalization of textiles has also been attained with the application of nanomaterials or nanocomposites. In conventional methods, resin-based finishings are commonly used for imparting the wrinkle resistance property to a textile fiber. A simple technique was first developed to functionalize the durable-press all-cotton fabrics by grafting silver-anchored TiO_2 nanoconjugates through enediol ligand-metal oxide bonding and resin dehydration. The functionalization is incorporated into a conventional pad dry cure mechanism [54].

6.5.6 Antiodor Textiles

Tourmaline nanomaterial-based nanofinishing on textiles is responsible for an odor-resisting property and can split moisture, bacteria, and odor [55]. Including fragrant material in antiodor finishing by nanoencapsulation in synthetic fibers or by formulation might help release fragrance in the course of its utilization.

6.6 Conclusion

"Green synthesis" is a way to minimize the generation of unwanted or harmful by-products through the build-up of reliable, sustainable, and eco-friendly synthesis processes. Using ideal solvent systems and natural resources is significant to achieve this aim. The textile industry is growing worldwide at a fast pace and revealing its major effect on the environment. However, traditional methods such as chemical and processing methods used in the textile industry cause environmental pollution and toxicity. Consequently, to overcome these drawbacks, green chemistry plays a major role in protecting human health, textiles, and environmental risks in an economically better way. For this, green nanomaterials play a key role in technological evolution. Generally, the most common methods followed for nanoparticle synthesis include gel sol, solvothermal, hydrothermal, sonochemical methods, radiation chemical reduction, and microwave-assisted synthesis. These methods are time-consuming, costly, and use hazardous compounds that lead to harmful reactions. However, the green method or green synthesis utilizes natural materials or microorganisms to decrease the use and manufacture of poisonous materials produced by the synthesis of the nanoparticles. Using this new approach of green synthesis, one has allowed results to be obtained using natural materials, however, equivalent with the chemical nanoparticle synthesis methods.

The current chapter discusses the research on the "green synthesis of metal/metal oxide nanoparticles and their use in environmental remediation applications." Detailed synthesis mechanisms related to the textile functionalization techniques. Modernization-based nanotechnology-based textile industry has been discussed. Applications of nanotechnology with different functional textiles have been explored. In summary, future research and development of prospective "green" nanomaterials should be directed toward extending laboratory-based work to a commercial level by considering traditional issues, especially health and environmental effects.

Conflict of Interest

The authors declare no competing interest.

References

1 Hulla, J., Sahu, S., and Hayes, A. (2015). Nanotechnology: history and future. *Human & Experimental Toxicology.* 34 (12): 1318.
2 Taniguchi N. (1974). On the basic concept of nano-technology. *Proceedings of the International Conference on Production Engineering*, Tokyo. Part II Japan Society of Precision Engineering.
3 Saleem, H. and Zaidi, S.J. (2020). Sustainable use of nanomaterials in textiles and their environmental impact. *Materials* 13 (22): 5134.
4 Yang, G. and Park, S.J. (2019). Deformation of single crystals, polycrystalline materials, and thin films: a review. *Materials* 12 (12): 2003.
5 Shahid-ul-Islam, Shabbir, M., and Mohammad, F. (2017). Insights into the functional finishing of textile materials using nanotechnology. In: *Textiles and Clothing Sustainability: Nanotextiles and Sustainability* (ed. S.S. Muthu), 97–115. Springer.
6 Singh, J., Dutta, T., Kim, K.H. et al. (2018). 'Green' synthesis of metals and their oxide nanoparticles: applications for environmental remediation. *J. Nanobiotechnol.* 16 (1): 1–24.
7 Hasan, K.F., Pervez, M.N., Talukder, M.E. et al. (2019). A novel coloration of polyester fabric through green silver nanoparticles (G-AgNPs@ PET). *Nanomaterials* 9 (4): 569.
8 Ortelli, S., Costa, A.L., and Dondi, M. (2015). TiO_2 nanosols applied directly on textiles using different purification treatments. *Materials* 8 (11): 7988–7996.
9 Tomšič, B., Simončič, B., Orel, B. et al. (2009). Antimicrobial activity of AgCl embedded in a silica matrix on cotton fabric. *Carbohydr. Polym.* 75 (4): 618–626.
10 Shabbir, M. and Mohammad, F. (2018). Multifunctional AgNPs@ wool: colored, UV-protective and antioxidant functional textiles. *Appl. Nanosci.* 8: 545–555.

11 Yazdanshenas, M.E. and Shateri-Khalilabad, M. (2013). In situ synthesis of silver nanoparticles on alkali-treated cotton fabrics. *J. Ind. Text.* 42 (4): 459–474.

12 Dong, B.H. and Hinestroza, J.P. (2009). Metal nanoparticles on natural cellulose fibers: electrostatic assembly and in situ synthesis. *ACS Appl. Mater. Interfaces* 1 (4): 797–803.

13 Montazer, M., Alimohammadi, F., Shamei, A., and Rahimi, M.K. (2012). In situ synthesis of nano silver on cotton using Tollens' reagent. *Carbohydr. Polym.* 87 (2): 1706–1712.

14 Luque, P.A., Chinchillas-Chinchillas, M.J., Nava, O. et al. (2021). Green synthesis of tin dioxide nanoparticles using *Camellia sinensis* and its application in photocatalytic degradation of textile dyes. *Optik* 229: 166259.

15 Hiremath, L., Kumar, S.N., and Sukanya, P. (2018). Development of antimicrobial smart textiles fabricated with magnetite nano particles obtained through green synthesis. *Mater. Today Proc.* 5 (10): 21030–21039.

16 Joshi, M. and Adak, B. (2019). 6. Nanotechnology-based textiles: a solution for the emerging automotive sector. In: *Rubber Nanocomposites and Nanotextiles* (ed. B. Banerjee), 179–230. De Gruyter.

17 Haji, A., Mousavi Shoushtari, A., Mazaheri, F., and Tabatabaeyan, S.E. (2016). RSM optimized self-cleaning nano-finishing on polyester/wool fabric pretreated with oxygen plasma. *J. Text. Inst.* 107 (8): 985–994.

18 Ghosh, G., Sidpara, A., and Bandyopadhyay, P.P. (2018). High efficiency chemical assisted nanofinishing of HVOF sprayed WC-Co coating. *Surf. Coat. Technol.* 334: 204–214.

19 Osman, E. (2020). Nanofinished medical textiles and their potential impact to health and environment. In: *Nanoparticles and Their Biomedical Applications* (ed. A.K. Shukla), 127–145. Springer.

20 Ghosh, S., Smith, T., Rana, S., and Goswami, P. (2020). Nanofinishing of textiles for sportswear. *World Textile Information Network* 44–49.

21 Naeem, M.A., Alfred, M., Saba, H. et al. (2019). A preliminary study on the preparation of seamless tubular bacterial cellulose-electrospun nanofibers-based nanocomposite fabrics. *J. Compos. Mater.* 53 (26–27): 3715–3724.

22 Qi, K., Zhou, Y., Ou, K. et al. (2020). Weavable and stretchable piezoresistive carbon nanotubes-embedded nanofiber sensing yarns for highly sensitive and multimodal wearable textile sensor. *Carbon* 170: 464–476.

23 Tebyetekerwa, M., Xu, Z., Yang, S., and Ramakrishna, S. (2020). Electrospun nanofibers-based face masks. *Adv. Fiber Mater.* 2 (3): 161–166.

24 Nguyen-Tri, P., Nguyen, T.A., Carriere, P., and Ngo Xuan, C. (2018). Nanocomposite coatings: preparation, characterization, properties and applications. *Int. J. Corros.* 2018: 19.

25 Temesgen, A.G., Turşucular, Ö.F., Eren, R., and Ulcay, Y. (2018). Novel applications of nanotechnology in modification of textile fabrics properties and apparel. *Int. J. Adv. Multidiscip. Res.* 5 (12): 49–58.

26 Malhotra, B.D. and Ali, M.A. (2018). Nanomaterials in biosensors: fundamentals and applications. In: *Nanomaterials for Biosensors*, 1–74. Elsevier.

27 Attia, N.F. and Morsy, M.S. (2016). Facile synthesis of novel nanocomposite as antibacterial and flame retardant material for textile fabrics. *Mater. Chem. Phys.* 180: 364–372.

28 Shubha, J.P., Kavalli, K., Adil, S.F. et al. (2022). Facile green synthesis of semiconductive ZnO nanoparticles for photocatalytic degradation of dyes from the textile industry: a kinetic approach. *J. King Saud Univ. Sci.* 34 (5): 102047–102059.

29 Darwesh, O.M., Li, H., and Matter, I.A. (2023). Nano-bioremediation of textile industry wastewater using immobilized CuO-NPs myco-synthesized by a novel Cu-resistant *Fusarium oxysporum*. *Environ. Sci. Pollut. Res.* 30: 16694–16706.

30 Elbadawy, H.A., Elhusseiny, A.F., Hussein, S.M. et al. (2023). Sustainable and energy-efficient photocatalytic degradation of textile dye assisted by ecofriendly synthesized silver nanoparticles. *Sci. Rep.* 13: 2302.

31 Zeebaree, A.Y.S., Rashid, R.F., Zebari, O.I.H. et al. (2022). Sustainable engineering of plant-synthesized TiO_2 nanocatalysts: diagnosis, properties and their photocatalytic performance in removing of methylene blue dye from effluent. A review. *Curr. Res. Green Sustainable Chem.* 5: 100312.

32 Sreelekshmi, P.B., Pillai, R.R., and Meera, A.P. (2022). Controlled synthesis of novel graphene oxide nanoparticles for the photodegradation of organic dyes. *Top. Catal.* 65: 1659–1668.

33 Panel, Y., Shahadat, M., Ali, S.W. et al. (2022). Chitosan-based green nanomaterials for treatment of textile industry dyes. In: *Green Nanomaterials for Industrial Applications: Micro and Nano Technologies* (ed. U. Shanker, C.M. Hussain, and M. Rani), 153–176. Elsevier.

34 Vigneshwaran, N. and Arputharaj, A. (2020). Functional finishing of cotton textiles using nanomaterials. In: *Advances in Functional Finishing of Textiles* (ed. M. Shahid and R. Adivarekar), 43–56. Springer.

35 Dhineshbabu, N.R. and Bose, S. (2019). UV resistant and fire retardant properties in fabrics coated with polymer based nanocomposites derived from sustainable and natural resources for protective clothing application. *Compos. Part B Eng.* 172: 555–563.

36 Fouda, A., Saad, E.L., Salem, S.S., and Shaheen, T.I. (2018). In-vitro cytotoxicity, antibacterial, and UV protection properties of the biosynthesized zinc oxide nanoparticles for medical textile applications. *Microb. Pathogen.* 125: 252–261.

37 Tsuzuki, T. and Wang, X. (2010). Nanoparticle coatings for UV protective textiles. *Res. J. Text. Apparel* 14: 9–20.

38 Radetić, M. (2013). Functionalization of textile materials with TiO_2 nanoparticles. *J. Photochem. Photobiol. C Photchem. Rev.* 16: 62–76.
39 Attia, N.F., Moussa, M., Sheta, A.M. et al. (2017). Synthesis of effective multifunctional textile based on silica nanoparticles. *Prog. Org. Coat.* 106: 41–49.
40 Attia, N.F., Moussa, M., Sheta, A.M. et al. (2017). Effect of different nanoparticles based coating on the performance of textile properties. *Prog. Org. Coat.* 104: 72–80.
41 Dhineshbabu, N.R. and Bose, S. (2018). Smart textiles coated with eco-friendly UV-blocking nanoparticles derived from natural resources. *ACS Omega* 3 (7): 7454–7465.
42 Noorian, S.A., Hemmatinejad, N., and Navarro, J.A. (2020). Ligand modified cellulose fabrics as support of zinc oxide nanoparticles for UV protection and antimicrobial activities. *Int. J. Biol. Macromol.* 154: 1215–1226.
43 Saleemi, S., Naveed, T., Riaz, T. et al. (2020). Surface functionalization of cotton and PC fabrics using SiO_2 and ZnO nanoparticles for durable flame retardant properties. *Coatings* 10 (2): 124.
44 Ali, S.W., Basak, S., and Shukla, A. (2020). Nanoparticles: a potential alternative to classical fire retardants for textile substrates. In: *Handbook of Nanomaterials for Manufacturing Applications* (ed. C.M. Hussain), 265–278. Elsevier.
45 Norouzi, M., Zare, Y., and Kiany, P. (2015). Nanoparticles as effective flame retardants for natural and synthetic textile polymers: application, mechanism, and optimization. *Polym. Rev.* 55 (3): 531–560.
46 Yazhini, K.B. and Prabu, H.G. (2015). Study on flame-retardant and UV-protection properties of cotton fabric functionalized with ppy–ZnO–CNT nanocomposite. *RSC Adv.* 5 (61): 49062–49069.
47 Fanglong, Z., Qun, X., Qianqian, F. et al. (2016). Influence of nano-silica on flame resistance behavior of intumescent flame retardant cellulosic textiles: remarkable synergistic effect? *Surf. Coat. Technol.* 294: 90–94.
48 Asif, A.K. and Hasan, M.Z. (2018). Application of nanotechnology in modern textiles: a review. *Int. J. Curr. Eng. Technol.* 8 (2): 227–231.
49 Talebi, S. and Montazer, M. (2020). Denim fabric with flame retardant, hydrophilic and self-cleaning properties conferring by in-situ synthesis of silica nanoparticles. *Cellulose* 27: 6643–6661.
50 Debener, S., Emkes, R., De Vos, M., and Bleichner, M. (2015). Unobtrusive ambulatory EEG using a smartphone and flexible printed electrodes around the ear. *Sci. Rep.* 5 (1): 16743.
51 Wang, R., Wang, X., and Xin, J.H. (2010). Advanced visible-light-driven self-cleaning cotton by $Au/TiO_2/SiO_2$ photocatalysts. *ACS Appl. Mater. Interfaces* 2 (1): 82–85.

52 Xu, B., Ding, J., Feng, L. et al. (2015). Self-cleaning cotton fabrics via combination of photocatalytic TiO_2 and superhydrophobic SiO_2. *Surf. Coat. Technol.* 262: 70–76.

53 Ullah, N., Yasin, S., Abro, Z. et al. (2014). Mechanically robust and antimicrobial cotton fibers loaded with silver nanoparticles: synthesized via Chinese holly plant leaves. *Int. J. Text. Sci.* 3: 1–5.

54 Tripathi, R., Narayan, A., Bramhecha, I., and Sheikh, J. (2019). Development of multifunctional linen fabric using chitosan film as a template for immobilization of in-situ generated CeO_2 nanoparticles. *Int. J. Biol. Macromol.* 121: 1154–1159.

55 Joshi, M. and Adak, B. (2019). Advances in nanotechnology based functional, smart and intelligent textiles: a review. In: *Comprehensive Nanoscience and Nanotechnology* (ed. D.L. Andrews, R.H. Lipson, and T. Nann), 253–290. Elsevier.

7

Drug-delivery, Antimicrobial, Anticancerous Applications of Green Synthesized Nanomaterials

Sivasubramanian Murugappan, Monika Pebam, Sri Amruthaa Sankaranarayanan, and Aravind Kumar Rengan

Department of Biomedical Engineering, Indian Institute of Technology, Hyderabad, Sangareddy, TS, India

Abstract

Nanomaterials have been extensively used in various applications owing to the unique physicochemical properties exhibited by these materials in the size of 1–100 nm. Nanomaterials are extensively used in biomedical applications owing to their easy surface functionalization for targeted therapeutics, integration of multiple modalities within a single nanoplatform, size, shape-dependent tuning of physicochemical properties, etc. Despite the technical advancements in the development of nanomaterials, they are still limited in their clinical use due to their toxic side effects. Green synthesis of nanomaterials, typically using plant-based products/extracts or microbial extracts, has emerged as an affordable, simple yet safe alternative for the synthesis and development of several nanosystems that can be used for biomedical applications. Hence, this chapter focuses on the methods and use of green synthesis of various metallic nanosystems and their subsequent use in various biomedical applications such as drug delivery for antimicrobial and anticancer activity in detail using the supporting literature.

Keywords *green synthesis; nanomaterials; antimicrobial; anticancer; drug delivery*

Green Synthesis of Nanomaterials: Biological and Environmental Applications, First Edition.
Edited by Archana Chakravarty, Preeti Singh, Saiqa Ikram, and R.N. Yadava.

7.1 Introduction

Nanotechnology is an emerging alternative to conventional therapeutic modalities for the treatment of various diseases. It basically deals with nanoparticles in size ranging from 1 to 100 nm in any one of its dimensions [1]. When compared with their bulk counterpart, nano-sized materials exhibit unique physical, chemical, and optical characteristics, which make them interesting to explore for a wide range of applications. The significant properties of nano-sized materials include a high surface area to volume ratio owing to the small size, high surface energy, and aspect ratio, which influence the role of nanomaterials in biomedical applications [2]. The three most important requisites for the synthesis of any nanosystem include (i) the selection of an optimum eco-friendly solvent, (ii) a reducing agent, and (iii) a stabilizing agent. Conventional methods of nanoparticle synthesis involve physical and chemical routes. Chemical methods involve the use of expensive and toxic precursors or harmful by-products, causing potential risk. Physical methods are laborious and not easy to reproduce [3].

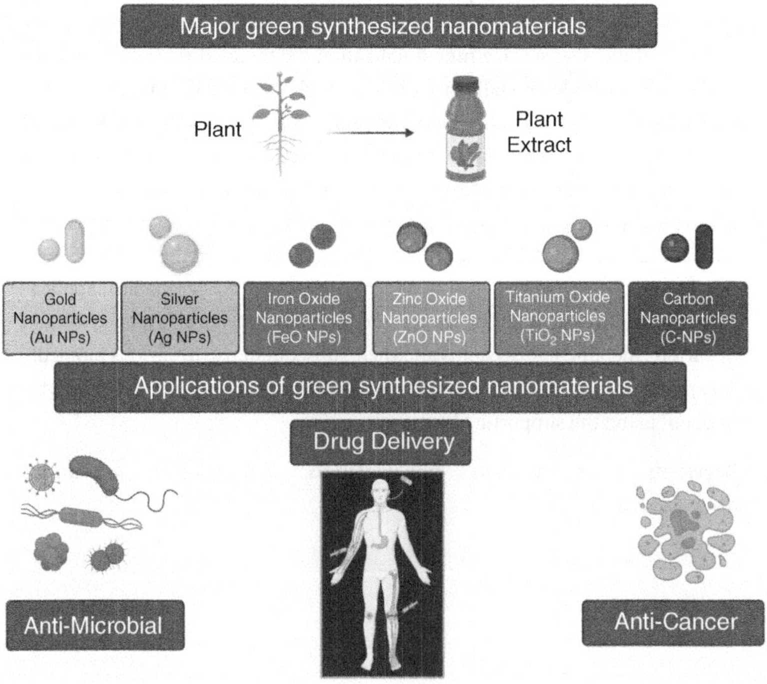

However, green methods of nanoparticle synthesis have emerged as a cost-effective alternative to conventional methods. Green synthesis involves the use of plant/microbial-based extracts for reducing and stabilizing the nanoparticles. The green route is known to be safe and biocompatible, easy to reproduce, eco-friendly approach, and extensively used for biomedical applications. In the case of plants, extracts of various parts of the plants such as leaves, stems, roots, fruits, etc. are used owing to the presence of bioactive molecules, which play an important role in stabilizing and reducing the nanoparticles [4]. Additionally, physical and chemical methods yield NPs that are uncertain in terms of their size, shape, and dispersity despite the expensive products involved. This chapter focuses on the green synthesis of six different nanomaterials, namely, gold, silver, titanium oxide, zinc oxide, iron oxide, and carbon-based nanosystems, and their potential applications in drug delivery and as anticancer and antimicrobial agents [5].

7.2 Gold Nanoparticles

7.2.1 Synthesis of AuNPs

Gold nanoparticles (AuNPs) have been explored for a long time in the field of medicine due to their unique physicochemical properties. They are majorly synthesized by chemical reduction methods though there are other physical and biological methods due to convenience in synthesis. However, with growing research in AuNPs, researchers are proposing new convenient green syntheses. Khan et al. have used *Clerodendrum inerme* for the leaf extracts for both reduction and stabilization. They were able to synthesize AuNPs of an average size of 5.82 nm with a spherical morphology [6]. Similarly, *Croton caudatus* Geisel leaf extracts were used by Kumar et al. to reduce $HAuCl_4$ to AuNPs using a reduction process, following the Pearson acid–base concept. When those soft metals (Au) are bound by hard ligands (leaf extract), then there is no complex formation; however, it is reduced to form NPs. The particles synthesized were of sizes 20–50 nm [7]. Ismail et al. used the leaf extract of *Corchorus olitorius* to synthesize AuNPs. Their results showed the relationship between concentration and particle morphology as higher concentration yielded quasi-spherical particles while low concentration gave a combination of hexagon and triangle-shaped particles [8]. Borady and group developed AuNPs with four different concentrations of Parsley leaf extracts giving different morphologies like spherical, semi-rod, and flower for each concentration. Both of these were quantitatively found easily through the color change in each set of AuNPs [9]. The work by Bogireddy et al. used *Coffea arabica* seed extracts to reduce and cap gold precursors to form AuNPs. They altered the pH to know their effect on the size of AuNPs formed. Though both low and high pH groups agglomerated, low pH groups formed bigger particles, which was attributed mainly to the limited availability of capping agent, the phenolic group (OH groups), as the high pH group produced particles of comparatively smaller sizes [10]. Another group that worked on Cinnamon bark extract was able to produce spherical AuNPs with sizes of around 35 nm. The carbohydrate helps in capping, while terpenoids and proteins are the reason behind the reduction of Au ions into AuNPs [11]. Singh et al. developed AuNPs using the cortex and xylem parts of *Cannabis sativa*, commonly called Industrial hemp. The xylem extract, containing cannabinoids and alkaloids (phenolic compounds), produced core AuNPs (C-AuNPs) that were majorly spherical; however, it also contained some triangular, hexagonal, and rod-shaped particles. The size distribution range was around 12–20 nm, with a polydispersity index of 0.48 for F-AuNPs and 0.24 for C-AuNPs. This shows the electrostatic stability of the particles contributing to the prevention of aggregation [12]. *Scutellaria barbata* plant extract was used by Wang et al. for the reduction of gold. The average size of the AuNPs produced was 154 nm and was spherical in shape. However, the aggregation of particles was also seen in the system [13].

7.2.2 AuNPs in Drug Delivery

Kumar and colleagues developed a gold-based drug delivery system with *Punica granatum* (Pomegranate, P) peel extract as the reducing and stabilizing agent for the formation of AuNPs (PAuNPs). Then, it was functionalized with folic acid for tumor targeting of the drug fluorouracil (5-Fu). Furthermore, the *in vitro* release kinetics reveals a burst release of the drug initially in the first hour, followed by a slow release of about 22.92% over 48 hours [14]. A similar study by Devi et al. used *Vitex negundo* leaf extract as a reducing agent and Arabic gum as a capping agent for AuNP synthesis. While the targeting agent remained folic acid, the drug was decided to be epirubicin. In terms of its *in vitro* release, the free drug showed faster release while the Fa-E-AuNPs showed a sustained release with 5% in one hand and 72.32% in 12 hours [15]. In a different study, xanthum gum was utilized to reduce and cap gold. The so-formed AuNPs were loaded with doxorubicin (Dox). The high loading efficiency of $71.4 \pm 2.53\%$ was observed due to the interaction between pyruvic acid (negatively charged) in Xanthum gum and an amine group (positively charged) in Dox. At the end of 10 hours, the drug release was 98.1% in SAB and 83.6% in PBS [16]. Marine red algae-derived Carrageenan oligosaccharide reduced AuNPs (CAO-AuNPs) were used as a drug delivery system for the delivery of epirubicin. A total of 12.5% and 94.3% were the loading and encapsulation efficiencies of the drug in CAO-AuNPs, respectively. At the end of 72 hours, the release of epirubicin at a pH of 5.0 (96%) was three times higher than that of the release at a pH of 7.4 (30%), indicating tumor selective release of the drug [17].

7.2.3 Antimicrobial Activity of AuNPs

The antimicrobial activity of *Jasminum auriculatum* leaf-extracted AuNPs was tested against a number of human pathogenic bacterial and fungal species. The zone of inhibition (ZOI) of the positive control groups chloramphenicol (for bacteria) and streptomycin (for fungi) performed equally or better than the biogenic AuNPs in all the concentrations tested [18]. Similar was the case in Annona Muricata extracted AuNPs where the positive controls Streptomycin and Amphotericin B for the bacterial and fungal strains, respectively. Considering the positive controls had 100% ZOI, the AuNPs had only 30–66% against fungi and 40–54% against bacteria [19]. Hamelian et al. used cephalexin and kanamycin as positive controls for gram-negative and gram-positive bacteria, respectively. These positive controls overperformed the ZOI's of *Pistacia atlantica*-extracted AuNPs [20]. Mandhata and colleagues utilized the cyanobacterium *Anabaena spiroides* to reduce $HAuCl_4$ to AuNPs. It was found that they were able to produce significant antibacterial activity against various multi-drug resistant bacterial strains [21]. The pure *Origanum vulgare* (OVE) extract and AuNPs formed with the help of leaf extract of OVE were compared for their antimicrobial activity. And the results clearly showed the enhanced antimicrobial activity of AuNPs, especially against *S. aureus* and *C. albicans*, by around fourfold that of OVE (Table 7.1) [22].

7.2.4 Anticancer Activity of AuNPs

AuNPs obtained from *Abies spectabilis* were tested for their anticancer ability against T24 cell lines. Their IC50 values were between 15 and 20 μg/ml. Further, it was found that condensation and DNA fragmentation were the major mechanisms by which the AuNPs induce apoptosis in this study [24]. Similarly, Ke et al. synthesized AuNPs using *Catharanthus roseus* extract to check their potency against cervical cancer cells. The developed AuNPs were found to follow caspase-mediated apoptosis in HeLa cells with an IC50 value of 5 μg/ml [25]. The work of Barai and the group used *Nerium oleander* for

Table 7.1 Zone of inhibition (ZOI) of green synthesized AuNPs.

Reducing agent (green compound)	Microbial species	ZOI (mm)	Reference
Origanum vulgare – Herba	*Salmonella enteritidis*	10	[22]
	Escherichia coli	8	
	Listeria monocytogenes	10	
	Staphylococcus aureus	21	
	Candida albicans	28	
Uncaria gambir Roxb. leaf	*S. aureus*	7	[23]
	E. coli	9	
Pistacia atlantica	*Pseudomonas aeruginosa*	23	[20]
	E. coli	21	
	Bacillus subtilis	20	
	S. aureus	20	
Jasminum auriculatum leaf	*Streptococcus pyogenes*	12	[18]
	S. aureus	9	
	E. coli	12	
	Klebsiella pneumoniae	7	
	Aspergillus fumigatus	4	
	C. albicans	4	
	Trichoderma viride	5	
	Lecanicillium lecanii	5	

the reduction of Au to AuNPs. This stem extract and stem extract reduced AuNPs were treated against MCF-7 breast cancer cell lines, showing excellent cytotoxicity [26]. On a similar note, Red marine algae obtained carrageenan oligosaccharides (CAO), and CAO-reduced AuNPs were checked for their IC50 values against HCT-116 and MDA-MB-231 [27]. Jeyrani et al. also treated their *Gelidium pusillum* reduced AuNPs against MDA-MB-231 cell lines and found them to possess good cancerous activity while being compatible with the HEK-293 cells. Renal carcinoma cell lines A498 and SW156 were dosed with *Curcuma wenyujin* mediated AuNPs to know their IC50 values. It was found that A498 cells were comparatively more sensitive to the biogenic AuNPs than SW156 with an IC50 value of 25 μg/ml while SW156 had 40 μg/ml [28]. Then, Sun et al. used their green synthesized AuNPs (from *Marsdenia tenacissima*) against A549 cell lines, showing excellent anticancer activity with a less IC50 value of 15 μg/ml [29]. The AuNPs formed with the use of *Commiphora wightii* extract as a reducing agent, when evaluated for cancer activity *in vitro*, revealed that the G2/M phase against the MCF-7 cell lines by apoptosis (Table 7.2) [32].

Table 7.2 Anticancer activity of green synthesized AuNPs.

Reducing agent (green compound)	Cell line	IC-50 (μg/ml)	Reference
Corchorus olitorius	HCT-116	12.2	[8]
	HepG-2	10.3	
	MCF-7	11.2	
Parsley leaves	CO-II	56.83	[9]
Anacardium occidentale leaves	PBMC	600	[30]
	MCF-7	6	
Jasminum auriculatum leaf	HeLa	104	[18]
Abies spectabilis	T24	20	[24]
Catharanthus roseus	HeLa	5	[25]
Nerium oleander	MCF-7	74.04	[26]
Carrageenan oligosaccharide	HCT-116	34.4 ± 1.7	[27]
	MDA-MB-231	129.2 ± 1.7	
Gelidium pusillum	MDA-MB-231	43.09 ± 1.6	[31]
	HEK-293	>150	
Commiphora wightii	MCF-7	66.11	[32]
Marsdenia tenacissima	A549	15	[29]

7.3 Silver Nanoparticles

7.3.1 Synthesis of AgNPs

Silver (Ag) nanoparticles are mainly used in batteries, glass, ceramic, or for the treatment of antimicrobial infections. Metallic nanoparticles can be synthesized using different approaches, such as electrochemical methods, decomposition, microwave-assisted techniques, and wet chemical procedures [33]. The lotus leaves were crushed into powder and heated in deionized water at 100 °C, followed by filtration using Whatman paper and kept at 4 °C for further tests, and phytochemical screening tests were performed. Lotus leaves and silver nitrate were mixed in a 1 : 6 ratio to form nanoparticles (NPs) as the solution appeared brownish-red. The size of NP was found to be in the range of 15–20 nm according to dynamic light scattering (DLS) and transmission electron microscopy (TEM). The concentration of silver ions was less than 1 mM [34]. Garlic was weighed and crushed with deionized water and maintained at room temperature for 24 hours.

The presence of metal salts reduces its oxidation state promotes a fast reaction and forms smaller size NPs [35]. The use of biomolecules (tannins, terpenoids, alkaloids, proteins, etc.) prevents the agglomeration of NPs thus enhancing antimicrobial activities. Rhizome of *Zingerbie officinale and C. longa* after air-dried and macerated was added into 250 ml of distilled water and boiled for 30 minutes. The phytochemical constituents were filtered using Whatman paper. AgNPs were prepared using 1 M $AgNO_3$ solution turning brownish color and showing a reduction from Ag^+ to Ag^0 nanoparticles, size ranging from 49 to 69 nm [36]. Another group explored *Tectona grandis* seed extract for the reduction of silver nitrate ($AgNO_3$) to AgNPs and did its characterization using visible spectrum, scanning electron microscopy (SEM), energy-dispersive X-ray spectroscopy (EDX), X-ray diffraction (XRD). $AgNO_3$ to seed extract was used 1 : 9 ratio and kept at constant stirring at 800 rpm, the resultant color turned brownish due to surface plasmon vibration. TEM images of AgNPs showed a spherical shape in aggregation in the 10–30 nm range and were further used to check antimicrobial activity in Gram-negative bacteria (*Escherichia coli*) showing a higher inhibition zone. This may be due to a thin peptidoglycan cell wall layer [37]. AgNPs enhance membrane permeability due to the expedited leakage of sugar and proteins from the cytosolic region of bacteria. AgNPs synthesized by green chemistry expedite a novel alternate therapy against chemically synthesized AgNPs to overcome the multidrug resistance concerned.

7.3.2 AgNPs in Drug Delivery

AgNPs have gained attention for drug delivery due to their intrinsic properties. Various biosynthesis techniques of AgNPs like emulsion, solvent-evaporation, and reduction methods using reducing agents like citrate or sodium borohydride. Green synthesis is highly encouraged because of its eco-friendly nature, and economical, large-scale production possibility. The phytochemical agent products obtained from plant extracts like *Azadirachta indica*, *Crocus sativus* L., *Calliandra haematocephala*, green tea, *Curcuma longa*, *Withania somnifera*, etc. act as reducing, stabilizing, or capping agents. AgNPs generally have a spherical shape of about 20–50 nm size due to the reduction of Ag^+ to Ag^0, metallic clusters agglomerate into oligomeric clusters. Different surfactants were used to prevent the AgNPs from agglomeration. Surface functionalization of AgNPs helps in increasing stability, enhanced catalytic activity, and sustained drug release for applications in therapy [38]. Green synthesis of AgNPs has the ability to alter signaling pathways. AgNPs being toxic in free form were biosynthesized using natural products has been explored. Hyaluronic acid (HA) is a polysaccharide that has surface cell receptors for CD44 on cancer cells. So a based green synthesis of AgNPs utilizes HA as a reducing and capping agent for sustained drug release in the acidic microenvironment of tumor cells. AgNPs enhanced passive and active targeting when administered as NPs. The high intracellular uptake of NPs increases lipophilicity interacting with intrinsic mechanisms leading to apoptosis due to increased expression of p53, caspase 3, and Bcl-2 [39]. Shandiz et al. showed that imatinib-loaded silver nanoparticles (IMAB-AgNPs) also show dose-dependent cytotoxicity against MCF-7 cells (IC50: 9.63 μM, 3.02 μM). Another study led by Muhammad et al. showed that methotrexate-AgNPs (AgMTX) coated with polyethylene glycol (PEG) denoted as (PEG-Ag-MTX) has significant anticancer activity against MCF-7 cells (IC50 258.6 μg/ml) when compared with MTX alone (512.7 μg/ml). Benyettou et al. demonstrated that aldronate (Ald@AgNPs) has a significantly higher toxicity at IC50 10.1 μM, potentially due to the higher lipophilicity and increased cellular uptake [39]. AgNPs have received high commercialization, with 55.4% of the total nanomaterial-based products in the market. These include silver-based biocomposites like Acticoat™ and Bactigras™ (Smith & Nephew), Aquacel™ (ConvaTec), PolyMem Silver™ (Aspen), and Tegaderm™ (3M) approved by the United States (US) FDA for wound-dressing applications [40].

7.3.3 Antimicrobial Activity of AgNPs

Biosynthesized AgNPs showed excellent dose-dependent against gram-negative *E. coli* showing a minimum inhibitory concentration of 62.5 μg/ml. In the disk-diffusion method, the zone of inhibition increased with an increasing dose of silver nanoparticles. Ag releases from NP cause cell wall break penetrating to bacteria leading to DNA-damage, protein denaturation, and production of oxygen species such as hydroxide and superoxide radicals, collectively leading to cell death [34, 41]. Applicability of biosynthesized AgNPs showed more antimicrobial when compared to chemically synthesized AgNPs. In the disk-diffusion method, using the highest concentration of AgNPs showed more antimicrobial activity when compared to the standard drug streptomycin having a 21–26 mm zone of inhibition. AgNPs exhibit broad-spectrum antibacterial activity against Gram-negative bacteria due to the presence of proteoglycan in the cell wall content [42]. Green AgNPs showed more antimicrobial activity than treatment done alone using plant extracts of *Azadirachta indica* [43], *Coffea arabica* [44], *Fagonia cretica, Tinospora cordifolia* where it intercalates the DNA leading to the rupture of sulfur and phosphorus bases break leading to cell death [45].

Green synthesized AgNPs using *Elephantopus scaber* having spherical shape of 37 nm were tested against the *B. subtilis*, *L. lactis*, *P. fluorescens*, *P. aeruginosa*, *A. flavus*, and *A. penicillioides* strains using agar diffusion method showing 16–24 and 11–12 mm inhibition zones against tested bacterial and fungal strains. Green AgNPs synthesized using green and black tea leaf extract having 10–20 nm (spherical shape) were testes against Methicillin- and vancomycin-resistant *S. aureus* by disk diffusion methods and exhibited a 19–21 mm zone of inhibition. Green synthesized *Phyllanthus amarus* – AgNPs having flower shapes of 32–42 nm size was used to check its antimicrobial properties using disk diffusion methods against strains of *E. coli*, *Pseudomonas* spp., *Bacillus* spp., *Staphylococcus* spp., *A. niger*, *A. flavus*, and *Penicillium* spp. showed maximum inhibition zones [41].

7.3.4 Anticancer Activity of AgNPs

AgNPs remarkably work against tumor angiogenesis, inhibiting the growth and migration of endothelial cells. It leads to apoptotic cell death enhancing caspase expression [46]. Medicinal plants have future prospects for cancer treatment with the new approaches for using green synthesis of metallic NPs. AgNPs have a site-specific target that enhances the drug delivery efficiency across the impermeable membrane and restricts immune-reactivity. Studies show that inherent change in the cell morphology, plasma membrane, and cytoplasm disintegration. AgNPs synthesized from *Siberian ginseng* have shown anticancer potential against colon and lung cancer against the cisplatin drug. It showed the downregulation of the caspase/p-38 MAPK pathway along with reactive oxygen species (ROS) generation [47]. Another group has shown the anticancer potential for the green synthesis of AgNPs using *Dendropanax morbifera* leaves against HepG2 and A549 cells enhancing ROS production and induction of apoptosis by downregulation of EGFR/p38 MAPK pathway [48]. In another study, green synthesized AgNPs using *Azadirachta indica* (neem) leaf extract exhibited a size of 90 nm approximately, induced dose-dependent cytotoxicity against lung cancer (NCI-H460) showing nuclear morphological changes and cell death by apoptosis when compared to chemically synthesized AgNPs. Similarly, studies have shown the extract from onion leaf, *Aloe vera, Embilica officinalis* also act as reducing agents for metal nanoparticle synthesis [49]. Green synthesized AgNPs using *Pinus roxburghii* exhibit cytotoxic effect against prostate and lung cancer via intrinsic pathway causing mitochondrial membrane potential disruption and DNA damage, thereby inducing cell cycle arrest at the G_2/M phase and apoptosis. *Cynara scolymus* extract used as a reducing agent in AgNPs also showed antitumor activity employed with photodynamic therapy (PDT) through intracellular ROS production and mitochondrial apoptosis, regulating apoptotic proteins elevation [50]. AgNPs majorly act on increasing free radical species. ROS generation, in turn, causes DNA, protein, and lipid damage. It further activates apoptotic proteins like Bak/Bax, double-stranded DNA break, disintegration of mitochondrial membrane, and finally cell shrinkage. So, AgNPs can provide a promising drug delivery target for efficient cancer therapeutics.

7.4 Zinc Oxide Nanoparticles

7.4.1 Synthesis of ZnO NPs

Few studies indicate that ZnO is synthesized from precursors like zinc acetate, zinc nitrate by chemical reduction have shown adverse effects. For safety ZnO preparation, the green synthesized ZnO NPs formed by the Zn ion's reaction with polyphenols followed by hydrolysis, forming $Zn(OH)_2$. Further, calcination and decomposition of $Zn(OH)_2$ leads to the formation of ZnO NPs possessing good anticancer, UV protection, and antimicrobial abilities [51]. *Ulva lactuca* seaweed extract reduced zinc acetate to ZnO NPs and cap their size in the range of 10–50 nm. The asymmetrical shapes and agglomeration are attributed majorly to oxides which undergo a solution combustion mechanism to form NPs [52]. The ZnO NPs formed using Oak Fruit Hull's aqueous extract produced uniform spherical nanoparticles having an average diameter of 34 nm [53]. Another study used *Deverra tortuosa* to reduce and cap the ZnO NPs with sizes of 9.26 and 31.18 nm [54]. *Garcinia mangostana* pericarp extract reduced ZnO NPs was a spherical shape in a 5–45 nm size range [55]. The work of Chikanna and team uses novel, cheap, eco-friendly components as reducing agents to synthesize ZnO NPs. The utilized goat fecal matter (GFM) and sheep fecal matter (SFM) formed ZnO NPs of sizes 40–120 and 60–130 nm, respectively, with spongy and flower-like structures found from SEM [56]. The root extract of *Scutellaria baicalensis* acted as a reducing agent to form ZnO NPs from zinc nitrate. The particles formed were spherical and were around 50 nm in size [57]. Similar was the elemental composition in *Laurus nobilis* leaf extract reduced ZnO NPs. They had a composition of about 81.3% zinc and 18.7% oxygen, 76.9% zinc, and 23.1% oxygen for the zinc acetate and zinc nitrate precursors, respectively. They were spherical in shape with an average diameter of 21.49 and 25.26 nm [58]. Ogunyemi et al. synthesized ZnO NPs from three different green component extracts namely *Olea europaea* (olive leave), *Matricaria chamomilla* L. (chamomile flower), and *Lycopersicon esculentum* M.

(red tomato fruit). The overall size range of these particles lies between 40.5 and 191 nm and elemental composition data also lies very close to that of the ideal composition results of *Scutellaria* and *Laurus* [51].

7.4.2 Role of ZnO NPs in Drug Delivery

The usage of ZnO NPs as nanocarriers is less explored compared to its counterparts like AuNPs, AgNPs, etc. Here we present those few papers (to the best of our knowledge) that have delved into this idea. Vimala et al. developed a ZnO NPs carrier system using *Borassus flabellifer* fruit extract. The system was loaded with Dox and was tested both *in vitro* and *in vivo*. The rod-shaped ZnO NPs had high encapsulation and loading efficiencies. The release was followed for 80 hours, an initial of 40 hours, a burst release and the remaining had a sustained slow release [59]. The same group has also loaded the drug Erbitux in a green synthesized ZnO NPs system modified with folic acid and polyethylene glycol to target breast cancer [60]. Wang and colleagues followed ultrasound and coffee leaf extract-mediated ZnO NP carrier for loading mangiferin. The system showed low encapsulation and loading efficiencies of $19.01 \pm 1.90\%$ and $3.66 \pm 0.35\%$, respectively. In terms of release kinetics, more than 50% of the drug was released in the initial 12 hours and around 97% at the end of 132 hours strongly contributing to the delivery system for sustained release [61].

7.4.3 Antimicrobial Activity of ZnO NPs

ZnO NPs prepared using the leaf extracts of *Euphorbia hirta* showed increased ZOI against both bacterial and fungal microorganisms with an increase in ZnO NPs concentration leading to increased H_2O_2 levels [62]. Similar results were obtained on *Punica granatum* leaf and flower extract-mediated ZnO NPs when tested against a dozen bacterial species [63]. ZnO NPs reduced using Orange peel extract were checked for their antibactericidal activity by Thi et al., the results showed significant antibactericidal activity [64]. The ZOI of the *Atalantia monophylla* leaf extract, the extract reduced ZnO NPs, ZnO, and positive control were compared for both bacterial and fungal activity [65]. Ansari et al., extracted *Cinnamomum verum* bark using hexane as a solvent and utilized the obtained extract to Zinc nitrate to ZnO NPs. The ZnO NPs were further tested against *E. coli* and *S. aureus* for their antimicrobial activity and streptomycin as positive control. ZnO NPs were able to inhibit the bacterial groups, however, could not overperform streptomycin [66] similar to the work of Mallikarjunaswamy et al. (Table 7.3).

Table 7.3 Zone of inhibition of green synthesized ZnO NPs.

Reducing agent (green compound)	Microbial species	ZOI (mm)	Reference
Olea europaea	*Xanthomonas oryzae* pv. *oryzae*	22	[51]
Atalantia monophylla leaf	*Staphylococcus aureus*	28 ± 2.91	[65]
	Escherichia coli	23 ± 1.76	
	Bacillus cereus	21 ± 1.22	
	Pseudomonas aeruginosa	22 ± 1.34	
	Bacillus subtilis	20 ± 1.13	
	Klebsiella pneumoniae	19 ± 1.01	
	Candida albicans	24 ± 1.97	
	Aspergillius niger	18 ± 0.99	
Mussaenda frondose (stem extract)	*E. coli*	28.64	[67]
	P. aeruginosa	20.31	
	B. subtilis	19.13	
Callus extract	*S. aureus*	21.51	
Goat fecal matter (GFM)	*Salmonella typhimurium*	3.23 ± 0.15	[56]
	B. subtilis	2.46 ± 0.06	
Sheep fecal matter (SFM)	*S. typhimurium*	2.93 ± 0.15	
	B. subtilis	2.1 ± 0.10	
Ulva lactuca seaweed	*Bacillus licheniformis*	26.3 ± 1.6	[52]
	Bacillus pumilus	21.2 ± 0.9	
	E. coli	24.0 ± 1.0	
	Proteus vulgaris	20.3 ± 0.7	
Aloe socotrina leaf	*E. coli*	25.3 ± 1.7	[68]
	K. pneumoniae	19.2 ± 0.8	
	P. vulgaris	25.0 ± 0.9	
	P. aeruginosa	19.4 ± 0.8	
Vitis vinifera (Grape)	*K. pneumonia*	20	[69]
	S. aureus	23.85	
K-carrageenan	*MRSA*	15.5	[70]
Tecoma castanifolia leaf	*E. coli*	17	[71]
	P. aeruginosa	15	
	B. subtilis	15	
	S. aureus	17	

(Continued)

Table 7.3 (Continued)

Reducing agent (green compound)	Microbial species	ZOI (mm)	Reference
Cissus quadrangularis	*P. aeruginosa*	16 ± 1.00	[72]
	Acinetobacter baumannii	16 ± 0.54	
	B. cereus	19 ± 1.00	
	E. coli	19 ± 0.54	
	Mycobacterium smegmatis	22 ± 1.50	
Bergenia ciliata rhizome	*Yersenia enterocolitica*	16	[73]
	P. aeruginosa	17	
	Salmonella typhi	13	
	E. coli	15	
	B. subtilis	7	
	S. aureus	7	
Euphorbia hirta leaf	*Streptococcus mutans*	28	[62]
	S. aureus	29	
	Clostridium absonum	27	
	E. coli	24	
	Arthogrophis cuboida	29	
	Aspergillius fumigates	27	
	A. niger	20	
Prunus dulcis (almond gum)	*S. aureus*	18	[74]
	E. coli	32	
	Salmonella paratyphi	25	
	C. albicans	18	
Pomegranate (*Punica granatum*)	*S. aureus*	10.50 ± 0.87	[63]
	S. typhi	10.00	
	E. coli	13.83 ± 0.76	
	Aeromonas hydrophila	12.33 ± 0.58	
	B. cereus	13.67 ± 0.58	
	Listeria monocytogenes	12.00 ± 1.00	
	Enterococcus faecalis	14.33 ± 0.58	
	P. aeruginosa	11.33 ± 0.58	
	Enterococcus faecium	15.17 ± 0.29	
	Salmonella diarizonae	12.00 ± 1.73	

Table 7.3 (Continued)

Reducing agent (green compound)	Microbial species	ZOI (mm)	Reference
	K. pneumoniae	11.67 ± 1.53	
	Moraxella catarrhalis	9.67 ± 1.53	
	Streptococcus pneumoniae	12.67 ± 0.58	
Beta vulgaris	*E. coli*	10	[75]
	A. niger	9	
Cinnamomum verum	*E. coli*	8	
	S. aureus	8	
	Candida albanicans	8	
Brassica oleracea var. *italica*	*E. coli*	8	
	S. aureus	8	
Cinnamomum tamala	*E. coli*	10	
	S. aureus	8	
	A. niger	8	
	C. albanicans	8	
Indian bael (*Aegle marmelos*)	*B. cereus*	12.10 ± 1.42	[76]
	Micrococcus luteus	25.66 ± 0.81	
	S. aureus	12.98 ± 0.02	
	K. pneumoniae	31.03 ± 3.00	
	Enterobacter aerogenes	12.00 ± 0.06	
	E. coli	12.12 ± 1.44	
	Pseudomonas fuorescens	13.48 ± 1.54	
	P. aeruginosa	22.55 ± 0.11	
	Salmonella enteritidis	13.35 ± 0.67	
Cinnamomum verum Bark	*S. aureus*	16.75 ± 0.47	[66]
	E. coli	13.25 ± 0.75	

7.4.4 Anticancer Activity of ZnO NPs

Ruangtong et al. produced ZnO nanosheets by making use of banana peel extract as the reducing agent. These ZnO nanosheets were checked for their anticancer efficiency against the skin (A431), colorectal (SW620), liver (HepG2) cancer cells, and normal cell line (Vero). The result demonstrated an inhibitory effect only on cancer cells promoting the green synthesized ZnO nanosheet's potency to be an anticancer drug [77]. Green synthesized ZnO NPs from the leaf extracts of *Tecoma castanifolia* and *Raphanus sativus var. Longipinnatus* were treated against A549 lung cancer cell lines [71, 78]. On the same note, Jayappan and group used callus and stem extracts of *Mussaenda frondosa* to get their ZnO NPs. The 50% cell inhibition against A549 cancer cell lines required more than 65 μg/ml for both callus and stem-extracted ZnO NPs [67]. Further, Selim et al. also tested their *D. tortuosa* reduced ZnO NPs against A549 cell lines along with WI38 lung fibroblast cell lines and Caco-2 colon cancer cell lines with Dox as the positive control [54]. A similar study against MIA PaCa-2 revealed *Cissus quadrangularis* reduced ZnO NPs to have a significant cancer activity even at low drug concentrations (Table 7.4) [72]. Dulta et al.'s work used the rhizome extract of *Bergenia ciliata* to synthesize ZnO NPs and tested against HT-29 and HeLa and the results suggested IC50 values over 100 μg/ml [73].

Table 7.4 Anticancer activity of green synthesized ZnO NPs.

Reducing agent (green compound)	Cell line	IC-50 (μg/ml)	Reference
Deverra tortuosa	A549	83.47	[54]
	Caco-2	50.81	
	WI38	434.60	
	PANC-1	40 ± 5.6	
	AsPC-1	30 ± 4.6	
	Hu02	80 ± 2.1	
Banana peel	SW-620	31.92	[77]
	A431	54.63	
	HepG2	102.7	
Tecoma castanifolia leaf	A549	65	[71]
Raphanus sativus var. *longipinnatus*	A549	40	[78]
Cissus quadrangularis	MIA PaCa-2	7.02 ± 0.02	[72]
Bergenia ciliata rhizome	HeLa	101.7	[73]
	HT-29	124.3	

7.5 Titanium Dioxide Nanoparticles

7.5.1 Synthesis of Titanium Dioxide NPs (TiO_2NPs)

Titanium dioxide nanoparticles (TiO_2NPs) were synthesized using leaf extract of *A. vera* at pH 9 wherein 100 ml of leaf extract was added drop-by-drop to 100 ml 1N of $TiCl_4$ under continuous stirring. TEM images showed TiO_2NPs were in the range of 13.3 ± 0.35 nm having spherical shape without aggregation. The NPs exhibit sustained drug release. Green synthesized TiO_2NPs were non-toxic to normal human embryonic kidney cell lines up to 100 μg/ml, and leukemia cell lines [79]. TiO_2NPs were synthesized using facile green synthesis from jasmine flower extract as a reducing agent. Using the aqueous extract of jasmine, a 20 ml solution of titanium tetra isopropoxide (TTIP) was added dropwise that showed a color change from pale white to yellowish–gray color, confirming TiO_2NPs. SEM reveals uniform morphology having a spherical shape which was collated with the crystallization observed during XRD [80]. Another group explored the root extract of *W. somnifera* as a stabilizing and reducing agent. TiO_2NPs when examined using TEM and SEM-EDS showed size (50–90 nm) with sphere and square shape NPs, further studied for antibiofilm and anticancer activity [81]. TiO_2NPs were synthesized using the facile green synthesis method using $TiCl_3$ hydrolysis in an aqueous solution under acidic conditions for 24 hours using mango-peel extract. The rutile to anatase ratio of the formed NPs exhibits a shape like rice grain (27 nm) and anatase phase NPs showed a spherical shape (17 nm) [82]. TiO_2NPs were also synthesized by another group using *Moringa oleifera* ethanolic extract leaves using titanium tetraisopropoxide under continuous stirring at 50 °C. The anatase phase of TiO_2 particles showed a tetragonal structure when observed using XRD analysis [83].

7.5.2 TiO_2NPs in Drug Delivery

TiO_2NPs can be a choice for drug delivery carriers using established drugs like Dox, cisplatin, and paclitaxel. The electrostatic interaction increases between drugs and the NPs which can overcome multidrug resistance enhancing more encapsulation of the drug. The nanostructured TiO_2NPs mediated drug target via cell internalization through a glycoprotein pumping system [84]. Another interesting approach established by Lui et al. developed TiO_2NPs loaded with cisplatin by modifying using hyaluronic acid (HA) in pH-responsive for non-adjuvant chemotherapy of ovarian cancer. The drug loading capacity in A2780 ovarian cancer cells increased through endocytosis showing significant anticancer activity [85]. TiO_2NPs can be used as biosensors for drug delivery systems in photodynamic (PDT) and photothermal therapy (PTT) as they are hydrophilic, with thermal conductivity and stability that convert photons to heat energy [86]. Photocatalytic properties of TiO_2NPs can generate hydroxyl radicals and anionic oxygen radicals with high redox actions that can eradicate cancer cells, bacteria, and viruses [84]. Photoactivated TiO_2NPs increased Bax protein expression which further activates caspase 3 and caspase 9 leading to apoptosis when studied using MDA-MB-468 cells. Hence nanosized TiO_2NPs could be promising against the treatment of multidrug resistance tumor cells [84]. TiO_2NPs have wide biomedical applications as a choice for bone implants, prosthetics, vascular stents, pharmaceuticals, cosmetics, etc. Nitrogen doping of TiO_2NPs was used in PDT for passive targeting of tumors that induce singlet oxygen species and hydroxyl anions radical production. TiO_2NPs having photocatalytic properties were used as photosensitizing agents for cancer therapy as well as inactivation of photodynamic against antibiotic resistance bacteria [87].

7.5.3 Antibacterial Activities of TiO_2NPs

Synthesized TiO_2NPs using root extracts of *W. somnifera* exhibited significant antibiofilm activity and destruction of cell walls among various pathogenic microbes with minimum inhibitory concentration (MIC) values over *E. coli* (32 μg/ml), *Psuedomonas aeruginosa* (32 μg/ml), *Candida albicans* (64 μg/ml), *S. marcescens* (8 μg/ml) [81]. Green synthesis of *Trigonella foenum-graecum* (TF) extract showed a dose-dependent antibacterial activity showing zone of inhibition for the following *S. aureus* (11.2 mm), *S. faecalis*(11.6 mm), *E. coli* (10.8 mm), *E. faecalis* (11.4 mm), and *Y. enterocolitica* (10.6 mm). TF-TiO_2NPs were in the anatase phase having spherical shape morphology with size ranging between 20 and 90 nm [88]. Antibacterial studies of TiO_2NPs were performed against *E. coli* (14 mm), *Staphylococcus aureus* (8 mm), and *Klebsiella pneumonia* (12 mm) as a zone of inhibition. This led to the rupture of the peptidoglycan cell wall of bacteria probably due to the electrostatic force of interaction between positive charge TiO_2NPs and gram negative bacteria strains. It results in the induction of ROS generation as hydroxyl and superoxide ions help in rupturing the cell wall and cytoplasm membrane due to the increased surface area of NPs [80]. Studies have reported that TiO_2NPs trigger the onset of ROS production by disrupting phospholipid content, reducing cell adhesion, inhibition respiratory cytosolic rate, and attenuation macromolecules producing substantial effects on cellular levels and gene expression [89]. Another group synthesized TiO_2NPs using *Piper betel* leaf extract and chemogenic source nitric acid as capping and reducing agents by hydrothermal method. TEM analysis showed the spherical shape of NPs with an average size of 8 nm. Further, it was observed that it showed antimicrobial properties against *S. aureus* and *E. coli* having MIC of 25 µg/ml [90]. TiO_2NPs showed cell cytotoxicity when tested against breast cancer cells like MDA-MB-468 cells, and MCF-7 leading to apoptosis executed by caspase-3 cleavage. Commercially available PEG-TiO_2NPs (anatase phase) induce cell death in melanoma cells (B16F10) induced by hyperthermia, leading to solid tumor destruction and necrotic cell death pathway and helping in cell cycle arrest. Photoactivated nitrogen-doped/TiO_2NPs with visible showed significant PARP cleavage in MDA-MB-231 which reports DNA mutations (double-strand DNA/single-strand DNA breakage) [91].

7.5.4 Anticancer Activities of TiO_2NPs

Greenly synthesized TiO_2NPs loaded with Dox showed enhanced anticancer activity in Ehrlich tumor-bearing mice. The release of Dox from TiO_2NPs is due to electrostatic force and attributed to the sustained release of Dox for intra-tumoral injection. Dox-TiO_2NPs having a spherical shape with a size of 14.53 ± 4.68 nm helps in passive accumulation in tumor sites by enhanced permeability effect as cellular uptake is obtained by endocytosis [79]. TiO_2NPs prepared from methanolic Guava (*Psidium guajava*) leaf extracts that contain aromatic amines and alcohols help in the TiO_2NPs formation. This exhibited significant anticancer activity due to oxidative stress metabolism generating free radicals, superoxide dismutase, peroxidase that eventually leads to lipid bilayer disruption, and DNA enzymes [92]. Another group synthesized TiO_2 nanoparticles modified with *Withania somnifera, Eclipta prostrata,* and *Glycyrrhiza glabra* having a size range of 50 nm. They showed anticancer potential against oral cancer cells in a dose-dependent manner [93]. Palladium-doped Pd@TiO_2NPs have been synthesized using *A. vera* as a capping and reducing agent and showed cytotoxicity against A549 cells (165 µg/ml). Low bandgap metal produced excess electrons resulting in ROS generation that accumulates in the tumor region as Pd TiO_2NPs show a red-shift in the visible region [94]. Anatase TiO_2 nanoparticles were modified with bio-agents such as turmeric, ginger, and garlic, exhibiting crystalline structure in XRD that showed red-shift absorbance spectra. The anticancer property was executed using oral cancer cells using various concentrations of turmeric, ginger, and garlic-doped TiO_2NPs and its toxicity profile increased with increased NPs concentrations. Dopped

TiO_2NPs exhibits more antioxidant when compared with pure TiO_2NPs leading to intracellular ROS production which in turn damage DNA and leads to apoptosis [95]. Nanostructured TiO_2NPs have various biomedical applications due to their biocompatibility nature, but being photo-excited exhibits generations of free electrons which in turn react with water and oxygen to yield reactive oxygen species, which ultimately trigger cell death owing to altered signaling pathways. Biosynthesized TiO_2NPs using *Ledebouria revolute* plant extracts, having an average crystalline size of 47 nm in a spherical shape, were checked for their cell inhibition against concentrations range of 6.25–100 μg/ml in A549 cells. Upon increasing concentrations cell death increased which showed anticancer potential [96].

7.6 Iron Oxide Nanoparticles

7.6.1 Synthesis of IONPs

Recently, iron oxide nanoparticles were synthesized using *Ficus carica* leaf extracts and found 43–57 nm sized nanoparticles exerted potential antioxidant properties due to the presence of flavonoids, phenolic compounds, fatty acids, etc., in the extract used [97]. Another study reports the green synthesis of IONPs using an aqueous extract of *Cassia auriculata* flower. The method yielded spherical nanoparticles with crystalline nature and magnetic properties and was successfully used for environmental remediation (photocatalytic degradation) and potent larvicidal activity [98]. Another recent study reported a bio-inspired route of using *Terminalia chebula* dried seed extracts for the synthesis of IONPs. The as synthesized nanoparticles exhibited superior electro-chemical properties that could be used for biosensing and better photocatalytic activity over methylene blue for removing MB from MB-contaminated water bodies [99]. Apart from plants, microbial extracts can be utilized for the biosynthesis of IONPs. Recently, extracellular metabolic extracts of the manglicolous fungi *Aspergillus niger* BSC-1 strain were used to synthesize IONPs, resulting in needle-shaped NPs of 20–40 nm size. It is revealed that the hydrocarbons, amide bonds carboxyl groups NADH-dependent hydrolase enzymes, etc. in the fungal extract help in the reduction of the ferric ions and form NPs [100]. Similarly, IONPs have also been synthesized using extracts of three filamentous fungi, *Phialemoniopsis ocularis*, *Trichoderma asperellum*, and *Fusarium incarnatum*. The type of fungal extract and its components strongly influence the size and properties of the synthesized IONPs. The three extracts yielded spherical NPs of size 25, 13.13, and 30.56 nm, respectively, and it is revealed through FTIR analysis emphasized amide I and II bonds corresponding to membrane proteins aid as reducing agents in the synthesis [101]. Bacterial extracts of *Bacillus cereus* HMH1 strains have also been used to synthesize IONPs, yielding spherical NPs of 30 nm size [102].

7.6.2 IONPs in Drug Delivery

The magnetic properties of IONPs can be exploited for the targeted delivery of therapeutic moieties to the desired cells using an external magnetic stimulus. Such a stimulus-responsive targeting reduced the burden of cost and dependency on expensive targeting ligands and antibodies. IONPs also offer additional advantages of biocompatibility, stability, water solubility, and ease of surface modification. It is well reported that modification of IONPs surface with biopolymers such as chitosan and PEG has shown promising improvement in drug loading and releasing efficiency. However, using green techniques such as plant extracts and microbial extracts as potential alternatives reduces the overall cost of production and improves biocompatibility and therapeutic effects owing to the phenolic/flavonoids present in these extracts.

One of the recent reports on IONPs using *M. oleifera* leaves extract synthesized through an economical and eco-friendly approach yielded NPs of uniform shape and size of 16 nm, which makes it a potential candidate for drug delivery [103]. A low-cost synthesis method of SPIONs was reported recently using coconut water and the study analyzed the ability of macrophages to internalize these 4 nm-sized NPs. These SPIONs could efficiently internalized into the membrane component of macrophages via an endocytic/phagocytic pathway which could be exploited for targeted drug delivery [104]. IONPs synthesized using a 2% weight percentage of pomegranate extract were successfully used to load an anticancer chemo drug, 5-fluorouracil, and yielded a maximum release of 79% at physiological pH showing a potential anticancer activity in HCT116 colon cancer cells. It is evident from the above-reported studies that opting for a greener route for the synthesis of IONPs can yield stable and biocompatible IONPs of varying shapes and sizes which could be further tuned as per the required biomedical applications such as MRI contrast, hyperthermia, drug delivery, antimicrobial property, etc. [105].

7.6.3 Antibacterial Activity of IONPs

The antimicrobial properties of IONPs have been known to exist since centuries. Several studies have reported the synthesis of microbial and plant extract-reduced IONPs and proved their effect against several pathogenic bacterial strains. Recently, biosynthesized IONPs using Piper betel leaves were reported to exhibit potential antibacterial activity against both gram-positive and gram-negative bacteria. The highest activity was found against *P. aeruginosa* and the lowest for *S. aureus* [106]. A recent study reported the use of *Penicillium* sp. isolated from soil in the synthesis of IONPs. FTIR analysis revealed the involvement of fungal proteins in the formation and stability of the NPs, yielding spherical NPs of 3–10 nm size with a high stability (zeta potential of +33.9 mV). The as-synthesized IONPs exerted superior bactericidal activity when tested against strains of highly infectious *S. aureus*, *K. pneumonia*, *E. coli*, *P. aeroginosa*, and *S. sonnie*. These NPs also exhibit significant antioxidant activity on par with ascorbic acid with an IC50 of 12 μg/ml. The excellent antioxidant activity can be extended and utilized for a potential anticancer application [107]. Another study reported the synthesis of IONPs using leaf extract of *Celosia argenta*. The IONPs exhibit a broad range of potential biomedical applications such as biofilm inhibition, antiinflammatory, antioxidant, and larvicidal. The as synthesized particles exhibited antimicrobial activity against infectious *E. coli* and *S. aureus* [108]. Recently, an unconventional, eco-friendly green synthesis of IONPs was reported using extracts of *Hibiscus rosa sinensis* flower as both reducing and stabilizing agents. Additionally, microwave-assisted method was used to synthesize NPS of an average diameter of 50 nm with significant inhibition effect against various bacterial activity against *S. aureus*, *P. aeruginosa*, *K. pneumonia*, and *E. coli* [109]. Previously, IONPs synthesized using *Tridax procumbens* were reported to exhibit significant activity against *P. aeroginosa*. Leaf extract of *Dadonaea viscosa* was used for the synthesis of ZVI, recording an instant color change from yellow to greenish-black solution at RT. The as synthesized ZVI exhibited spherical morphology of an average size of 27 nm and FTIR analysis revealed the role of flavonoids, saponins, and other biomolecules in the extract for stabilizing and acting as a capping agent. It is evident from the above studies that the use of plant biomolecules as reducing/capping agents offers two important advantages, the former being the increased stability and narrow size distribution of the NPs and the latter, being the most important, preventing of oxidation of NPs to their oxide forms. It is also very clear that green synthesized IONPs.

7.6.4 Anticancer Activity of IONPs

The type and concentration of extracts used in the synthesis of iron oxide NPs have a profound influence on the type of anticancer activity exhibited by these NPs. One of the studies reported the influence of the weight percentage of fruit peel extract of pomegranate on the synthesis of iron oxide

NPs and its subsequent anticancer activity. It was reported that the addition of fruit peel extract acted as a green stabilizer yielding NPs of size less than 11 nm. However, a potential anticancer activity was observed only when the weight percentage of the extract was maintained at 2% and 4% against specific nasopharyngeal carcinoma cell line, HONE1 [110]. Another recent study reported the use of piper betel leaves for the synthesis of IONPs, in specific α-Fe_2O_3. The technique yielded cubic-shaped NPs with an average size of 25 nm showing a potential anticancer activity in a concentration-dependent manner with IC50 of 100 μg/ml. Microscopic analysis revealed shrunk cells with condensed and fragmented nuclei owing to the cytotoxic effect of green synthesized IONPs. *Psoralea corylifolia* reduced IONPs were reported to exhibit anticancer activity against Caki-2 and MDCK cell lines via a capase-3-dependent apoptosis pathway [111].

IONPs are mainly used in cancer treatment through magnetic hyperthermia, a modality wherein the incident external high-energy magnetic field is converted into thermal energy, causing the ablation of cancer cells. IONPs, most importantly being magnetic in nature, aid in the magneto-sensitization of cancer cells for a localized therapeutic effect. Recently, NIR-responsive IONPs were green synthesized *Pimenta dioica* leaf extract. With excellent biocompatibility, the as synthesized IONPs also possessed superior OCT contrast and produced significant heat due to the external magnetic field on HeLa cell lines, thereby qualifying as a potential candidate for cancer theranostics [112]. Gardenia, an ancient medicinal plant was recently utilized to synthesize eco-friendly hematite NPs. It was revealed that the phenolic groups in the extract reduce the ferric ions and the hydroxyl –OH groups act as capping by bonding over the NPs surface. The method yielded uniform spherical-shaped NPs of 5 nm size with excellent hyperthermia properties at very low concentrations proving as a potential candidate for anticancer applications [113]. Another group reported the IONPs synthesis using pomegranate extract and revealed that at 4% of the weight percentage of the extract, the as synthesized NPs were stable with a size of 14.38 nm, with a remarkable heating capacity under the influence of an external magnetic field and acceptable MRI contrast.

7.7 Carbon Based Nanomaterials

7.7.1 Synthesis of Carbon-Based Nanomaterials

One of the critical challenges in large-scale industrial production of carbon nanotubes despite its various uses, is its low dispersibility in the solvents. The use of bio-based alternatives for carbon sources such as olive oil or coconut oil has been reported to overcome this challenge. Several materials have been used as bio-alternatives of carbon sources for a large-scale synthesis of carbon nanotubes [114]. Along similar lines, the use of $KClO_3$ and Mn_2O_7 in the synthesis of single or multi-layer graphene sheets is associated with the toxicity of Mn^{2+} and the instable intermediates with potential explosive risk. Recently, biopolymers such as chitosan, alginate, lignin, Clove extract, Industrial soot, carrageenan, vitamin C, etc. have been used as biological precursors for the synthesis of graphene sheets [115]. Thus, green synthesis of carbon dots has emerged and several studies report the synthesis of carbon dots from plant or microbial extracts for various biomedical applications [116, 117]. A recent study reports the use of *Malus floribunda* as an eco-friendly precursor for the synthesis of C-dots, yielding particles of 3.5 nm with a yield of 18%. Similarly, an easily scalable hydrothermal-based method was reported by using grape extract as a precursor for C-dot synthesis, with a yield of 18.67% of 4 nm sized NPs [118]. In a recent study, carbon dots, were for the first time synthesized using ryegrass-type agro-industrial residues such as malt bagasse and *Lolium perenne* [119].

7.7.2 Carbon Based Nanomaterials in Drug Delivery

Owing to their unique and novel surface properties, carbon-based nanomaterials are promising for drug delivery to the targeted site. Owing to its very small size, in terms of 2–10 nm, CDs possess a very large surface area for flexible modification/conjugation with therapeutic moieties. CDs are potential candidates for theranostics, involving both bioimaging and targeted drug delivery. A threefold symmetric polyphenolic compound, phloroglucinol was used to synthesize green emission CDs, which could be used for loading Dox via electrostatic interaction with an efficiency of 54.62%. The synthesized CDs exhibited potential cytotoxicity at both pH of 5.0 and 6.5 mimicking the tumoral environment, in addition to enhanced ROS generation. The synthesized CDs also exhibited significant membrane-targeting potential and could be used for tracking the cellular response without photobleaching [120]. Water-soluble CDs have been synthesized using root extracts of red Korean ginseng through a microwave-assisted method. The excitation-dependent emission properties were used effectively for bioimaging. The efficacy of the synthesized CDs as a drug delivery system was tested by loading Rutin, a natural flavanoid. The loading efficiency was 12.7%, however, the CDs possessed a sustained drug release behavior at physiological conditions. The loading with flavonoids improved the radical scavenging activity of the CDs significantly. A dose-dependent antibacterial activity was also reported against a broad spectrum of bacterial strains enhanced upon NIR exposure [121]. Leaf extracts of black pepper, a common household spice have also been used to synthesize CDs in an eco-friendly hydrothermal method. On par with other biosynthesized CDs, the pepper-derived CDs also exhibited excitation-dependent emission properties with an average size of 4 nm. The synthesized CDs also exhibited antioxidant, larvicidal, and potential Fe^{3+}-sensing properties [122]. Exploiting the versatility of CDs in terms of their excitation-dependent fluorescence properties, several studies reported the integration of these fluorescent carbon-based materials for a synergistic and additive effect [123]. Green synthesized CDs using sorbitol and Arabic gum have also been tested for delivery of Dox and antibiotic, ciprofloxacin towards respective targets [124, 125]. Furthermore, glycerol and green tea-derived CDs could be used as theranostic agents with the additional ability to control the drug release of Dox and quinine sulfate [126, 127]. Harmful cyanobacteria have also been used in the synthesis of CDs and chemically conjugated with Dox for targeted drug delivery into cancer cells [128].

7.7.3 Antimicrobial Activity of Carbon-Based Nanomaterials

The use of photodynamic killing of bacteria emerged as a potential alternative to overcome the multi drug resistance of various pathogenic strains. Photosensitizers that generate ROS upon NIR irradiation are used to induce oxidative stress and subsequent protein/DNA damage. Among the various classes of photosensitizers, carbon dots represent a potential material for both detection and killing of bacteria owing to their unique physicochemical properties [129]. Recently, carbon dots were prepared using turmeric leaves (*Curcuma longa*) with a spherical shape and size of 2.6 nm. The synthesized carbon dots were found to exhibit cytotoxicity against contagious gram-negative *E. coli* and *K. pneumoniae* and gram positive, *S. aureus* and *S. epidermis* at very low concentrations. The killing action was elucidated by ROS generation [130]. CDs synthesized using onion extracts were reported to possess excessive sulfur content compared to the dots synthesized using garlic, ginger, and fish extracts. The as synthesized onion carbon dots exhibited significant antibacterial activity against *E. coli*, *S. aureus*, and *P. fragi* over a wide range of pH [131]. Carbon dots synthesized using seaweed extracts were found to exert potential antifungal activity against cucumber downy at very low concentrations. The presence of large π-conjugated nanostructures and the abundance of hydroxyl and amino groups make CDs a good delivery platform. Flumorph, a fungicide was successfully loaded with an efficiency of 47.18% and upon laser irradiation, resulted in significant antifungal synergistic activity [132]. Therefore, it is evident from the above studies, that the type and nature of biosource used as a carbon source influences the physicochemical and optical properties of the yielded carbon nanomaterials, and has a strong influence over the antimicrobial activity against pathogenic strains to overcome multi-drug resistance.

7.7.4 Anticancer Activity of Carbon-Based Nanomaterials

Owing to their unique luminescence properties, ease of synthesis, and functionalization, graphene or carbon dots have been widely used in cancer theranostics. Recently, a study reported the synthesis of C-dots using honey, a one-pot method, yielding particles of 7 nm size, which are much smaller than the conventionally used NPs such as gold, copper, etc. for imaging. Due to the strong absorption in the NIR range, these NPs showed significant signal enhancement for sentinel lymph node imaging. This could be thereby extended to monitor the cancer progression and treatment efficacy [133]. Withered leaves of plants have been used to synthesize graphene dots in a one-step facile procedure and yield NPs with NIR-responsive properties. These dots were capable of self-assembly, and the exceptional fluorescence properties could be exploited for the biolabeling of both normal and cancer cells. Additionally, NIR responsive nature aids in increased cytotoxicity via ROS generation in cancer cells upon 808 nm laser irradiation. Tracking of cell death using the inherent fluorescence properties of these dots even pot-irradiation proves the photostability of such materials in comparison with organic dyes undergoing photobleaching. Several studies reported synthesis methods of carbon nanomaterials using biosources for improving the yield and subsequent bioapplications. A recent study reports the synthesis of C-dots using *Manihot esculenta* via a hydrothermal method with a size range of 3–5 nm showing high selectivity and sensitivity towards Fe^{3+} ions, and for selective bioimaging of breast cancer cell lines, thereby proving to be a potential candidate for sensing and theranostics [134].

7.8 Conclusion and Future Directions

Biosynthesized metallic NPs are used in biomedicine as it has the potential to replace conventional therapy for various ailments. They have the capability to induce photocatalytic pathway wherein the using of light energy can increase the radical generations leading to oxidative stress. Due to oxidative stress, the cell signaling pathway further alters its program due to cellular membrane/lipid bilayer disruption due to DNA damage, and mitochondrial membrane potential disruption. Green synthesized NPs gained importance in the medical field as it is less toxic, and increased sustained drug release profile to engulf the tumor cell. Nanomaterials can be an effective therapy as it compromises unfavorable conventional methods like chemotherapy, and radiotherapy that results in multiple health relapse. NPs due to their size range between 1 and 100 nm help in renal clearance and can be used in passive targeting using non-invasive materials for diagnostics and therapy. This review article provides promising insights into the nanomedicine field using green synthesized nanostructured or doped metallic nanoparticles for their roles in drug-delivery, antimicrobial, and anticancer directions. These nanostructured metallic nanoparticles have a better light absorption rate that paves the way for passive and active targeting.

Acknowledgment

The authors would like to thank MHRD IMPRINT (4291), ICMR (No.35/1/2020-GIA/Nano/BMS), DST-Inspire (DST/INSPIRE/04/2015/000377), DST-AMT(DST/TDT/AMT/2017/227), SERB-CRG (CRG/2020/005069) grants. Author MP would like to acknowledge DST INSPIRE (IF180306) for funding her fellowship. Author SAS and SM would like to gratefully acknowledge MoE-PMRF (ID: 2000832) and ICMR for the fellowship, respectively.

Conflicts of Interest

The authors declare no conflict of interest.

References

1 Hood, L. (2013). Systems biology and P4 medicine: past, present, and future. *Rambam Maimonides Med. J.* 4 (2): e0012.
2 Muthu, M.S., Leong, D.T., Mei, L., and Feng, S.S. (2014). Nanotheranostics – application and further development of nanomedicine strategies for advanced theranostics. *Theranostics* 4 (6): 660–677.
3 Park, Y. (2014). A new paradigm shift for the green synthesis of antibacterial silver nanoparticles utilizing plant extracts. *Toxicol. Res.* 30 (3): 169–178.

4 Purohit, J., Chattopadhyay, A., and Singh, N.K. (2019). Green synthesis of microbial nanoparticle: approaches to application. In: *Microbial Nanobionics*, Nanotechnology in the Life Sciences (ed. R. Prasad), 35–60. Springer.

5 Narayanan, K.B. and Sakthivel, N. (2011). Green synthesis of biogenic metal nanoparticles by terrestrial and aquatic phototrophic and heterotrophic eukaryotes and biocompatible agents. *Adv. Colloid Interface Sci.* 169 (2): 59–79.

6 Khan, S.A., Shahid, S., and Lee, C.S. (2020). Green synthesis of gold and silver nanoparticles using leaf extract of *Clerodendrum inerme*; characterization, antimicrobial, and antioxidant activities. *Biomolecules* 10: 835.

7 Vijaya Kumar, P., Mary Jelastin Kala, S., and Prakash, K.S. (2019). Green synthesis of gold nanoparticles using *Croton caudatus* Geisel leaf extract and their biological studies. *Mater. Lett.* 236: 19–22.

8 Ismail, E.H., Saqer, A.M.A., Assirey, E. et al. (2018). Successful green synthesis of gold nanoparticles using a *Corchorus olitorius* extract and their antiproliferative effect in cancer cells. *Int. J. Mol. Sci.* 19: 2612.

9 El-Borady, O.M., Ayat, M.S., Shabrawy, M.A., and Millet, P. (2020). Green synthesis of gold nanoparticles using Parsley leaves extract and their applications as an alternative catalytic, antioxidant, anticancer, and antibacterial agents. *Adv. Powder Technol.* 31: 4390–4400.

10 Bogireddy, N.K.R., Pal, U., Gomez, L.M., and Agarwal, V. (2018). Size controlled green synthesis of gold nanoparticles using *Coffea arabica* seed extract and their catalytic performance in 4-nitrophenol reduction. *RSC Adv.* 8: 24819–24826.

11 ElMitwalli, O.S., Barakat, O.A., Daoud, R.M. et al. (2020). Green synthesis of gold nanoparticles using cinnamon bark extract, characterization, and fluorescence activity in Au/eosin Y assemblies. *J. Nanopart. Res.* 22: 309.

12 Singh, P., Pandit, S., Garnæs, J. et al. (2018). Green synthesis of gold and silver nanoparticles from *Cannabis sativa* (Industrial hemp) and their capacity for biofilm inhibition. *Int. J. Nanomed.* 13: 3571–3591.

13 Wang, L., Xu, J., Yan, Y. et al. (2019). Green synthesis of gold nanoparticles from *Scutellaria barbata* and its anticancer activity in pancreatic cancer cell (PANC-1). *Artif. Cells Nanomed. Biotechnol.* 47: 1617–1627.

14 Ganeshkumar, M., Sathishkumar, M., Ponrasu, T. et al. (2013). Spontaneous ultra fast synthesis of gold nanoparticles using *Punica granatum* for cancer targeted drug delivery. *Colloids Surf., B* 106: 208–216.

15 Devi, P.R., Kumar, C.S., Selvamani, P. et al. (2014). Synthesis and characterization of Arabic gum capped gold nanoparticles for tumor-targeted drug delivery. *Mater. Lett.* 139: 241–244.

16 Pooja, D., Panyaram, S., Kulhari, H. et al. (2014). Xanthan gum stabilized gold nanoparticles: characterization, biocompatibility, stability and cytotoxicity. *Carbohydr. Polym.* 110: 1–9.
17 Chen, X., Han, W., Zhao, X. et al. (2019). Epirubicin-loaded marine carrageenan oligosaccharide capped gold nanoparticle system for pH-triggered anticancer drug release. *Sci. Rep.* 9 (1): 1–10.
18 Balasubramanian, S., Kala, S.M.J., and Pushparaj, T.L. (2020). Biogenic synthesis of gold nanoparticles using *Jasminum auriculatum* leaf extract and their catalytic, antimicrobial and anticancer activities. *J. Drug Delivery Sci. Technol.* 57: 101620.
19 Folorunso, A., Akintelu, S., Oyebamiji, A.K. et al. (2019). Biosynthesis, characterization and antimicrobial activity of gold nanoparticles from leaf extracts of *Annona muricata*. *J. Nanostruct. Chem.* 9: 111–117.
20 Hamelian, M., Hemmati, S., Varmira, K., and Veisi, H. (2018). Green synthesis, antibacterial, antioxidant and cytotoxic effect of gold nanoparticles using *Pistacia atlantica* extract. *J. Taiwan Inst. Chem. Eng.* 93: 21–30.
21 Mandhata, C.P., Sahoo, C.R., Mahanta, C.S., and Padhy, R.N. (2021). Isolation, biosynthesis and antimicrobial activity of gold nanoparticles produced with extracts of *Anabaena spiroides*. *Bioprocess. Biosyst. Eng.* 44: 1617–1626.
22 Benedec, D., Oniga, I., Cuibus, F. et al. (2018). *Origanum vulgare* mediated green synthesis of biocompatible gold nanoparticles simultaneously possessing plasmonic, antioxidant and antimicrobial properties. *Int. J. Nanomed.* 13: 1041–1058.
23 Arief, S., Nasution, F.W., and Zulhadjri, L.A. (2020). High antibacterial properties of green synthesized gold nanoparticles using *Uncaria gambir* Roxb. leaf extract and triethanolamine. *J. Appl. Pharm Sci.* 10 (8): 124–130.
24 Wu, T., Duan, X., Hu, C. et al. (2019). Synthesis and characterization of gold nanoparticles from *Abies spectabilis* extract and its anticancer activity on bladder cancer T24 cells. *Artif. Cells Nanomed. Biotechnol.* 47: 512–523.
25 Ke, Y., Al Aboody, M.S., Alturaiki, W. et al. (2019). Photosynthesized gold nanoparticles from *Catharanthus roseus* induces caspase-mediated apoptosis in cervical cancer cells (HeLa). *Artif. Cells Nanomed. Biotechnol.* 47: 1938–1946.
26 Barai, A.C., Paul, K., Dey, A. et al. (2018). Green synthesis of *Nerium oleander*-conjugated gold nanoparticles and study of its in vitro anticancer activity on MCF-7 cell lines and catalytic activity. *Nano Convergence* 5: 10.
27 Chen, X., Zhao, X., Gao, Y. et al. (2018). Green synthesis of gold nanoparticles using carrageenan oligosaccharide and their in vitro antitumor activity. *Mar. Drugs* 16: 277.

28 Li, W., Liu, R., Pei, Q. et al. (2019). Apoptotic effect of green synthesized gold nanoparticles from *Curcuma wenyujin* extract against human renal cell carcinoma a498 cells. *Int. J. Nanomed.* 14: 4091–4103.

29 Sun, B., Hu, N., Han, L. et al. (2019). Anticancer activity of green synthesised gold nanoparticles from *Marsdenia tenacissima* inhibits A549 cell proliferation through the apoptotic pathway. *Artif. Cells Nanomed. Biotechnol.* 47: 4012–4019.

30 Sunderam, V., Thiyagarajan, D., Lawrence, A.V. et al. (2019). In-vitro antimicrobial and anticancer properties of green synthesized gold nanoparticles using *Anacardium occidentale* leaves extract. *Saudi J. Biol. Sci.* 26: 455–459.

31 Jeyarani, S., Vinita, N.M., Puja, P. et al. (2020). Biomimetic gold nanoparticles for its cytotoxicity and biocompatibility evidenced by fluorescence-based assays in cancer (MDA-MB-231) and non-cancerous (HEK-293) cells. *J. Photochem. Photobiol., B* 202: 111715.

32 Uzma, M., Sunayana, N., Raghavendra, V.B. et al. (2020). Biogenic synthesis of gold nanoparticles using *Commiphora wightii* and their cytotoxic effects on breast cancer cell line (MCF-7). *Process Biochem.* 92: 269–276.

33 Alsammarraie, F.K., Wang, W., Zhou, P. et al. (2018). Green synthesis of silver nanoparticles using turmeric extracts and investigation of their antibacterial activities. *Colloids Surf., B* 171 (April): 398–405. https://doi.org/10.1016/j.colsurfb.2018.07.059.

34 Hamedi, S. and Shojaosadati, S.A. (2019). Rapid and green synthesis of silver nanoparticles using *Diospyros lotus* extract: evaluation of their biological and catalytic activities. *Polyhedron* 171: 172–180. https://doi.org/10.1016/j.poly.2019.07.010.

35 Arumai Selvan, D., Mahendiran, D., Senthil Kumar, R., and Kalilur Rahiman, A. (2018). Garlic, green tea and turmeric extracts-mediated green synthesis of silver nanoparticles: phytochemical, antioxidant and in vitro cytotoxicity studies. *J. Photochem. Photobiol., B* 180 (February): 243–252. https://doi.org/10.1016/j.jphotobiol.2018.02.014.

36 Venkatadri, B., Shanparvish, E., Rameshkumar, M.R. et al. (2020). Green synthesis of silver nanoparticles using aqueous rhizome extract of *Zingiber officinale* and *Curcuma longa*: in-vitro anti-cancer potential on human colon carcinoma HT-29 cells. *Saudi J. Biol. Sci.* 27 (11): 2980–2986.

37 Rautela, A., Rani, J., and Debnath (Das), M. (2019). Green synthesis of silver nanoparticles from *Tectona grandis* seeds extract: characterization and mechanism of antimicrobial action on different microorganisms. *J. Anal. Sci. Technol.* 10 (1): 1–10. https://jast-journal.springeropen.com/articles/10.1186/s40543-018-0163-z.

38 Ivanova, N., Gugleva, V., Dobreva, M. et al. (2018). Silver nanoparticles as multi-functional drug delivery systems. In: *Nanomedicines* (ed. M.A. Farrukh), 71–92. IntechOpen.

39 Hussein, H.A. and Abdullah, M.A. (2021). Novel drug delivery systems based on silver nanoparticles, hyaluronic acid, lipid nanoparticles and liposomes for cancer treatment. *Appl. Nanosci.* 12: 3071–3096.

40 Paladini, F. and Pollini, M. (2019). Antimicrobial silver nanoparticles for wound healing application: progress and future trends. *Materials* 12: 2540. www.mdpi.com/journal/materials.

41 Roy, A., Bulut, O., Some, S. et al. (2019). Green synthesis of silver nanoparticles: biomolecule-nanoparticle organizations targeting antimicrobial activity. *RSC Adv.* 9 (5): 2673–2702.

42 Ahmed, R.H. and Mustafa, D.E. (2020). Green synthesis of silver nanoparticles mediated by traditionally used medicinal plants in Sudan. *Int. Nano Lett.* 10 (1): 1–14. https://doi.org/10.1007/s40089-019-00291-9.

43 Ahmed, S., Saifullah, Ahmad, M. et al. (2016). Green synthesis of silver nanoparticles using *Azadirachta indica* aqueous leaf extract. *J. Radiat. Res. Appl. Sci.* 9 (1): 1–7.

44 Dhand, V., Soumya, L., Bharadwaj, S. et al. (2016). Green synthesis of silver nanoparticles using *Coffea arabica* seed extract and its antibacterial activity. *Mater. Sci. Eng., C* 58: 36–43.

45 Garibo, D., Borbón-Nuñez, H.A., de León, J.N.D. et al. (2020). Green synthesis of silver nanoparticles using *Lysiloma acapulcensis* exhibit high-antimicrobial activity. *Sci. Rep.* 10 (1): 1–11. https://doi.org/10.1038/s41598-020-69606-7.

46 Jabeen, S., Qureshi, R., Munazir, M. et al. (2021). Application of green synthesized silver nanoparticles in cancer treatment – a critical review. *Mater. Res. Express* 8: 092001. https://doi.org/10.1088/2053-1591/ac1de3.

47 Kim, C.G., Castro-Aceituno, V., Abbai, R. et al. (2018). Caspase-3/MAPK pathways as main regulators of the apoptotic effect of the phyto-mediated synthesized silver nanoparticle from dried stem of *Eleutherococcus senticosus* in human cancer cells. *Biomed. Pharmacother.* 99: 128–133.

48 Castro Aceituno, V., Ahn, S., Simu, S.Y. et al. (2016). Silver nanoparticles from *Dendropanax morbifera* Léveille inhibit cell migration, induce apoptosis, and increase generation of reactive oxygen species in A549 lung cancer cells. *In Vitro Cell. Dev. Biol. Anim.* 52 (10): 1012–1019. https://link.springer.com/article/10.1007/s11626-016-0057-6.

49 Kummara, S., Patil, M.B., and Uriah, T. (2016). Synthesis, characterization, biocompatible and anticancer activity of green and chemically synthesized silver nanoparticles – a comparative study. *Biomed. Pharmacother.* 84: 10–21.

50 Jain, N., Jain, P., Rajput, D., and Patil, U.K. (2021). Green synthesized plant-based silver nanoparticles: therapeutic prospective for anticancer and antiviral activity. *Micro Nano Syst. Lett.* 9 (1): 1–24. https://mnsl-journal.springeropen.com/articles/10.1186/s40486-021-00131-6.

51 Ogunyemi, S.O., Abdallah, Y., Zhang, M. et al. (2019). Green synthesis of zinc oxide nanoparticles using different plant extracts and their antibacterial activity against *Xanthomonas oryzae* pv. *oryzae*. *Artif. Cells Nanomed. Biotechnol.* 47: 341–352.

52 Ishwarya, R., Vaseeharan, B., Kalyani, S. et al. (2018). Facile green synthesis of zinc oxide nanoparticles using *Ulva lactuca* seaweed extract and evaluation of their photocatalytic, antibiofilm and insecticidal activity. *J. Photochem. Photobiol., B* 178: 249–258.

53 Sorbiun, M., Shayegan Mehr, E., Ramazani, A., and Taghavi Fardood, S. (2018). Green synthesis of zinc oxide and copper oxide nanoparticles using aqueous extract of oak fruit hull (jaft) and comparing their photocatalytic degradation of basic violet 3. *Int. J. Environ. Res.* 12: 29–37.

54 Selim, Y.A., Azb, M.A., Ragab, I., and H. M. Abd El-Azim, M. (2020). Green synthesis of zinc oxide nanoparticles using aqueous extract of *Deverra tortuosa* and their cytotoxic activities. *Sci. Rep.* 10: 3445.

55 Aminuzzaman, M., Ying, L.P., Goh, W.S., and Watanabe, A. (2018). Green synthesis of zinc oxide nanoparticles using aqueous extract of *Garcinia mangostana* fruit pericarp and their photocatalytic activity. *Bull. Mater. Sci.* 41: 50.

56 Chikkanna, M.M., Neelagund, S.E., and Rajashekarappa, K.K. (2019). Green synthesis of zinc oxide nanoparticles (ZnO NPs) and their biological activity. *SN Appl. Sci.* 1: 117.

57 Chen, L., Batjikh, I., Hurh, J. et al. (2019). Green synthesis of zinc oxide nanoparticles from root extract of *Scutellaria baicalensis* and its photocatalytic degradation activity using methylene blue. *Optik* 184: 324–329.

58 Fakhari, S., Jamzad, M., and Kabiri Fard, H. (2019). Green synthesis of zinc oxide nanoparticles: a comparison. *Green Chem. Lett. Rev.* 12 (1): 19–24.

59 Vimala, K., Sundarraj, S., Paulpandi, M. et al. (2014). Green synthesized doxorubicin loaded zinc oxide nanoparticles regulates the Bax and Bcl-2 expression in breast and colon carcinoma. *Process Biochem.* 49 (1): 160–172.

60 Vimala, K. and Soundarapandian, K. (2017). Erbitux conjugated zinc oxide nanoparticles to enhance antitumor efficiency via targeted drug delivery system for breast cancer therapy. *Ann. Oncol.* 28: x41.

61 Wang, Q., Mei, S., Manivel, P. et al. (2022). Current research in food science zinc oxide nanoparticles synthesized using coffee leaf extract assisted with ultrasound as nanocarriers for mangiferin. *Curr. Res. Food Sci.* 5 (May): 868–877.

62 Ahmad, W. and Kalra, D. (2020). Green synthesis, characterization and anti microbial activities of ZnO nanoparticles using *Euphorbia hirta* leaf extract. *J. King Saud Univ. Sci.* 32 (4): 2358–2364.
63 Ifeanyichukwu, U.L., Fayemi, O.E., and Ateba, C.N. (2020). Green synthesis of zinc oxide nanoparticles from pomegranate (*Punica granatum*) extracts and characterization of their antibacterial activity. *Molecules* 25 (19): 4521.
64 Doan Thi, T.U., Nguyen, T.T., Thi, Y.D. et al. (2020). Green synthesis of ZnO nanoparticles using orange fruit peel extract for antibacterial activities. *RSC Adv.* 10 (40): 23899–23907.
65 Vijayakumar, S., Mahadevan, S., Arulmozhi, P. et al. (2018). Green synthesis of zinc oxide nanoparticles using *Atalantia monophylla* leaf extracts: characterization and antimicrobial analysis. *Mater. Sci. Semicond. Process.* 82: 39–45.
66 Ansari, M.A., Murali, M., Prasad, D. et al. (2020). *Cinnamomum verum* bark extract mediated green synthesis of ZnO nanoparticles and their antibacterial potentiality. *Biomolecules* 10 (2): 336.
67 Jayappa, M.D., Ramaiah, C.K., Kumar, M.A.P. et al. (2020). Green synthesis of zinc oxide nanoparticles from the leaf, stem and in vitro grown callus of *Mussaenda frondosa* L.: characterization and their applications. *Appl. Nanosci.* 10: 3057–3074.
68 Fahimmnisha, B., Ishwarya, R., AlSalhi, M.S. et al. (2019). Green fabrication, characterization and antibacterial potential of zinc oxide nanoparticles using *Aloe socotrina* leaf extract – a novel drug delivery approach. *J. Drug Delivery Sci. Technol.* 55: 101465.
69 Hussein, B.Y. and Mohammed, A.M. (2021). Green synthesis of ZnO nanoparticles in grape extract: their application as anti-cancer and anti-bacterial. *Mater. Today Proc.* 42: A18–A26.
70 Vijayakumar, S., Saravanakumar, K., Malaikozhundan, B. et al. (2020). Biopolymer K-carrageenan wrapped ZnO nanoparticles as drug delivery vehicles for anti MRSA therapy. *Int. J. Biol. Macromol.* 144: 9–18.
71 Sharmila, G., Thirumarimurugan, M., and Muthukumaran, C. (2019). Green synthesis of ZnO nanoparticles using *Tecoma castanifolia* leaf extract: characterization and evaluation of its antioxidant, bactericidal and anticancer activities. *Microchem. J.* 145: 578–587.
72 Sathappan, S., Kirubakaran, N., Gunasekaran, D. et al. (2021). Green synthesis of zinc oxide nanoparticles (ZnO NPs) using *Cissus quadrangularis*: characterization, antimicrobial and anticancer studies. *Proc. Natl. Acad. Sci. India Sect. B - Biol. Sci.* 91 (2): 289–296.
73 Dulta, K., Koşarsoy Ağçeli, G., Chauhan, P. et al. (2021). A novel approach of synthesis zinc oxide nanoparticles by *Bergenia ciliata* rhizome extract: antibacterial and anticancer potential. *J. Inorg. Organomet. Polym. Mater.* 31 (1): 180–190.

74 Theophil Anand, G., Renuka, D., Ramesh, R. et al. (2019). Green synthesis of ZnO nanoparticle using *Prunus dulcis* (Almond Gum) for antimicrobial and supercapacitor applications. *Surf. Interfaces* 17 (June): 100376.

75 Pillai, A.M., Sivasankarapillai, V.S., Rahdar, A. et al. (2020). Green synthesis and characterization of zinc oxide nanoparticles with antibacterial and antifungal activity. *J. Mol. Struct.* 1211: 128107.

76 Mallikarjunaswamy, C., Lakshmi Ranganatha, V., Ramu, R. et al. (2020). Facile microwave-assisted green synthesis of ZnO nanoparticles: application to photodegradation, antibacterial and antioxidant. *J. Mater. Sci.: Mater. Electron.* 31 (2): 1004–1021.

77 Ruangtong, J., T-Thienprasert, J., and T-Thienprasert, N.P. (2020). Green synthesized ZnO nanosheets from banana peel extract possess anti-bacterial activity and anti-cancer activity. *Mater. Today Commun.* 24: 101224.

78 Umamaheswari, A., Prabu, S.L., John, S.A., and Puratchikody, A. (2021). Green synthesis of zinc oxide nanoparticles using leaf extracts of *Raphanus sativus* var. *longipinnatus* and evaluation of their anticancer property in A549 cell lines. *Biotechnol. Rep.* 29: e00595.

79 Abdel Fadeel, D.A., Hanafy, M.S., Kelany, N.A., and Elywa, M.A. (2021). Novel greenly synthesized titanium dioxide nanoparticles compared to liposomes in drug delivery: in vivo investigation on Ehrlich solid tumor model. *Heliyon* 7 (6): e07370. https://doi.org/10.1016/j.heliyon.2021.e07370.

80 Aravind, M., Amalanathan, M., and Mary, M.S.M. (2021). Synthesis of TiO_2 nanoparticles by chemical and green synthesis methods and their multifaceted properties. *SN Appl. Sci.* 3: 409.

81 Al-Shabib, N.A., Husain, F.M., Qais, F.A. et al. (2020). Phyto-mediated synthesis of porous titanium dioxide nanoparticles from *Withania somnifera* root extract: broad-spectrum attenuation of biofilm and cytotoxic properties against HepG2 cell lines. *Front. Microbiol.* 11: 1680.

82 Isnaeni, I.N., Indriyati, D., Sumiarsa, D., and Primadona, I. (2021). Green synthesis of different TiO_2 nanoparticle phases using mango-peel extract. *Mater. Lett.* 294: 129792.

83 Patidar, V. and Jain, P. (2017). Green synthesis of TiO_2 nanoparticle using *Moringa oleifera* leaf extract. *Int. Res. J. Eng. Technol.* 4: 470–473.

84 Lagopati, N., Evangelou, K., Falaras, P. et al. (2021). Nanomedicine: photo-activated nanostructured titanium dioxide, as a promising anticancer agent. *Pharmacol. Ther.* 222: 107795.

85 Liu, E., Zhou, Y., Liu, Z. et al. (2015). Cisplatin loaded hyaluronic acid modified TiO_2 nanoparticles for neoadjuvant chemotherapy of ovarian cancer. *J. Nanomater.* 16: 275.

86 He, C.F., Wang, S.H., Yu, Y.J. et al. (2016). Advances in biodegradable nanomaterials for photothermal therapy of cancer. *Cancer Biol. Med.* 13 (3): 299.

87 Ziental, D., Czarczynska-Goslinska, B., Mlynarczyk, D.T. et al. (2020). Titanium dioxide nanoparticles: prospects and applications in medicine. *Nanomaterials* 10 (2): 387.

88 Subhapriya, S. and Gomathipriya, P. (2018). Green synthesis of titanium dioxide (TiO_2) nanoparticles by *Trigonella foenum-graecum* extract and its antimicrobial properties. *Microb. Pathogen.* 116: 215–220.

89 Nadeem, M., Tungmunnithum, D., Hano, C. et al. (2018). The current trends in the green syntheses of titanium oxide nanoparticles and their applications. *Green Chem. Lett. Rev.* 11 (4): 492–502. https://www.tandfonline.com/doi/abs/10.1080/17518253.2018.1538430.

90 Hunagund, S.M., Desai, V.R., Kadadevarmath, J.S. et al. (2016). Biogenic and chemogenic synthesis of TiO_2NPs via hydrothermal route and their antibacterial activities. *RSC Adv.* 6 (99): 97438–97444. https://pubs.rsc.org/en/content/articlehtml/2016/ra/c6ra22163g.

91 Galata, E., Georgakopoulou, E.A., Kassalia, M.E. et al. (2019). Development of smart composites based on doped-TiO_2 nanoparticles with visible light anticancer properties. *Materials (Basel, Switzerland)* 12 (16): 2589. https://pubmed.ncbi.nlm.nih.gov/31416238/.

92 Ilyas, M., Waris, A., Khan, A.U. et al. (2021). Biological synthesis of titanium dioxide nanoparticles from plants and microorganisms and their potential biomedical applications. *Inorg. Chem. Commun.* 133: 108968.

93 Maheswari, P., Harish, S., Navaneethan, M. et al. (2020). Bio-modified TiO_2 nanoparticles with *Withania somnifera*, *Eclipta prostrata* and *Glycyrrhiza glabra* for anticancer and antibacterial applications. *Mater. Sci. Eng. C Mater. Biol. Appl.* 108: 110457. https://pubmed.ncbi.nlm.nih.gov/31924033/.

94 Hariharan, D., Thangamuniyandi, P., Selvakumar, P. et al. (2019). Green approach synthesis of Pd@TiO_2 nanoparticles: characterization, visible light active picric acid degradation and anticancer activity. *Process Biochem.* 87: 83–88.

95 Maheswari, P., Ponnusamy, S., Harish, S. et al. (2020). Hydrothermal synthesis of pure and bio modified TiO_2: characterization, evaluation of antibacterial activity against gram positive and gram negative bacteria and anticancer activity against KB oral cancer cell line. *Arabian J. Chem.* 13 (1): 3484–3497.

96 Shiva Samhitha, S., Raghavendra, G., Quezada, C., and Hima Bindu, P. (2022). Green synthesized TiO_2 nanoparticles for anticancer applications: mini review. *Mater. Today Proc.* 54: 765–770.

97 Üstün, E., Önbaş, S.C., Çelik, S.K. et al. (2022). Green synthesis of iron oxide nanoparticles by using *Ficus carica* leaf extract and its antioxidant activity. *Biointerface Res. Appl. Chem.* 12 (2): 2108–2116.

98 Hemalatha, S. and Venkatesan, R.S. (2021). Fabrication of iron oxide nano particle using *Cassia auriculata* leaf extract and its characterization. *Ann. Rom. Soc. Cell Biol.* 25 (3): 1683–1689.

99 Singh, P., Singh, K.R., Verma, R. et al. (2022). Efficient electro-optical characteristics of bioinspired iron oxide nanoparticles synthesized by *Terminalia chebula* dried seed extract. *Mater. Lett.* 307: 131053.

100 Chatterjee, S., Mahanty, S., Das, P. et al. (2020). Biofabrication of iron oxide nanoparticles using manglicolous fungus *Aspergillus niger* BSC-1 and removal of Cr(VI) from aqueous solution. *Chem. Eng. J.* 385: 123790.

101 Mahanty, S., Bakshi, M., Ghosh, S. et al. (2019). Green synthesis of iron oxide nanoparticles mediated by filamentous fungi isolated from Sundarban Mangrove Ecosystem, India. *Bionanoscience* 9 (3): 637–651.

102 Fatemi, M., Mollania, N., Momeni-Moghaddam, M., and Sadeghifar, F. (2018). Extracellular biosynthesis of magnetic iron oxide nanoparticles by *Bacillus cereus* strain HMH1: characterization and in vitro cytotoxicity analysis on MCF-7 and 3T3 cell lines. *J. Biotechnol.* 270 (January): 1–11.

103 Kiwumulo, H.F., Muwonge, H., Ibingira, C. et al. (2022). Green synthesis and characterization of iron-oxide nanoparticles using *Moringa oleifera*: a potential protocol for use in low and middle income countries. *BMC Res. Notes* 15 (1): 1–8.

104 Verçoza, B.R.F., Bernardo, R.R., Pentón-Madrigal, A. et al. (2019). Therapeutic potential of low-cost nanocarriers produced by green synthesis: macrophage uptake of superparamagnetic iron oxide nanoparticles. *Nanomedicine* 14 (17): 2293–2313.

105 Yusefi, M., Shameli, K., Hedayatnasab, Z. et al. (2021). Green synthesis of Fe_3O_4 nanoparticles for hyperthermia, magnetic resonance imaging and 5-fluorouracil carrier in potential colorectal cancer treatment. *Res. Chem. Intermed.* 47 (5): 1789–1808.

106 Yoonus, J., Resmi, R., and Beena, B. (2020). Evaluation of antibacterial and anticancer activity of green synthesized iron oxide (α-Fe_2O_3) nanoparticles. *Mater. Today Proc.* 46: 2969–2974.

107 Zakariya, N.A., Majeed, S., and Jusof, W.H.W. (2022). Investigation of antioxidant and antibacterial activity of iron oxide nanoparticles (IONPS) synthesized from the aqueous extract of Penicillium spp. *Sens. Int.* 3 (February): 1–9.

108 Velsankar, K., Parvathy, G., Mohandoss, S. et al. (2021). *Celosia argentea* leaf extract-mediated green synthesized iron oxide nanoparticles for bio-applications. *J. Nanostruct. Chem.* 12: 625–640.

109 Buarki, F., AbuHassan, H., Al Hannan, F., and Henari, F.Z. (2022). Green synthesis of iron oxide nanoparticles using *Hibiscus rosa sinensis* flowers and their antibacterial activity. *J. Nanotechnol.* 2022: 1–6.

110 Yusefi, M., Shameli, K., Ali, R.R. et al. (2020). Evaluating anticancer activity of plant-mediated synthesized iron oxide nanoparticles using *Punica granatum* fruit peel extract. *J. Mol. Struct.* 1204: 127539.

111 Nagajyothi, P.C., Pandurangan, M., Kim, D.H. et al. (2017). Green synthesis of iron oxide nanoparticles and their catalytic and in vitro anticancer activities. *J. Cluster Sci.* 28 (1): 245–257.

112 Kharey, P., Indoliya, A., Gupta, R. et al. (2022). Near-infrared active superparamagnetic iron oxide nanoparticles for magnetomotive optical coherence tomography imaging and magnetic hyperthermia therapeutic applications. *J. Magn. Magn. Mater.* 549 (December 2021): 169038.

113 Karade, V.C., Parit, S.B., Dawkar, V.V. et al. (2019). A green approach for the synthesis of α-Fe_2O_3 nanoparticles from *Gardenia resinifera* plant and it's in vitro hyperthermia application. *Heliyon* 5 (7): e02044.

114 Chufa, B.M., Murthy, H.C.A., Gonfa, B.A., and Anshebo, T.Y. (2021). Carbon nanotubes: a review on green synthesis, growth mechanism and application as a membrane filter for fluoride remediation. *Green Chem. Lett. Rev.* 14 (4): 640–657.

115 Primo, A., Atienzar, P., Sanchez, E. et al. (2012). From biomass wastes to large-area, high-quality, N-doped graphene: catalyst-free carbonization of chitosan coatings on arbitrary substrates. *Chem. Commun.* 48 (74): 9254–9256.

116 Yuan, X., Wei, Y., Chen, S. et al. (2016). Bio-based graphene/sodium alginate aerogels for strain sensors. *RSC Adv.* 6 (68): 64056–64064.

117 Sharma, A. and Das, J. (2019). Small molecules derived carbon dots: synthesis and applications in sensing, catalysis, imaging, and biomedicine. *J. Nanobiotechnol.* 17 (1): 1–24.

118 Huang, H., Xu, Y., Tang, C.J. et al. (2014). Facile and green synthesis of photoluminescent carbon nanoparticles for cellular imaging. *New J. Chem.* 38 (2): 784–789.

119 Afonso, A.J.G., Aquino, F.T., Dalmônico, G.M.L. et al. (2022). Green synthesis of carbon nanodots from agro-industrial residues. *Carbon Lett.* 32 (1): 131–141.

120 Dutta, S.D., Hexiu, J., Kim, J. et al. (2022). Two-photon excitable membrane targeting polyphenolic carbon dots for long-term imaging and pH-responsive chemotherapeutic drug delivery for synergistic tumor therapy. *Biomater. Sci.* 10 (7): 1680–1696.

121 Tejwan, N., Kundu, M., Ghosh, N. et al. (2022). Synthesis of green carbon dots as bioimaging agent and drug delivery system for enhanced antioxidant and antibacterial efficacy. *Inorg. Chem. Commun.* 139 (February): 109317.

122 Vasimalai, N., Vilas-Boas, V., Gallo, J. et al. (2018). Green synthesis of fluorescent carbon dots from spices for in vitro imaging and tumour cell growth inhibition. *Beilstein J. Nanotechnol.* 9 (1): 530–544.

123 Flak, D., Przysiecka, Ł., Nowaczyk, G. et al. (2018). GQDs-MSNs nanocomposite nanoparticles for simultaneous intracellular drug delivery and fluorescent imaging. *J. Nanopart. Res.* 20 (11): 306.

124 Phadke, C., Mewada, A., Dharmatti, R. et al. (2015). Biogenic synthesis of fluorescent carbon dots at ambient temperature using *Azadirachta indica* (Neem) gum. *J. Fluoresc.* 25 (4): 1103–1107.

125 Thakur, M., Pandey, S., Mewada, A. et al. (2014). Antibiotic conjugated fluorescent carbon dots as a theranostic agent for controlled drug release, bioimaging, and enhanced antimicrobial activity. *J. Drug Delivery* 2014: 1–9.

126 Irmania, N., Dehvari, K., Gedda, G. et al. (2020). Manganese-doped green tea-derived carbon quantum dots as a targeted dual imaging and photodynamic therapy platform. *J. Biomed. Mater. Res. - Part B Appl. Biomater.* 108 (4): 1616–1625.

127 Lai, C.W., Hsiao, Y.H., Peng, Y.K., and Chou, P.T. (2012). Facile synthesis of highly emissive carbon dots from pyrolysis of glycerol; Gram scale production of carbon dots/mSiO_2 for cell imaging and drug release. *J. Mater. Chem.* 22 (29): 14403–14409.

128 Lee, H.U., Park, S.Y., Park, E.S. et al. (2014). Photoluminescent carbon nanotags from harmful cyanobacteria for drug delivery and imaging in cancer cells. *Sci. Rep.* 4: 1–7.

129 Ghirardello, M., Ramos-Soriano, J., and Galan, M.C. (2021). Carbon dots as an emergent class of antimicrobial agents. *Nanomaterials* 11 (8): 1–24.

130 Saravanan, A., Maruthapandi, M., Das, P. et al. (2021). Green synthesis of multifunctional carbon dots with antibacterial activities. *Nanomaterials* 11 (2): 1–11.

131 Lin, R., Cheng, S., and Tan, M. (2022). Green synthesis of fluorescent carbon dots with antibacterial activity and their application in Atlantic mackerel (*Scomber scombrus*) storage. *Food Funct.* 13 (4): 2098–2108.

132 Zhao, S., Huang, L., Xie, Y. et al. (2021). Green synthesis of multifunctional carbon dots for anti-cancer and anti-fungal applications. *Chin. J. Chem. Eng.* 37: 97–104.

133 Wu, L., Cai, X., Nelson, K. et al. (2013). A green synthesis of carbon nanoparticles from honey and their use in real-time photoacoustic imaging. *Nano Res.* 6 (5): 312–325.

134 Sivanandhan, M., Parasuraman, A., Surya, C. et al. (2022). Facile approach for green synthesis of fluorescent carbon dots from *Manihot esculenta* and their potential applications as sensor and bio-imaging agents. *Inorg. Chem. Commun.* 137 (October 2021): 109219.

8

How Eco-friendly Nanomaterials are Effective for the Sustainability of the Environment

Manoj Kumar[1], *Preeti Sharma*[2], *Archana Chakravarty*[3], *Sikandar Paswan*[4], *and Deepak Kumar Bhartiya*[5]

[1] *Department of Chemistry, Government Degree College Dhadha Bujurg-Hata, Kushinagar, UP, India*
[2] *Department of Natural Sciences, University of Maryland Eastern Shore, Princess Anne, MD, USA*
[3] *Department of Chemistry, Jamia Millia Islamia, New Delhi, DL, India*
[4] *Department of Chemistry, Baba Raghav Das PG College, Deoria, UP, India*
[5] *Department of Zoology, Government Degree College Dhadha Bujurg-Hata, Kushinagar, UP, India*

Abstract

Recent developments in nanoscience have significantly attraction of researchers to grow novel nanomaterials to mitigate human and environmental threats. In light of the sustainability of the environment, the green synthetic route of green nanotechnology is a valuable method because hazardous compounds are not used for the synthetic protocol. Using naturally found plant-based extracts, microorganisms, biopolymers, etc. attains various benefits as ecofriendly and biocompatible for different therapeutic and pharmacological applications. Metallic nanoparticles that are energy-efficient, affordable and nontoxic have been produced in recent years using algae, plants, bacteria, fungi and viruses. Nanotechnology plays a significant role in nano-adsorbents, nano-photocatalyst, nano-filters and magnetic nanomaterials, for removing and identifying pollutants from water, air and soil environment nanosensors and plays a significant tool in attaining sustainable development. It can considerably influence the emerging of green and clean technology for environmental and health welfare. The eco-friendly nanomaterial has the potential to give solutions to clear up soil and air pollution, mitigate water contamination, enhance the efficiency of old-fashioned technology employed in environmental clean-up and for the sustainable development of society.

Green Synthesis of Nanomaterials: Biological and Environmental Applications, First Edition.
Edited by Archana Chakravarty, Preeti Singh, Saiqa Ikram, and R.N. Yadava.

Keywords *eco friendly; sustainable; nanomaterials; water contamination; soil pollution; biomass*

8.1 Introduction

Now-a-days, nanotechnology has become one of the most scientific and significant industrial revolutions because its applications cross the scientific frontiers due to the advanced manufacturing of electronic appliances to pharmacological products. Many materials and methods are used for the synthesis that produces hazardous by-products. Therefore, using green resources, such as natural and renewable materials, in synthesizing and designing nanomaterials and green nanotechnology has taken on new significance [1]. The green synthetic method received considerable attention over the conventional method because it has become significant toward plentiful advantages like economic and nonhazards and has many uses in nanomedicine and pharmacy. As the population increases and the stock of resources comes to be limited, thus, there is a need for pollution-free technology for clean energy supply and environmental remediation for sustainable development.

According to Buzea et al., nanomaterials typically range in size (in at least one dimension) among 1–100 nm [2]. Nanoscale substances have electrical, machine-driven, optical and magnetic properties that differ dramatically than those of conventional materials. Nanomaterials can take on a variety of different shapes in addition to catalysis, intense reactivity and adsorption. In recent years, a variety of industries have successfully used nanomaterials, including biological science [3], catalytic activity [4], healthcare [5] and sensory sectors [6]. Nanosafety is essential due to the risks posed by man-made nanomaterials (MNMs), whose effects might be greater than the benefits they provide for ecological usage. Even though they are incremental and unquestionably beneficial for human health and Earth's natural ecosystems, the potential hazards that arise from MNMs' safety for people and the planet are impeding the implementation of pollution control, remediation and prevention strategies using nanotechnology [7].

Although nanotechnology attracted much attention in the last decades, clean development and sustainability of nanoparticles is an encouraging job for researchers. The most often performed task by scientists is green nanoparticle production using plant extracts. However, they become valueless by using organic and inorganic bases. Using human urine as a reducing agent,

Dabhane et al. stated the biogenic production of copper oxide nanoparticles (CuO NPs) [8]. In the presence of solar light, the CuO NPs break down phenol in an aqueous medium in an environmentally benign manner. The removal's effectiveness was assessed with a device called a spectrophotometer. CuO NPs have reportedly been demonstrated to have specific antibacterial capabilities when tested on gram-positive and gram-negative pathogenic bacteria, according to Dabhane et al. [8]. The ability to kill fungi was also demonstrated in several distinct fungus strains [8].

According to Anindita De et al., water pollution poses a serious hazard in poor or emerging nations and is one of the prime causes of human fatalities globally [9]. Additionally, the issue of water contamination is creating an acute risk to health and nutrition due to rapid industrial development and urbanization. Water carries a variety of contaminants, depending on the source of the water bodies. It needs various methods or technology to treat them and prevent contamination. However, these methods have many drawbacks and disadvantages restricted to large-scale, long-term uses and sustainability [9]. Nanomaterial synthesized by green technology has been used for water treatment and making pure. Green synthesized nanomaterials are used as an adsorbent, catalyst and membrane for water refining [9]. Another nanomaterial harms the environment because different chemicals are used for fabrication.

Regarding this, green nanomaterial can make a sustainable and effective tool for the environment. Due to their smaller size and opportunity to target higher efficacy in wastewater treatment, nanomaterials play a significant role in removing pollutants from wastewater. In order to avoid water pollution, biosynthesized nanomaterials are used, such as silver-impregnated cyclodextrin nanocomposites, silver nanoparticles in *Aloe vera* plant extract, magnetic, zinc oxide; and metal nanomaterials coated with biopolymer [10]. One efficient method to remove dangerous compounds from water is to use green nanoparticles. Sustainable nanoparticles provide a straightforward, economical; and ecofriendly wastewater treatment option, according to Neeti et al. [10]. The ecosystem and public health are seriously endangered by contaminated water, soil and air, which makes them essential for sustainable development. Older methods like thermal processing, pump-and-treat and oxidization by chemicals are included in the list of proven remediation procedures, as are more recent ones like "nano remediation." In this chapter, we review and explain various routes for using green nanomaterials for water treatment, as well as consolidate recent advancements in this vital research area.

8.2 Eco-friendly Nanomaterials

Green chemistry is receiving greater prominence due to global environmental challenges, and as a result, sustainable methods or green manufacturing of nanoparticles are expanding quickly. The engineered nanomaterials are now used for remediation of the environment and sustainable development, which is less expensive and more efficient than the majority of conventional approaches [11, 12]. The word "nanoparticles" refers to atom clusters with a size between 1 and 100 nm (10^{-9} nm). Because of its exponential potential, materials with nanoscale dimensions have garnered significant interest from scientists; this is a popular study topic in the physical, chemical and biological sciences as well as in the disciplines of materials and medicine throughout time. Nowadays, many scientists have synthesized nanoparticles using physicochemical processes. Although ablation with lasers and evaporation–condensation are the most prominent physical approaches for the production of nanoparticles, numerous researchers are currently adopting physicochemical processes [13]. These techniques have a low yield and are too expensive. When several chemical processes (such as hydrothermal, sonochemical, solvothermal, sol–gel and mechanical chemistries) are used to create nanoparticles, These procedures generate toxic by-products that are harmful to the environment [14–16]. Nowadays, the focus of the study has shifted to creating nanoparticles using bio-based techniques, which are affordable, nontoxic and environmentally benign [17–20]. Due to their distinctive physicochemical and therapeutic efficacies, algae-mediated nanoparticle production has entered a variety of prospective applications in research, technology, and medicine. Biological nanomaterials have tremendous applications in the field of biotechnology, physics, quantum technology, chemistry and material science. These nanoparticles are critical in the elimination of environmental toxins because of their extremely reactive outer layer, which operates as adsorbents, biosensors, and catalysts for the removal of different contaminants [21]. In the twenty-first century, nanotechnology has emerged as a prominent area of unconventional advancement and research across a number of sectors, including those of engineering, agriculture, cosmetics, medicine and food production [22]. Biosynthesized nanomaterials as efficient diagnostic or medication delivery systems are possible and developed by different scientists [23]. These nanoparticles are crucial for the treatment of wastewater. Water sources can get contaminated by various ions, heavy metals, petroleum hydrocarbons, pesticides, radioactive elements and novel contaminants, including pharmaceuticals and personal care products. Several pollutants, including pathogenic bacteria, viruses, radionuclides and heavy metal ions, are discharged into water supplies as a result of increasing industrialization, endangering human health. Wastewater is created when industrial effluents, organic contaminants or other substances damage the quality of water [24]. Activated carbons, carbon nanotubes, graphene and its oxide are ideal for water purification processes to remove impurities such as heavy metals, fluorides, textile colors or medicines due to their physicochemical features. Due to their characteristics, which include selectivity to specific contaminants and their absorption capacity, many methods based on nanoparticles are being employed in the cleaning of water. Metallic nanoparticles, carbon-based materials and biopolymeric membranes are the most common forms of nanomaterials employed in water treatment. [25–27]. Because of their excellent reactivity, photolytic and adsorbent properties which are brought about by their enormous surface area and attraction to various chemical groups, a variety of metal oxide nanoparticles, including iron oxide (Fe_2O_3/Fe_3O_4), zinc oxide (ZnO) and titanium dioxide (TiO_2), have been utilized for water treatment [28]. It is crucial to choose nanomaterials for water purification based on their potential for reuse, environmental friendliness and economic viability [29, 30]. Nanotechnology is increasingly being applied in various environmental protection domains today. It is renowned for its distinctive physiochemical features of nanomaterials. The elimination of organic toxins, hazardous metals and inorganic pollutants is being studied by utilizing a range of nanomaterials. Economically, nanotechnology promotes the efficient use of water sources and lower energy consumption [31]. Nanostructured adsorbents additionally serve in sewage treatment because of their significantly higher effectiveness and faster reactivity. Magnetic nanoparticles are important and designed for the sorption of metals and organic pollutants. Silver nanomaterials play a vital role as antibacterial agents against various microorganisms present in nature and also in the purification of the community water in hospitals. They have been utilized in place of chlorine in purifying methods. Silver nanoparticles have the potential to be employed in sewage and wastewater treatment as biofouling-reducing agents and also as an effective disinfectant for the removal of *E. coli* and other bacteria [32–34]. Most frequently, carbon nanoparticles are employed to filter water. Carbon nanoparticles can be used to assess the cost-effectiveness of using activated carbon as well as to recover the adsorption capacity of used carbon. Furthermore, iron nanoparticles are the next most commonly used in the sewage treatment process. Reductive dehalogenation reaction is the primary basis for this treatment by utilizing iron nanoparticles. Iron nanoparticles are inexpensive and form metal hydroxide by the reaction with pollutants, which act as a flocculant to remove both inorganic and organic pollution [35]. Lignocellulosic biomass is a term for plant-based materials having ecologically friendly qualities that may be used to produce biofuels, biochemicals, bioethanol and other high-value commodities sustainably [36, 37]. Biomass is a great source of nano-sized cellulose, which is highly prized as a reinforcing material due to its high cellulosic content. Nanocellulose formed from agro-industrial biomass, or specifically from plant fibers, is favored because of its greater biocompatibility, ability to decompose, sustainability and high hydrophilic properties. Aside from being lightweight, nanocellulose materials have an excellent aspect ratio, an extensive surface area, the ability to modify the surface, a particular permeation and resistance, a high Young's modulus and a low thermal expansion coefficient [38–40]. Soil degradation and pollution have escalated as a result of the presence and growth of industrialization and excessive urbanization. Nanomaterials have lately become more popular for soil remediation because of their intense reactivity, large surface-to-volume ratio, functionalization on the surface and ability to alter physical characteristics like size, shape, permeability and chemical structure [41]. Additionally, using nanoscale zero-valent iron, high cleaning percentages for the treatment of trichloroethene, Dichlorodiphenyltrichloroethane, hexavalent chromium, nitrate, lead and cadmium have been demonstrated [42]. Air pollution, which affects both public health and climate change, is one of the largest problems the world is currently facing. The six prominent and harmful airborne contaminants are particulate matter, nitrogen oxides (NO_X), sulfur dioxide (SO_2), carbon monoxide (CO), lead and ozone at the ground level, which are formed by chemical interactions among NO_X and organic molecules that are volatile [43]. Nanomaterials are widely regarded in bio- and nanomedicine as a promising class of medicines useful for enhancing the efficacy of theranostics for serious disorders. Many researchers worked to create nanoparticles with customizable form and composition to enable high-contrast imaging using MRI, CT scans or near-infrared imaging, coupled with an effective remote therapy like hyperthermia and/or photo-ablation [44–46]. Researchers also looked at constructing multifunctional organic coatings or macromolecular vesicles with a variety of activities in precise magnetic nanoparticles, as well as approving furtivity, bio-distribution, targeting and drug delivery, as well as *in vitro* and *in vivo* validations of their productivity (Figure 8.1; Table 8.1) [59, 60].

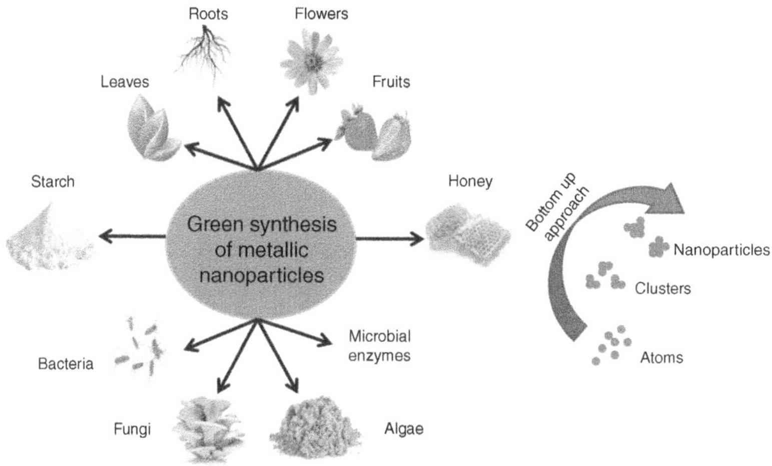

Figure 8.1 Synthetic route of eco-friendly nanomaterials. *Source:* Kumar et al. [47]/Reproduced from MDPI/CC BY 4.0.

8.3 Green Nanomaterial for Removal of Water Contamination

Currently, rapid industrialization and urbanization generate a large amount of wastewater, which have heavy metal and toxic chemicals. This leads to environmental threats and affects the health and life of living well-being, if not treated properly [61]. The organic pollutant in the water supply continually attracts a significant universal consideration. These organic pollutants are determined or can be converted into unruly products or biodegradable. The long-term consumption of toxic organic contaminants is associated with serious environmental issues and adverse health problems [62]. In this section, we cover the recent advancement and application of green nanoparticles synthesized from agricultural waste and/or biochars in removing pollutants or contamination from water bodies. Biochar is synthesized from waste biomass by thermal conversion in an anaerobic environment [62]. It is carbon residue and porous in nature (Figure 8.2).

Presently, there is much literature on water treatment, particularly the elimination of persistent organic toxins by the potential application of biochar. Biological and chemical modifications in biochar using nanoparticles (NPs) have generated a novel type of hybrid chars, which contain a large potential for the treatment of water to mitigate organic pollutants [62].

Table 8.1 Some eco-friendly nanomaterials and their applications in water treatment.

Serial number	Eco-friendly materials	Source	Application	Reference
1.	Silver nanoparticles	Vaccinium macrocarpon fruit	Reduction of methylene blue, methyl orange, rhodamine	[48]
2.	Ag–Au nanoparticles	*Albizia saman* leaf	Reduction of toxic Cr(VI)	[49]
3.	Mg nanoparticles	*Acacia* sp.	Removal of Ni(II), Pb(II), Cd(II), Cu(II), Zn(II), Co(II), Mn(II)	[50]
4.	Pd nanoparticles	*Lagerstroemia speciosa*	Reduction of organic pollutant	[51]
5.	Pd nanoparticles	*Anogeissus latifolia* gum	Reduction of dyes	[52]
6.	CuO/ZnO nanocomposite	*Melissa officinalis* L. leaf	Reduction of 4-NP and RhB	[53]
7.	Cu nanoparticles	*Tridax procumbens* leaf	Degradation of Bismarck brown	[54]
8.	CuO nanoparticles	*Tinospora cordifolia*	Photodegradation of methyl blue	[55]
9.	Cu nanoparticles	Extract of *Hibiscus sabdariffa* flowers	Removal of nitrate	[56]
10.	CdS	*P. aeruginosa* JP-11	Removal of Cd(II)	[57]
11.	Mn nanoparticles	*Pseudomonas putida* MnB1	Removal of Pb(II), Cd(II) and Zn(II)	[58]

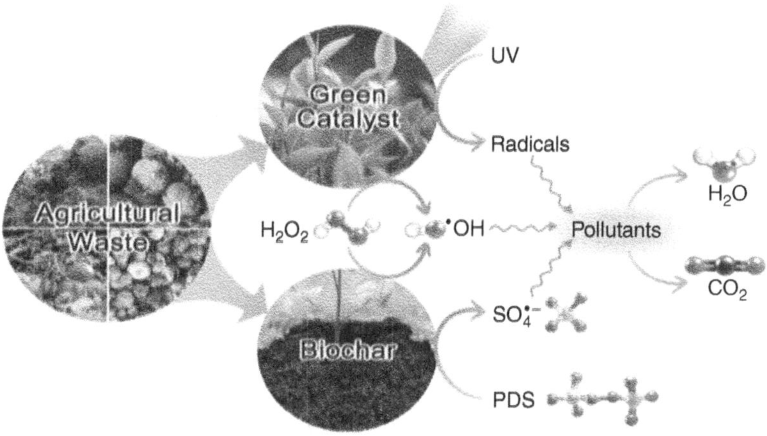

Figure 8.2 Synthetic route of eco-friendly biochar for sustainable development. *Source:* Kumar et al. [61]/Reproduced with permission from Elsevier.

Widespread interest has been generated by the use of nanomaterials in the treatment of water and wastewater. Nanomaterials have great adsorption capabilities and reactivity because of their small sizes, which result in huge specific surface areas [63, 64]. It has been shown that different types of nanomaterials may effectively remove heavy metals [5], organic pollutants [65], inorganic anions [66] and microorganisms [67, 68]. Recent years have seen successful applications of photocatalytic degradation by metal oxide nanoparticles, such as TiO_2, in the degradation of contaminants in water and wastewater. Due to its nontoxicity, chemical stability, economic viability, strong photoactivity, etc., TiO_2 has attracted a lot of research [69, 70]. Contaminants can gradually oxidize into low molecular weight intermediate products in the presence of light and a catalyst, and then they can be converted into CO_2, H_2O and anions such as NO_3^-, PO_4^{3-} and Cl^- [63]. Guo et al. prepared and used a new adsorbent, bead cellulose loaded with iron oxyhydroxide (BCF), for the adsorption and removal of arsenate and arsenite from aqueous systems [70]. Mahmoodi et al. studied the effect of immobilized titanium dioxide nanoparticles on the removal of Butachlor (*N*-butoxymethyl-2-chloro-2, 6-diethylacetanilide), which is one of the organic pollutants in agricultural soil and wastewater [71]. Because of the high cost of manufacture and the harmful effects of nanoparticles in the environment, the author employed the immobilized version of TiO_2, which is easily retrieved from water-based media. In this work, the optimum factors evaluated were inorganic SO^{2-} and, Cl^-, NO_3^- anions, hydrogen peroxide concentration and pH.

Co-precipitation is utilized to form a quaternary nano-photocatalyst, which is subsequently used to destroy colored contaminants entirely from water. By utilizing solar light, a low-cost natural resource, in conjunction with a reusable photocatalyst, pollutants such as color-giving dyes may be destroyed and mineralized on a huge scale, leaving behind harmless compounds. The technique might cleanse the water, allowing it to be used for cleaning, cooling, irrigation and other purposes. Further investigation demonstrates that the created unique quaternary photocatalyst ZrCdPbO4 outperforms binary or ternary analogues [72].

Through a batch and continuous flow column study, Xu and Zhao reduced chromium (Cr) (VI) in both water and soil media using carboxy methyl cellulose (CMC) stabilized ZVI nanoparticles [73]. They discovered that compared to the unstabilized ZVI nanoparticle, the stabilized ZVI nanoparticle precisely eliminates Cr(VI). In the batch trials, the decrease in levels of Cr(VI) went from 24% to 90% as the amount of ZVI raised from 0.04 to 0.12 g/l [73].

In order to remove Ni(II) from an aqueous solution, Li and Zhang exploited the principal-shell structure of iron nanoparticles as a sorbent and reductant [74]. The outcomes showed that 4.43 meq Ni(II)/g or 0.13 g Ni/g Fe was the sorption capacity for Ni(II) elimination. For the dechlorination of p-chlorophenol from water, Chen et al. also utilized ZVI nanoparticles and a commercial type of Fe0 powder with various mesh sizes [75]. When those particles were compared, it became clear that for the reduction process, Fe0 at the nanoscale was greater in effectiveness. Zero-valent iron (nZVI) nanoparticles were created by Celebi et al. and utilized to take the Ba^{2+} ion out of aqueous solutions [76]. Colorimetric and surface-enhanced Raman scattering dual-mode amoxicillin sensing was developed by Anh et al. using low-cost, ecologically friendly copper nanoparticles (CuNPs) and copper-graphene oxide (Cu-GO) nanocomposites [77]. The aggregation of CuNPs was responsible for the color shift from wine-red to light-red and brick red to brown in the CuNPs and Cu-GO nanocomposite solutions in the colorimetric sensors. This device provided a linear concentration range of 5–50 M with a limit of detection values as low as 2.17 and 1.71 M, allowing for on-site visual detection of amoxicillin in water samples.

8.4 Green Nanomaterial for Removal of Soil Pollution

To recover soil that heavy metals have contaminated, numerous remediation techniques have been developed. These are the main mechanical or physiochemical method-based measures, like soil incineration, excavation and landfilling, soil washing, solidification, and electric field application [78]. A wide range of adjustment agents have been used to manipulate the heavy metal bioavailability and to impede their diffusions in soil by inducing a number of sorption procedures, such as mineral surface adsorption, the creation of stable organic ligand complexes, ion exchanges and surface precipitation. Heavy metal activity in soils is controlled by desorption and sorption reactions with other components of soil [79].

Amendment mediators can be divided into two main categories: mobilizing agents that increase the mobility, bioavailability and elimination of heavy metals via soil washing and phytoextraction, as well as restraining amendment mediators that reduce mobility, bioavailability and transportation to food by halting groundwater leaching of heavy metals by phytostabilization. Both phytostabilization and phytoextraction developments are part of the phytoremediation method, which is employed to cope with polluted soils [69]. There are still some limitations to phytoremediation using these natural hyperaccumulators because it is a labor-intensive process that takes a very long time to remove heavy metal-affected soil, especially in moderately and highly contaminated areas. This may be largely explained by the slow growth rate and limited biomass production of these hyperaccumulators [80].

A wide-ranging decrease in production results from damage to agriculture that has not been repaired over many years due to a variety of factors like fungus, weeds, and insects. Additionally, it was shown that the SiO_2/QD/Au colloid particle-based *in vivo* (IVIS) nanoparticles that were implanted into a rate's chest wall were suitable for pesticide diagnostics. By infusing biochemical insecticides into agricultural areas, silicon matter has been considerably addressed. However, the strict regulation of chemical pesticides and the eventual degeneration of such applications have justified the use of alternatives as a primary remedy for environmental and public health concerns. Degraded soils and contaminated groundwater have resulted in nutritionally unbalanced and unproductive areas [81].

By enabling improved control and conservation of plant and animal production inputs, nanotechnology technologies have the potential to alter agricultural productivity. Comparing nano-fertilizers to traditional ones reveals advantages. In order to boost the solubility of Zn fertilizers, zinc oxide nanoparticles are used [82]. According to De la Rosa et al., alfalfa, tomatoes and cucumbers can grow and produce more biomass when zinc oxide nanoparticles are sprayed onto their leaves [83].

8.5 Conclusion

The ecofriendly nanomaterials are synthesized by clean, nontoxic plant extract and biomaterials and deal with economical and sustainable development. Therefore, green synthesized nanomaterials are upgraded as innovative and fabricated into different aspects. Many nanoparticles and their use and potential application have been discussed considering the aspect of increasing threat to the environment as well as human health in this review. In the upcoming years, these nanomaterials will continue to be used and put into practice for sustainable development. The introduction of nanotechnology will be helpful in the next few years to replace the risky methods currently in use since it will eliminate harmful contaminants from the soil and maintain the integrity of the environment. The use of nanotechnology in remediation has shown remarkable promise for reducing overall costs and cutting the time needed for refilling and completely repairing the damaged niche. The on-site clean-up reduces the overall cost of shipping, processing and soil dumping when the area has been replenished. The long-term implications of using nanoparticles are still a hot topic since appropriate analysis must be done before moving forward with widespread deployment to limit any potential environmental repercussions in the future.

References

1 Nakum, J. and Bhattacharya, D. (2022). Various green nanomaterials used for wastewater and soil treatment: a mini-review. *Front. Environ. Sci.* 9 (January): 1–9.
2 Buzea, C., Pacheco, I.I., and Robbie, K. (2007). Nanomaterials and nanoparticles: sources and toxicity. *Biointerphases* 2 (4): MR17–MR71. https://doi.org/10.1116/1.2815690.
3 Peter, B., Gauthier, O., Laïb, S. et al. (2006). Local delivery of bisphosphonate from coated orthopedic implants increases implants mechanical stability in osteoporotic rats. *J. Biomed. Mater. Res. A* 76A (1): 133–143. https://doi.org/10.1002/jbm.a.30456.
4 Ana-Alexandra S, Ion R-M, Nuta A, Bekaroğlu Ö. 2018. Nanomaterials for the detection and removal of pollutants from wastewater. *The 6th International Virtual Conference on Advanced Scientific Results.*
5 Tang, W., Zeng, G.-M., Gong, J.-L. et al. (2013). Impact of humic/fulvic acid on the removal of heavy metals from aqueous solutions using nanomaterials: a review. *Sci. Total Environ.* 468-469C: 1014–1027.
6 Kusior, A., Klich-Kafel, J., Trenczek-Zając, A. et al. (2013). TiO_2–SnO_2 nanomaterials for gas sensing and photocatalysis. *J. Eur. Ceram. Soc.* 33: 2285–2290.
7 Corsi, I., Venditti, I., Trotta, F., and Punta, C. (2022). Environmental safety of nanotechnologies: the eco-design of manufactured nanomaterials for environmental remediation. *Sci. Total Environ.* 864: 161181.

8 Dabhane, H., Ghotekar, S., Zate, M. et al. (2023). A novel approach toward the bio-inspired synthesis of CuO nanoparticles for phenol degradation and antimicrobial applications. *Biomass Convers. Biorefin.* https://doi.org/10.1007/s13399-023-03954-y.

9 Anindita, D., Singh, N.B., Guin, M., and Barthwal, S. (2023). Water purification by green synthesized nanomaterials. *Current Pharm. Biotechnol.* 24: 101–117. http://www.eurekaselect.com/article/123174.

10 Neeti, K., Singh, R., and Ahmad, S. (2023). The role of green nanomaterials as effective adsorbents and applications in wastewater treatment. *Mater. Today Proc.* 77: 269–276.

11 Ganie, A.S., Bano, S., Khan, N. et al. (2021). Nanoremediation technologies for sustainable remediation of contaminated environments: recent advances and challenges. *Chemosphere* 275: 130065.

12 Mukhopadhyay, R., Sarkar, B., Khan, E. et al. (2022). Nanomaterials for sustainable remediation of chemical contaminants in water and soil. *Crit. Rev. Env. Sci. Technol.* 52 (15): 2611–2660. https://doi.org/10.1080/10643389.2021.1886891.

13 Bharimalla, A., Patil, P.G., Mukherjee, S. et al. (2019). Nanocellulose-polymer composites: novel materials for food packaging applications. In: *Polymers for Agri-Food Applications* (ed. T.J. Gutiérrez), 553–599. Springer.

14 Nakada, T., Hirabayashi, Y., Tokado, T. et al. (2004). Novel device structure for Cu(In,Ga)Se_2 thin film solar cells using transparent conducting oxide back and front contacts. *Sol. Energy* 77: 739–747.

15 Fan, Z., Chang, P., Lu, J. et al. (2004). Photoluminescence and polarized photodetection of single ZnO NWs. *Appl. Phys. Lett.* 85: 6128–6130.

16 Aneesh, P.M., Vanaja, K.A., and Jayaraj, M.K. (2007). Synthesis of ZnO nanoparticles by hydrothermal method. *Proc. SPIE 6639, Nanophotonic Materials IV*, San Diego, CA.

17 Dalai, S., Pakrashi, S., Bhuvaneshwari, M. et al. (2014). Toxic effect of Cr(VI) in presence of n-TiO_2 and n-Al_2O_3 particles towards freshwater microalgae. *Aquat. Toxicol.* 146: 28–37.

18 Bhuvaneshwari, M., Velu, I., Raghavan, A. et al. (2015). Cytotoxicity of ZnO NPs towards fresh water algae *Scenedesmus obliquus* at low exposure concentrations in UV-C, visible and dark conditions. *Aquat. Toxicol.* 162: 29–38.

19 Basnet, P., Inakhunbi Chanu, T., Samanta, D., and Chatterjee, S. (2018). A review on bio-synthesized zinc oxide nanoparticles using plant extracts as reductants and stabilizing agents. *J. Photochem. Photobiol., B* 183: 201–221. https://doi.org/10.1016/j.jphotobiol.2018.04.036.

20 Ovais, M., Khalil, A., Islam, N.U. et al. (2018). Role of plant phytochemicals and microbial enzymes in biosynthesis of metallic nanoparticles. *Appl. Microbiol. Biotechnol.* 102: 6799–6814.

21 Khin, M., Nair, S., Veluru, J.B. et al. (2012). A review on nanomaterials for environmental remediation. *Energy Environ. Sci.* 5: 8075.

22 Ahmed, S., Ahmad, M., Swami, B., and Ikram, S. (2015). A review on plants extract mediated synthesis of silver nanoparticles for antimicrobial applications: a green expertise. *J. Adv. Res.* 7: 17–28.

23 Blum, A.P., Kammeyer, J.K., Rush, A.M. et al. (2015). Stimuli-responsive nanomaterials for biomedical applications. *JACS* 137 (6): 2140–2154. https://doi.org/10.1021/ja510147n.

24 Bellona, C. and Drewes, J.E. (2007). Viability of a low-pressure nanofilter in treating recycled water for water reuse applications: a pilot-scale study. *Water Res.* 41 (17): 3948–3958.

25 Jadhav, S.V., Bringas, E., Yadav, G.D. et al. (2015). Arsenic and fluoride contaminated groundwaters: a review of current technologies for contaminants removal. *J. Environ. Manage.* 162: 306–325. https://doi.org/10.1016/j.jenvman.2015.07.020.

26 Zamora-Ledezma, C., Negrete-Bolagay, D., Figueroa, F. et al. (2021). Heavy metal water pollution: a fresh look about hazards, novel and conventional remediation methods. *Environ. Technol. Innovation* 22: 101504.

27 Saikia, J., Gogoi, A., and Baruah, S. (2019). Nanotechnology for water remediation. In: *Environmental Nanotechnology* (ed. N. Dasgupta, S. Ranjan, and E. Lichtfouse), 195–211. Springer.

28 Aragaw, T.A., Bogale, F.M., and Aragaw, B. (2021). Iron-based nanoparticles in wastewater treatment: a review on synthesis methods, applications, and removal mechanisms. *J. Saudi Chem. Soc.* 25: 101280.

29 Oller, I., Malato, S., and Sánchez Pérez, J.A. (2010). Combination of advanced oxidation processes and biological treatments for wastewater decontamination – a review. *Sci. Total Environ.* 409: 4141–4166.

30 Zhang, L. and Fang, M. (2010). Nanomaterials in pollution trace detection and environmental improvement. *Nano Today* 5: 128–142.

31 Bora, T. and Dutta, J. (2014). Applications of nanotechnology in wastewater treatment – a review. *J. Nanosci. Nanotechnol.* 14 (1): 613–626.

32 Reidy, B., Haase, A., Luch, A. et al. (2013). Mechanisms of silver nanoparticle release, transformation and toxicity: a critical review of current knowledge and recommendations for future studies and applications. *Materials (Basel)* 6: 2295–2350.

33 Parthasarathi, V. and Thilagavathi, G. (2009). Synthesis and characterization of titanium dioxide nano-particles and their applications to textiles for microbe resistance. *J. Text Apparel, Technol. Manag.* 6: 1–8.

34 Amin, M., Alazba, P., and Manzoor, U. (2014). A review of removal of pollutants from water/wastewater using different types of nanomaterials. *Adv. Mater. Sci. Eng.* 2014: 1–24.

35 Klačanová, K., Fodran, P., Simon, P. et al. (2013). Formation of Fe(0)-nanoparticles via reduction of Fe(II) compounds by amino acids and their subsequent oxidation to iron oxides. *J. Chem.* 2013: 961629.
36 Nasir, M., Hashim, R., Sulaiman, O., and Asim, M. (2017). Nanocellulose: preparation method and applications. In: *Cellulose-Reinforced Nanofibre Composites* (ed. M. Jawaid, S. Boufi, and H.P.S. Abdul Khalil), 261–276. Elsevier.
37 Phanthong, P., Reubroycharoen, P., Hao, X. et al. (2018). Nanocellulose: extraction and application. *Carbon Resour. Convers.* 1: 32–43.
38 Sharma, A., Thakur, M., Bhattacharya, M. et al. (2019). Commercial application of cellulose nano-composites – a review. *Biotechnol. Rep.* 21: e00316.
39 Rol, F., Belgacem, N., Gandini, A., and Bras, J. (2018). Recent advances in surface-modified cellulose nanofibrils. *Prog. Polym. Sci.* 88: 241–264.
40 Wang, J., Liu, X., Jin, T. et al. (2019). Preparation of nanocellulose and its potential in reinforced composites: a review. *J. Biomater. Sci. Polym. Ed.* 30 (11): 919–946. https://doi.org/10.1080/09205063.2019.1612726.
41 Kumar, L., Ragunathan, V., Chugh, M., and Bharadvaja, N. (2021). Nanomaterials for remediation of contaminants: a review. *Environ. Chem. Lett.* 19 (4): 3139–3163. https://doi.org/10.1007/s10311-021-01212-z.
42 Guerra, F.D., Attia, M.F., Whitehead, D.C., and Alexis, F. (2018). Nanotechnology for environmental remediation: materials and applications. *Molecules* 23 (7): 1–23.
43 Manisalidis, I., Stavropoulou, E., Stavropoulos, A., and Bezirtzoglou, E. (2020). Environmental and health impacts of air pollution: a review. *Front. Public Health* 8 (February): 1–13.
44 Uthaman, S., Kim, H.S., Revuri, V. et al. (2018). Green synthesis of bioactive polysaccharide-capped gold nanoparticles for lymph node CT imaging. *Carbohydr. Polym.* 181: 27–33.
45 Varma, R. (2012). Greener approach to nanomaterials and their sustainable applications. *Curr. Opin. Chem. Eng.* 1: 123–128.
46 Varma, R.S. (2014). Journey on greener pathways: from the use of alternate energy inputs and benign reaction media to sustainable applications of nano-catalysts in synthesis and environmental remediation. *Green Chem.* 16 (4): 2027–2041. https://doi.org/10.1039/C3GC42640H.
47 Kumar, H., Bhardwaj, K., Kuča, K. et al. (2020). Flower-based green synthesis of metallic nanoparticles: applications beyond fragrance. *Nanomaterials* 10 (4): 1–16.
48 Khodadadi, B., Bordbar, M., Yeganeh-Faal, A., and Nasrollahzadeh, M. (2017). Green synthesis of Ag nanoparticles/clinoptilolite using *Vaccinium macrocarpon* fruit extract and its excellent catalytic activity for reduction of organic dyes. *J. Alloys Compd.* 719: 82–88.

49 Vellaichamy, B. and Periakaruppan, P. (2016). A facile, one-pot and eco-friendly synthesis of gold/silver nanobimetallics smartened rGO for enhanced catalytic reduction of hexavalent chromium. *RSC Adv.* 6 (62): 57380–57388. https://doi.org/10.1039/C6RA10544K.

50 Srivastava, V., Sharma, Y.C., and Sillanpää, M. (2015). Green synthesis of magnesium oxide nanoflower and its application for the removal of divalent metallic species from synthetic wastewater. *Ceram. Int.* 41: 6702–6709.

51 Garole, D., Garole, V., Choudhary, B., and Borse, S. (2018). Palladium nanocatalyst: green synthesis, characterization, and catalytic application. *Int. J. Environ. Sci. Technol.* 16: 7885–7892.

52 Kora, A.J. and Rastogi, L. (2015). Green synthesis of palladium nanoparticles using gum ghatti (*Anogeissus latifolia*) and its application as an antioxidant and catalyst. *Arabian J. Chem.* 11: 1097–1106.

53 Bordbar, M., Negahdar, N., and Nasrollahzadeh, M. (2017). *Melissa officinalis* L. leaf extract assisted green synthesis of CuO/ZnO nanocomposite for the reduction of 4-nitrophenol and Rhodamine B. *Sep. Purif. Technol.* 191: 295–300.

54 Nagarajan, K., Chakraborthy, P., Palanichamy, V., and Rajeswari, D. (2016). Synthesis and characterization of copper nanoparticles using *Tridax procumbens* and its application in degradation of bismarck brown. *Int. J. Chem. Tech. Res.* 9: 498–507.

55 Udayabhanu, Nethravathi, P.C., Kumar, M.A.P. et al. (2015). *Tinospora cordifolia* mediated facile green synthesis of cupric oxide nanoparticles and their photocatalytic, antioxidant and antibacterial properties. *Mater. Sci. Semicond. Process.* 33: 81–88.

56 Paixão, R., Reck, I., Bergamasco, R. et al. (2017). Activated carbon of babassu coconut impregnated with copper nanoparticles by green synthesis for the removal of nitrate in aqueous solution. *Environ. Technol.* 39: 1–37.

57 Raj, R., Dalei, K., Chakraborty, J., and Das, S. (2015). Extracellular polymeric substances of a marine bacterium mediated synthesis of CdS nanoparticles for removal of cadmium from aqueous solution. *J. Colloid Interface Sci.* 462: 166–175.

58 Zhou, D., Kim, D.-G., and Ko, S.-O. (2015). Heavy metal adsorption with biogenic manganese oxides generated by *Pseudomonas putida* strain MnB1. *J. Ind. Eng. Chem.* 24: 132–139.

59 Varma, R.S. (2016). Greener and sustainable trends in synthesis of organics and nanomaterials. *ACS Sustainable Chem. Eng.* 4 (11): 5866–5878. https://doi.org/10.1021/acssuschemeng.6b01623.

60 Varma, R.S. (2019). Biomass-derived renewable carbonaceous materials for sustainable chemical and ecological applications. *ACS Sustainable Chem. Eng.* 7 (7): 6458–6470. https://doi.org/10.1021/acssuschemeng.8b06550.

61 Kumar, M., Ambika, S., Hassani, A., and Nidheesh, P.V. (2023). Waste to catalyst: role of agricultural waste in water and wastewater treatment. *Sci. Total Environ.* 858: 159762. https://www.sciencedirect.com/science/article/pii/S0048969722068620.

62 Inyang, M. and Dickenson, E. (2015). The potential role of biochar in the removal of organic and microbial contaminants from potable and reuse water: a review. *Chemosphere* 134: 232–240. https://www.sciencedirect.com/science/article/pii/S0045653515003161.

63 Bratovcic, A. (2019). Photocatalytic degradation of organic compounds in wastewaters. *Technol. Acta* 11: 17–23.

64 Bratovcic, A. (2019). Different applications of nanomaterials and their impact on the environment. *Int. J. Mater. Sci. Eng.* 5 (1): 1–7.

65 Yan, J., Han, L., Gao, W. et al. (2015). Biochar supported nanoscale zerovalent iron composite used as persulfate activator for removing trichloroethylene. *Bioresour. Technol.* 175: 269–274. https://doi.org/10.1016/j.biortech.2014.10.103.

66 Liu, F., Yang, J.-H., Zuo, J. et al. (2014). Graphene-supported nanoscale zero-valent iron: removal of phosphorus from aqueous solution and mechanistic study. *J. Environ. Sci.* 26: 1751–1762.

67 Kalhapure, R.S., Sonawane, S.J., Sikwal, D.R. et al. (2015). Solid lipid nanoparticles of clotrimazole silver complex: an efficient nano antibacterial against *Staphylococcus aureus* and MRSA. *Colloids Surf., B* 136: 651–658. http://europepmc.org/abstract/MED/26492156.

68 Lu, H., Wang, J., Stoller, M. et al. (2016). An overview of nanomaterials for water and wastewater treatment. *Adv. Mater. Sci. Eng.* 2016: 1–10.

69 Li, J., Zhao, T., Chen, T. et al. (2015). Engineering noble metal nanomaterials for environmental applications. *Nanoscale* 7 (17): 7502–7519. https://doi.org/10.1039/C5NR00857C.

70 Guo, X. and Chen, F. (2005). Removal of arsenic by bead cellulose loaded with iron oxyhydroxide from groundwater. *Environ. Sci. Technol.* 39 (17): 6808–6818. https://doi.org/10.1021/es048080k.

71 Mahmoodi, N.M., Arami, M., Yousefi-Limaee, N. et al. (2007). Nanophotocatalysis using immobilized titanium dioxidenanoparticle. Degradation and mineralization of water containing organic pollutant: case study of Butachlor. *Mater. Res. Bull.* 42: 797–806.

72 Lohar, S., Prajapati, D.K., Sharma, A., and Bhardwaj, S. (2021). Wastewater treatment using nanomaterial quaternary photocatalyst ZrCdPbO4 – removal of colour pollutant by an eco friendly process. *J. Indian Chem. Soc.* 98 (11): 100194. https://www.sciencedirect.com/science/article/pii/S0019452221001941.

73 Xu, Y. and Zhao, D. (2007). Reductive immobilization of chromate in water and soil using stabilized iron nanoparticles. *Water Res.* 41 (10): 2101–2108. https://www.sciencedirect.com/science/article/pii/S0043135407001315.

74 Li, X. and Zhang, W. (2006). Iron nanoparticles: the core–shell structure and unique properties for Ni(II) sequestration. *Langmuir* 22: 4638–4642.

75 Su, J., Lin, S., Chen, Z. et al. (2011). Dechlorination of p-chlorophenol from aqueous solution using bentonite supported Fe/Pd nanoparticles: synthesis, characterization and kinetics. *Desalination* 280: 167–173.

76 Celebi, O., Üzüm, C., Shahwan, T., and Erten, H. (2007). A radiotracer study of the adsorption behavior of aqueous Ba2+ ions on nanoparticles of zero-valent iron. *J Hazard Mater.* 1 (148): 761–767.

77 Anh, N.T., Dinh, N.X., Van Tuan, H. et al. (2022). Eco-friendly copper nanomaterials-based dual-mode optical nanosensors for ultrasensitive trace determination of amoxicillin antibiotics residue in tap water samples. *Mater. Res. Bull. [Internet]* 147: 111649. https://www.sciencedirect.com/science/article/pii/S0025540821004463.

78 Yan, A., Wang, Y., Tan, S.N. et al. (2020). Phytoremediation: a promising approach for revegetation of heavy metal-polluted land. *Front. Plant Sci.* 11 (April): 1–15.

79 Socas-Rodríguez, B., González-Sálamo, J., Hernández-Borges, J., and Rodríguez-Delgado, M.Á. (2017). Recent applications of nanomaterials in food safety. *TrAC Trends Anal. Chem.* 96: 172–200. https://www.sciencedirect.com/science/article/pii/S0165993617301346.

80 Dou, X., Dai, H., Skuza, L., and Wei, S. (2022). Cadmium removal potential of hyperaccumulator *Solanum nigrum* L. under two planting modes in three years continuous phytoremediation. *Environ. Pollut.* 307: 119493. https://www.sciencedirect.com/science/article/pii/S0269749122007072.

81 Mazzaglia, A., Fortunati, E., Kenny, J. et al. (2017). Nanomaterials in plant protection. In: *Nanoscience and Nanotechnology for Agriculture and Food* (ed. M.A.V. Axelos and M. Van De Voorde), 115–134. Wiley. https://doi.org/10.1002/9783527697724.ch7.

82 Toscano, A., Marzo, A., Milani, M. et al. (2015). Comparison of removal efficiencies in Mediterranean pilot constructed wetlands vegetated with different plant species. *Ecol. Eng.* 75: 155–160. https://www.sciencedirect.com/science/article/pii/S0925857414006636.

83 de la Rosa, G., Lopez-Moreno, M., De Haro-Del Rio, D. et al. (2013). Effects of ZnO nanoparticles in alfalfa, tomato, and cucumber at the germination stage: root development and X-ray absorption spectroscopy studies. *Pure Appl. Chem.* 85: 2161–2174.

9

Magnetotactic Bacteria-Synthesized Nanoparticles and Their Applications

Juhi Gupta and Athar Adil Hashmi

Bioinorganic Research Lab, Department of Chemistry, Jamia Millia Islamia, New Delhi, DL, India

Abstract

Nanotechnology concentrates primarily on the study of nanoparticles. It is imperative that green nanotechnology be used to synthesize multifaceted sustainable nanomaterials with varying compositions, sizes and monodispersities. Research based on microbial-derived nanoparticle formation is an intriguing field with significant potential for exploration. Magnetotactic bacteria (MTB) synthesize magnetic nanocrystals with homogeneous conformations, configurations and forms at physiological functional states, which act as sources for various biomacromolecules that can subsequently be exploited for the formulation of diversified bio-inspired magnetic nanoparticles (MNPs) with high crystallinity, mono-dispersity, magnetism and bioengineerability. In this review, a slight overview of MTB and MNPs and the preparation, functionalization and physico-chemical properties of biocompatible MNPs will be discussed. The potential utility of these MNPs as next-generation materials over a broader range of functions including tackling real biomedical and environmental problems will also be emphasized. A brief comparison of the MNPs relative to chemically synthesized MNPs will be highlighted too.

Keywords *nanotechnology; magnetotactic bacteria; magnetosomes; MNPs; magnetic properties; magnetite*

Green Synthesis of Nanomaterials: Biological and Environmental Applications, First Edition.
Edited by Archana Chakravarty, Preeti Singh, Saiqa Ikram, and R.N. Yadava.

9.1 Introduction

Nanotechnology has transformed high-performance biomedical diagnostics and biotechnology devices [1, 2]. Owing to their superior and distinctive physico-chemical and biological attributes, nanoparticles (NPs) are widely studied [3]. The possibility of manipulating the characteristics of MNPs has sparked a lot of interest [4]. A prevalent misconception is that MNPs synthesized via co-precipitation contain the qualities necessary for several high-tech applications [5]. Nanotechnology's biggest challenge is controlling their structural, morphological and chemical characteristics [6]. Biogenic MNPs (magnetosomes) derived from magnetotactic bacteria meet all requirements for size, shape, biocompatibility and magnetism [7, 8].

9.1.1 Magnetotactic Bacteria (MTB)

MTB was originally presented in 1963 by S. Bellini while in 1975, bacterial magnetoreception was rediscovered by Richard P. Blakemore. Blakemore discovered that a vast population of water-borne coccoid bacteria was moving northward following the geomagnetic field, and that their movement could be reversed by applying magnetic field externally [8, 9]. Unprecedented magnetic-field-reliance cell motility sparked a lot of interest, which led to an extensive collection of microorganisms with comparable traits to be discovered [8, 10]. Subsequent Transmission Electron Microscopy (TEM) experiments demonstrated that these bacteria could take up iron and convert them to tiny, elongated (50–100 nm) magnetic nanocrystals *in vivo* [9]. An extensive variety of scientific and technical uses were made possible due to their peculiar coalition of magnetic, physical and optical properties [8, 11]. Figure 9.1 depicts a typical cell structure of MTB, comprising of magnetosomes (magnetic nanoparticles and proteins) [12].

MTB are morphologically, physiologically and phylogenetically variable group of aquatic and motile prokaryotes that move along the magnetic field lines of the Earth's magnetic field with the support of internal organelles that are termed magnetosomes [8]. The iron content of MTB is typically up to 3% by dry weight and an average of 10^{-13}–10^{-15} g of Fe per cell [12]. Different MTB forms have been discovered, ranging form the simpler spirilla to the more sophisticated multicellular bacteria, including the rod- and bean-shaped MTB forms, as well as the vibrioid and the ovoid [8]. As a result of MTBs wide range of phenotypic characteristics, all known MTB isolates possess a few common characteristics, such as Gram-negative cytoplasmic walls, flagella, sensitive to oxygen levels and biomineralization of magnetosomes [12, 13].

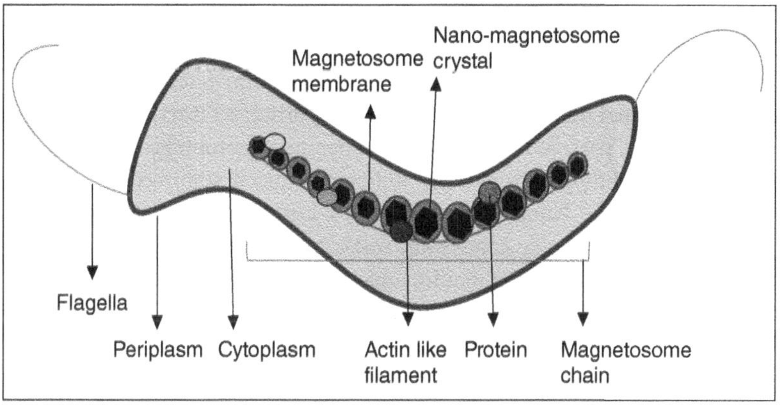

Figure 9.1 Schematic diagram of magnetotactic bacteria. *Source:* Adopted from Wang et al. [12].

MTBs may be found in virtually all aquatic forms, including freshwater sediment, marine biomes and soils. They have a propensity to survive in oxic-anoxic transition zone (OATZ) consisting of fluctuating oxygen levels in the native habitat in a vertical gradient system in the Northern and Southern Hemispheres, which is a zone where hypoxic water meets oxygenated water, which signifies the topmost portion of unstratified deposits dependent on the environment [8, 14]. When MTBs are below OATZ, the direction of their flagellar spinning is reversed, and they swim upward along the earth's geomagnetic field lines. Above OATZ, MTBs migrate in the other direction, which is down geomagnetic field lines. MTB has been isolated in numerous natural habitats; however, its finicky development and unique culture conditions impede its isolation and *in vitro* growth and development [8, 12].

9.1.2 Types of MTB

"Magnetotaxis" is a specialized ability of MTB that allows them to position themselves inside the geomagnetic field lines and then swim via flagella [15]. There are two forms of MTB: polar and axial. Depending on the magnetic crystal formations that occur within the MTB, polar MTB will either approach the south or the north [16]. MTB that are oriented toward the north magnetic pole are more prevalent in the northern hemisphere and migrate toward the south pole of a bar magnet. On the other hand, MTB that are oriented toward the south magnetic pole are more prevalent in the southern hemisphere and migrate toward the north pole [17]. They may co-exist almost equally around the geomagnetic equator from where they pick up their respective directions [18]. However, some inhabitants of south-seeking MTB were found in the northern hemisphere with direction of movement just reverse of all previously reported MTBs [19].

For the most part, MTB strains are either microaerophilic or anaerobic, growing chemolithoautotrophically utilizing reduced sulfur compounds as electron donors and being able to thrive in a range of environments such as from anaerobic to aerobic conditions, solids to liquids or from freshwater to seawater [8, 20]. All of them require anaerobic environment utilizing carbon sources in the form of short chain organic acids in their metabolism. Phylogenetic research of MTB has categorized them into the Alpha-, Gamma-, Delta classes of *Proteobacteria* and with the *Nitrospirae* phylum and candidate phylum *Omnitrophica* [8, 12]. Other MTB genera include *Magnetococcus*, *Magnetovibrio* and *Candidatus bavaricum*. *Magnetospirillum* is well-characterized based on genetic study. Some MTB species were obtained, grown and sequenced, including *Magnetospirillum magnetotacticum* strain MS-1 and *Magnetospirillium gryphiswaldense* strain MSR-1, *Magnetospirillum magneticum* strain AMB-1, *Magnetococcus marinus* strain MC-1 and *Magnetovibrio blakemorei* strain MV-1. Other MTB strains such *Candidatus Magnetobacterium bavaricum*, *Desulfovibrio magneticus* strain RS1 and *Magnetospira* sp.QH2 are accessible in axenic culture, full genome sequencing has not yet been established. Indian strains of *Magnetospirillum* sp. include VITRJS1, VITRJS2, VITRJS3, VITRJS4, VITRJS5, VITRJS6 and VITRJS7 [8].

9.2 Characteristics of Magnetosomes (MNPs)—Biogenic NPs and Their Physico-Chemical Properties

MTBs geomagnetic reactions are caused by the magnetic granules encased within its cells [12]. TEM has made it possible for us to see these embedded particles with uniform size and shape and chain-like formations [9, 21]. Each NP core is composed of organic phosphor-lipid bilayer membrane encapsulating several proteins arising from the cell's cytoplasmic membrane [9, 15]. This membrane is crucial for particle production and stability under physiological conditions. NPs with their membrane form a characteristic organelle termed as "Magnetosome," unique to MTB, which distinguishes them from other prokaryotes [9].

Thus, MTB synthesizes a lipid vesicle with magnetic iron-bearing nanocrystals embedded intracellularly, which is then attached to the bacterial membrane called magnetosomes [8, 12, 15]. Magnetosomes are grouped into single or multi-stranded chains within the cell, frequently along the long cellular axis if not spherical [14, 15]. Organic and inorganic components are combined in a regulated manner to form a magnetosome membrane. Magnetosome's inorganic core is magnetite (Fe_3O_4) or greigite (Fe_3S_4) [14]. Single-chain magnetosomes boost the bacterial cell's magnetic dipole moment. The magnetosome's organic phase was generated using vessels emanating from the inner membrane. Biomineralization controls the formation of organic magnetosome membrane and inorganic magnetic nanoparticle core under strict genetic control [8].

Magnetosomes are classified as either greigite (Fe_3S_4) or magnetite (Fe_3O_4) [22]. Greigite magnetosomes occur under sulfidic conditions in the absence of oxygen while magnetite magnetosomes require oxygen [13, 23]. However, the ratio of Fe(II) to Fe(III) in these magnetosome crystals might differ depending on the bacterial species or other conditions [12, 24].

Most MTB include 10–50 magnetosomes, each with MNPs [15]. The magnetite crystals seen in *M. bavaricum*, for example, are almost 600–1000 in number. Most MTB biomineralize one magnetosome mineral [16]. Only a few single-celled rod-shaped bacteria and multicellular magnetotactic prokaryotes can biomineralize Fe_3S_4 magnetosomes, whereas the majority of prokaryotes synthesize Fe_3O_4 magnetosomes. Some multicellular magnetotactic prokaryotes can produce both Fe_3O_4 and Fe_3S_4 minerals [12, 17].

Initially, magnetosome magnetite crystals were thought to be spherical, however, MTB biomineralized it in diverse sizes, shapes and orientations [8]. In addition, these crystals have a constant shape based on the MTB species/strain, a small range of crystal size, chemically pure and minimal crystallographic defects [25]. Magnetic nanocrystals in magnetosomes are controlled by species-specific chemical and genetic factors [12]. MTB has three typical morphologies of magnetite magnetosome crystals: nearly cubic (cubo-octahedral), elongated prismatic and bullet or tooth-shaped (anisotropic) [15].

The structure of magnetite crystal of magnetosome is based on the amalgamation of idealized morphologies of cube {100} and octahedron {111}, as a consequence of perfect biological control over production, according to High Resolution TEM (HRTEM) research. Extracts from *Magnetospirillum* sp. have a cubo-octahedral structure, with equal eight faces of an octahedron {111} and equal six faces of a cube {100}. This is a stable magnetite growth form with no deformities and excellent symmetry. Elongated prismatic crystals in *Magnetovibrio* sp. and Martian magnetite crystals comprise an elongated structure parallel to octahedron {111} to produce a {110} [8]. Greigite nanocrystals have frequently deformed wrinkled surfaces, however, cubo-octahedral and elongated prismatic crystals were found in unbred MTB [14, 15].

For magnetite and greigite crystals, the size of the stable single magnetic domain (SMD) varies from 35 to 120 nm [15]. The uncultured coccus MTB can also produce large magnetite magnetosome crystals with a length likely 250 nm [12]. Bacterial magnetosomes are the smallest permanent magnet crystals stable in SMD at ambient temperature without an external magnetic field. At room temperature, magnetic nanocrystals smaller than the SMD range (30 nm) are superparamagnetic and lack stable remnant magnetism. Initially, cells make these tiny particles, which eventually evolve into fully developed SMD crystals. Larger nanocrystals magnetize non-uniformly owing to domain barriers. There is a tendency for the domain walls to produce multifold magnetic states with opposite magnetic orientations, reducing the overall residual magnetic flux of the magnetosomes. MTB biomineralize SMD with optimal crystal size particles for maximal permanent magnetic dipole moment per magnetosome [12, 15].

Despite being versatile, these NPs show species-specific narrow and restricted size distribution [15]. MTB-MNPs have unique such as exceptionally truncated hexa-octahedral morphology; lesser defects in the crystal system; elongated pattern; smaller grain size, single domain; excellent purity; and aligned magnetosomal chains. Aligned magnetic crystals amplify the magnetization. The MTBs build up their internal magnetic dipole moments by aligning head-to-tail, making them more vulnerable to an external magnetism. Probably going by same rationale, if multiple magnetosomal chains occur in an MTB, they are invariably discovered parallelly. Single particles are ineffective, but such complexes make each MTB a miniature compass that can sense a geomagnetic field [9].

Magnetosomes direct MTB away from hypoxic condition and toward healthy ones to enhance growth and development [26]. Magnetosomes form a linear chain and hence, the maximum magnetic dipole moment likely be the summation of magnetic moments of the individual NPs. As a result, the bacterial cell rotates to align itself with the magnetic field [12, 13]. Magnetosome chains can be disrupted due to magnetic fields thereby reducing the stability. In previous studies, it was shown that a high magnetic field, i.e. 35 mT and above may annihilate the magnetosome and NP chains within MTB while the freeze-dried MTBs remain unscathed and do not spin in 1 T magnetic field [12].

Magnetite nanoparticles have stronger magnetism than greigite nanoparticles, making them more effective for hyperthermia, bio-separation and magnetic resonance imaging (MRI) contrast agents. And also, Greigite NP has not yet been successfully segregated in an uncultured medium [9]. In addition to these physical, chemical, magnetic features, the magnetosome membrane make them distinctive and important for biotechnology and nanotechnology applications [15].

Conductive minerals such as magnetite impart anaerobic marine sediments high conductivity, enhancing electrogen activity and allowing applications such as remote device powering and reductive dechlorination of contaminants [27]. Magnetite can facilitate electron transfer or act as extrinsic electrophile under acidic conditions from bacteria to external receptors [28]. Owing to the mixed valency of magnetite and potential to get oxidized by phototropic bacteria in light and reduced by electro-chemically active bacteria in dark, they might be used as battery [29].

9.3 Synthesis of Magnetosomes

As shown in Figure 9.2, magnetosome synthesis in most of the MTBs occurs via the following steps:- (i) invagination of the cytoplasmic membrane and subsequent magnetosome vesicle formation; (ii) absorption of iron and transport to the vesicle via enzymatic reduction of Fe(III) to Fe(II); (iii) formation of low-density ferric oxides analogous to ferrihydrite via partial oxidation of a redox enzyme; (iv) Reduction of about 33% of Fe(III) in hydrous oxides and subsequent dehydration to give Fe_3O_4. Co-crystallization of Fe(II) with iron oxidase leads to the formation of Fe(III) ferrihydrite, required for physiological biomineralization of magnetite or greigite in the magnetosome vesicle. This is a complex process that relies on several genes and proteins but has not been conclusively established [12].

Extracting/purifying magnetosomes is the next step in making magnetosome uses possible. To obtain magnetosomes, MTB cells must be lysed physically through ultrasonication, high-pressure homogenizer, or French press or

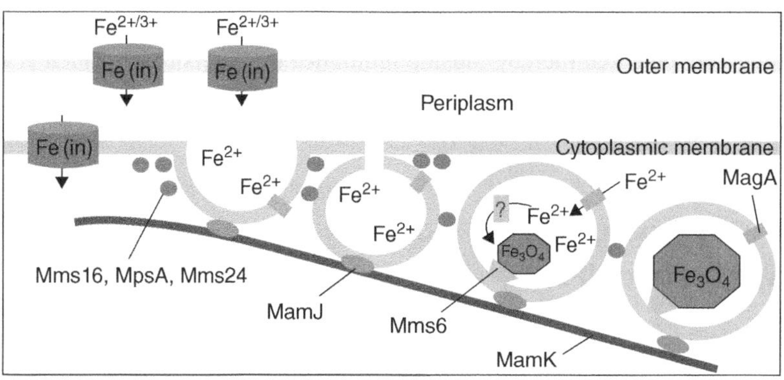

Figure 9.2 Scheme of the postulated magnetite biomineralization pathway. The graph was referenced and used with authorization. *Source:* Yan et al. [11]/with permission of Elsevier.

chemically via alkaline lysis. Aggressive physical or chemical lysis using strong acids or organic solvents/detergents might dissolve magnetite crystal or remove phospholipid membrane. Some organic solvents, such as phenol, chloroform, etc. and combustion reactions, have been recommended as further purification procedures to remove endotoxins. Recovering magnetosomes from this cell-free extract is commonly done using magnetic separation techniques [15, 30].

The structural features of the magnetosomes allow various molecules to bind to the surface which aids in functionalization of magnetosomes [5]. Broadly, two different functionalization technique via protein modification can be employed: Firstly, genetic engineering of genes encoding magnetosome surface protein(s); and secondly, simply by chemical modification [15]. Spider silk-coated magnetosomes have recently been created by expressing spider silk-inspired peptides linked to MamC in Ms. gryphiswaldense strain MSR-1, which enhanced the dispersibility and colloidal stability of NPs and since, spider silk is non-immunogenic, encapsulating magnetosomes with the polymer may improve their biocompatibility [31].

It is possible to create covalent linkages between amino groups ($-NH_2$) from magnetosome-surface proteins and certain functional groups in the structure of therapeutic agents using crosslinkers such as glutaraldehyde and genipin [30]. The total negative charge imparted by the phospholipids of the magnetosome membrane establishes a strong interaction with the positively charged coating macromolecules via amine crosslinking in majority of cases [15]. With the use of polyethylenimine (PEI) in conjunction with Doxorubicin, a link was established to bind the anticancer recombinant plasmid to the magnetosome surface [32]. Poly-amino acids, like poly-glutamic acid, can be used for conjugating functional molecules onto magnetosomes via charge interaction. Various studies have found that surface modification of magnetosomes for drug administration might alter their zeta potential thereby causing hindrance to dispersibility, which is a significant drawback. Incorporating poly-L-glutamic acid or other modifications can help minimize these alterations [15].

9.4 MNPs Relative to Chemically Synthesized NPs

Magnetosomes have several advantages over synthetic NPs in terms of:-

i) **Grain size distribution and range**
Synthetic MNPs may be synthesized in SMD and multi-magnetic domain with high dispersivity while the crystal size of magnetosome lies in the SMD range [33]. Among synthetic methods, thermal

decomposition at higher temperatures yields distinctly crystalline narrow size distribution NPs [34].

ii) **Magnetization**

Most magnetosomes are 40–50 nm in size. At ambient temperature, such SMD NPs orient themselves in such a way to generate static magnetization [35]. Thus, magnetosomes have superior magnetic characteristics at higher temperatures than MNPs [8]. Synthetic MNPs generated by chemical precipitation method are generally superparamagnetic with non-static magnetization having variable shapes and sizes [7]. Magnetosomes were found to be ferromagnetic, whereas synthesized MNPs were shown to be superparamagnetic [36].

Currently, MNP production focuses on 5–20 nm particles, which are superparamagnetic. Larger magnetite NPs having more per particle magnetism might benefit magnetic bio-separation and MRI. Contrary to the synthetic NPs, MTB-NPs often have a diameter of 35–120 nm, a size that is impossible to achieve with chemically produced NPs. Geomagnetic fields have little effect on superparamagnetic magnetite particles smaller than 20 nm. On the other hand, those bigger than 120 nm have several magnetic states that align themselves in various lines of force and may undermine the total magnetic moment [9].

iii) **Shape control**

Most synthetic magnetite NPs are spherical, polygonal or composite. The framework of MTB-NPs is species-specific, with parallelepipedal, cuboidal, and tooth, bullet, or arrowhead forms [9]. Thus, MTB-NPs show better shape control.

iv) **High crystallinity with fewer defects**

MNPs may be synthesized with excellent crystallinity, but this normally needs high temperatures; MTBs, on the other hand, can accomplish this at ambient temperature [9]. In contrast to synthetic NPs generated by the co-precipitation approach, biomineralization is thought to provide excellent purity and a perfect lattice for magnetosome [7, 8]. Studies have demonstrated that MTB-NPs are invariably made up of octahedral {111}, dodecahedral {110} and cubic {100} conformation [9]. Shape control in synthetic NP formulation is uncertain since organic molecules reportedly selectively suppress specific faces. MTB particles arrange along a peculiar axis, which is different for both greigite and magnetite NPs. Magnetite particles align along {111}, while greigite aligns along {100} [9].

v) **Better T_2 reducing effect**

Magnetic properties of small MNPs (>20 nm) may be compromised due to spin surface disorder and spin canting [37]. When it comes to decreasing T_2 levels, the MTB-NPs may be more effective than tiny synthetic equivalents since they have close-to-bulk magnetizations [38]. It was determined that the transverse relaxivity of the magnetite NP was 146/mmol/s, but the dextran-coated synthetic magnetite particles had a value of 62/mmol/s [9].

vi) **Ultrasonic velocity**
Magnetosome magnetite crystals also have a high ultrasonic velocity. It was discovered that the magnetosome suspension had a higher ultrasonic velocity while the synthetic MNPs had a velocity just slightly lower than the liquid transporter. The high ultrasonic velocity of magnetosomes is attributed to the magnetosomal membrane characteristics [6, 8].

vii) **Composition**
The MTB-NPs have a very conservative composition. In contrast to chemical synthesis, where the Fe precursor can be mixed with some other metal ions, attempts in synthesizing doped MTB-NPs by developing MTB with multiple cations are typically futile since the cells have the tendency to selectively entrap Fe even from restricted sources to grow pure magnetite nanocrystals [9].

viii) **Functionalization**
Magnetosomes are encased in a phospholipid bilayer polar membrane, which possesses a variety of chemical groups on the surface. Superficial amino and carboxyl groups on these membranes facilitate biomolecule functionalization while synthetic MNPs must be coated with PEG or glutaraldehyde before use. Magnetosome's membrane improves dispersivity and negative charge, making it a superior drug carrier [8]. The chain arrangement of particles enhances the magnetic dipole moment and decreases aggregation upon extraction. Optimized magnetosome synthesis provides great biocompatibility and minimal toxicity whereas synthetic MNPs are frequently made using hazardous ingredients and are less biocompatible. Following treatment with magnetosome (1 mg/ml), MTT tests on L929 cells indicated that 90% of the cells were viable, but only about 85% cells showed viability after treatment with proportionate synthetic MNPs [33].

ix) **Heating efficiency**
Magnetosomes have greater heating efficiency than chemically produced MNPs due to chain-like formulation and remarkable grain size [6]. They have good heating capabilities and generates better MRI than chemically produced NPs [39]. Both synthetic MNPs and bacterial magnetosomes perhaps be utilized as negative contrast agents [36].

NPs produced for use in drug delivery systems and bioimaging should satisfy three requirements: (i) size control, (ii) aggregation state and (iii) composition of viable liposome-encapsulated MNPs are all important considerations. Unlike synthetic MNPs, bacterial magnetosomes can be biomineralized as nanoscopic Fe minerals surrounded by phospholipid membranes, which match the aforementioned characteristics. Narrow size distribution, excellent dispersion and a naturally occurring biological covering make magnetosomes an enticing near-future option [8]. The lipid coating of MTB-NPs allows them to disperse in water, whereas manufactured particles must be water soluble, which can be challenging for particles larger than the superparamagnetic range [9].

9.5 Applications of Magnetosomes

Magnetosomes are membrane-bound, SMD crystals with magnetic properties, having a broader application spectrum considering their magnetic and physical properties. Nanocrystals of magnetite from the *Magnetospirillum* genus have so far been used in application. Magnetosome-inspired greigite NPs appear to have similar magnetic characteristics as magnetite magnetosome crystals, suggesting they are also promising candidates for biomedicinal applications [15].

The phospholipid coating makes the particles biocompatible, their suspensions stable and them easily modifiable [40]. MTB *Magnetospirillum* sp. AMB-1 has been genetically modified to produce a protein-magnetite complex [12]. MRI contrast agent, hyperthermia, drug delivery, and the immobilization of enzymes, antibodies, and oligonucleotides have all been achieved by using these MNPs. Here are a few examples of some of the applications:

9.5.1 Magnetosomes in Functionalization and Immobilization of Bio-active Molecules

Biologically active molecules such as oligonucleotides and enzymes can be immobilized in a variety of ways in bioscience and biotechnology. It is possible to employ the immobilized compounds to express their mode of action in a desired process or as affinity ligands to capture or change the target molecules or cells. External magnetic force could be employed to remove or target biologically active substances trapped on magnetic carriers. MTB-MNPs have been utilized to immobilize glucose oxidase and uricase, antibodies, oligonucleotides, etc. [41, 42]. Various antibodies, enzymes and protein A have been immobilized after curing with glutaraldehyde using MNPs catalyzed with 3-aminopropyltriethoxysilane [43].

The bipolar macroscopic molecule polyethylene glycol (PEG) can be used to chemically modify enzymes such that they become active and dissolvable in organic solvents. MNPs can then be coupled to PEG-enzyme conjugate. Or the magnetite-PEG conjugate can be synthesized first and then fused with the target enzyme. Magnetically modified lipase is stable in organic and aqueous solutions and catalyzes ester synthesis in organic solvents and may be retrieved by applying magnetic field. In addition, L-asparaginase and urokinase have also been altered in a similar manner. Similarly, antibodies can also be altered. Magnetoliposomes containing MNPs have been utilized to immobilize membrane-bound enzymes or antibodies and entrap pharmaceuticals, which improved the catalytic activity of lipid-depleted membrane-bound enzymes such cytochrome c-oxidase [41].

9.5.2 Magnetosomes in DNA, Xenobiotics and Antigen Detection Assays

Bio-active substances and xenobiotics can be detected using magnetic modifications of traditional immunoassays. Fine MNPs are used to covalently immobilize certain antibodies or antigens. The traditional microtitration plate-based assays are slower and less iterative than magnetically based assays, which have been shown to be simple, quick and sensitive. Magnetosomes have proved to be a powerful tool to discover single nucleotide polymorphism (SNP). SNP is a biological tool for detecting cancer, diabetes, and high blood pressure, as well as for separating cells and detecting DNA. DNA extraction method based on magnetosomes has been created through crosslinking with aminosilanes, which may bind to DNA through covalent bonds. Crosslinked DNA complexes are coupled to magnetic columns, and magnetic fields may separate DNA from debris [8, 21].

Magnetosomes were successfully employed in protein detection assays. The protein streptavidin was attached using biotin groups connected to the magnetosome membrane. Semisynthetic composites with biotin-binding ability might join bionics such as biotinylated DNA oligonucleotides or biotinylated antibodies. This method potentially improves immunological diagnostics and proteome research. Using this approach, an antibody-functionalized magnetosome surface-independent immunoassay (M-IPCR) was constructed to immobilize Hepatitis B antigen in human serum that is due to magnetic concentration of the sensor complex increased the output signal. It was found that M-IPCR was roughly 100-fold more efficient than Magneto-ELISA (Enzyme Linked Immunosorbent Assay). Magneto-ELISA employs synthetic NPs for increasing the antigen detection potential of ELISA and is always done parallelly to M-IPCR for comparative reasons [15]. Applying highly efficient, fully computerized and quick chemiluminescence immunoassay, antibody-conjugated bacterial MNPs were generated to assess the food allergen, i.e. lysozyme [44]. Recently, a sandwich immunoassay using antibody – protein A – MTB-MNP complexes and alkaline phosphatase-conjugated secondary antibody was reported for the detection of human insulin [45].

9.5.3 Treatment of Magnetic Hyperthermia

Magnetic hyperthermia promotes cell necrosis by heating MNPs. Magnetosome was a promising candidate for magnetic hyperthermia because of its superior magnetic properties. Hyperthermia destroys tumor cells by increasing intracellular heat stress to 41–46 °C. Non-selective heating can destroy healthy tissue and create serious negative effects [8]. According to the findings, the application of magnetosome-induced hyperthermia was found to be an effective method for killing tumor cells in mice. Using MDA-MB-231 breast cancer cells as a template, researchers evaluated the capacity of MTB cells, magnetosome chains and individual magnetosomes to produce heat [46–48].

9.5.4 Food Safety

Detecting pathogens in food with functionalized magnetosomes has shown to be a successful approach. Recombinant magnetosomes from *Magnetospirillum gryphiswaldense* strain MSR-1 cells were employed to build a capture system that combined the mamC gene with protein A, which was then linked to a particular immunoglobin in order to arrest *Vibrio parahaemolyticus*, a bacterium that is responsible for a wide variety of gastrointestinal disorders that are caused by consuming contaminated food [49].

9.5.5 Cell Separation

Separation of cells magnetically has also been accomplished by magnetosomes and particular antibodies. For example, synthesized recombinant magnetosomes with immunoglobin-linked protein A on their surface were, hence, used for cell separation. In *Magnetospirillum magneticum* strain AMB-1 cells, the ZZ domain of protein A was amalgamated to the gene encoding the magnetosome membrane protein MamC to give functionalized protein. Anti-CD14, CD19 and CD20 murine monoclonal antibodies were employed to pre-treat the peripheral blood cells with the functionalized magnetosomes, followed by coupling with anti-murine G immunoglobulins. The recovery efficiency of monocytes (CD14+) and B-lymphocytes (CD19+ and CD20+) was 95.7%, 97.2% and 98.8%, respectively [50].

9.5.6 Drug Delivery

Drug delivery systems utilizing MTB-MNPs have also been developed. Antitumor drug Doxorubicin (DOX) was covalently attached to MTB-MNPs to test tumor growth inhibition. On H22 tumor-impacted mice, these DOX-conjugated MTB-MNPs demonstrated an almost equivalent tumor suppression rate with respect to DOX alone (86.8% v/s 78.6%), with significantly decreased vascular damage. It is possible to magnetically guide them to aggregate and perform remedial actions just at diseased locations through subcutaneous administration into the solid tumor [51]. Using PEI-associated MTB-MNPs, galactosidase plasmids were distributed both *in vitro* and *in vivo*, in a similar manner to the one reported here. The MTB PEINP system was said to be more efficient and less hazardous than PEI alone [52]. MTB-MNP-based hyperthermia treatment has also been considered and similar investigations are under underway [9].

DOX and magnetosomes from *Magnetospirillum gryphyswaldense* strain MSR-1 were crosslinked together with glutaraldehyde to create a complex. The magnetosome-DOX combination looked to be quite stable, and the drug's release from the complex lasted longer than expected. The HL60 and EMT-6 cell lines of human leukemia and mouse breast cancer, respectively, were used to test the antitumor efficacy of the magnetosome-DOX complex. After 48 hours of incubation, 80% of the drug was still attached to the magnetosome. The complex remained intact during its passage through the systemic circulation, and a significant portion of the DOX was not appreciably released until the complex reached the target tissue for its intended impact. It also showed substantial anticancer efficacy as seen by a reduction in cancer cell proliferation. Coupling the medication to the magnetosome did not reduce anticancer activity due to any skeletal change [30]. The theorized applications associated with the magnetosomes in the literature have been cited in Table 9.1.

9.6 Conclusion and Future Perspective

Magnetosomes have been proven to be excellent nanocarriers for biomolecules in a variety of biotechnological and medicinal applications. The physiological composition of magnetosome membrane may be updated to create multifunctional magnetosomes, which has substantial potential for preclinical and clinical applications. Diagnostics and treatments may be performed simultaneously using multifunctional magnetosomes. A primary objective that has to be accomplished is the development of more efficient methods

Table 9.1 Some extra applications of magnetosomes cited in the literature.

Serial number	Applications	References
1.	Immobilization of protein	[53]
2.	Single nucleotide polymorphism detection	[54–57]
3.	Immobilization of enzymes	[58–61]
4.	Food safety	[62]
5.	Drug delivery	[63]
6.	Bio-catalysis	[64]
7.	Wastewater treatment	[65]
8.	Bio-templating	[66]
9.	MRI agents	[67–69]
10.	Hyperthermia	[70, 71]
11.	Immunoassay	[72, 73]

for the synthesis of magnetosomes from already existing strains or the identification of novel MTB strains that are capable of creating magnetosomes in large quantities. MTBs and magnetosomes have better qualities than chemically synthesized magnetic nanoparticles. The potential for magnetosomes to be used in a broader spectrum of new applications has considerably risen apparently because of recent developments in both the knowledge of biomineralization of magnetite nanocrystals and evolution of ways for functionalizing magnetosomes. The capability of magnetosomes to bind a wide variety of biomolecules has piqued the interest of researchers working in a variety of fields, including nanotechnology and biomedicine.

References

1 Parveen, S., Misra, R., and Sahoo, S.K. (2012). Nanoparticles: a boon to drug delivery, therapeutics, diagnostics and imaging. *Nanomed. Nanotechnol. Biol. Med.* 8 (2): 147–166.

2 Jha, Rakhi, J., Jha, Pradeep, K., Chaudhury, K. et al. (2014). An emerging interface between life science and nanotechnology: present status and prospects of reproductive healthcare aided by nano-biotechnology. *Nano Rev.* 5.

3 Mahmoudi, M., Sant, S., Wang, B. et al. (2011). Superparamagnetic iron oxide nanoparticles (SPIONs): development, surface modification and applications in chemotherapy. *Adv. Drug Delivery Rev.* 63 (1–2): 24–46.

4 Pankhurst, Q.A., Connoly, J., Jones, S.K., and Dobson, J. (2003). Applications of magnetic nanoparticles in biomedicine. *J. Phys. D: Appl. Phys.* 36 (13): R167.
5 Araujo, A.C.V., Abreu, F., Silva, K.T. et al. (2015). Magnetotactic bacteria as potential sources of bioproducts. *Mar. Drugs* 13: 389–430.
6 Józefczak, A., Leszczyński, B., Skumiel, A., and Hornowski, T. (2016). A comparison between acoustic properties and heat effects in biogenic (magnetosomes) and abiotic magnetite nanoparticle suspensions. *J. Magn. Magn. Mater.* 407: 92–100. https://doi.org/10.1016/j.jmmm.2016.01.054.
7 Alphandéry, E. (2014). Applications of magnetosomes synthesized by magnetotactic bacteria in medicine. *Front. Bioeng. Biotechnol.* 2 (MAR): 1–6.
8 Jacob, J.J. and Suthindhiran, K. (2016). Magnetotactic bacteria and magnetosomes – scope and challenges. *Mater. Sci. Eng., C* 68: 919–928. https://doi.org/10.1016/j.msec.2016.07.049.
9 Xie, J., Chen, K., and Chen, X. (2009). Production, modification and bio-applications of magnetic nanoparticles gestated by magnetotactic bacteria. *Nano Res.* 2 (4): 261–278.
10 Schüler, D. (2008). Genetics and cell biology of magnetosome formation in magnetotactic bacteria. *FEMS Microbiol. Rev.* 32 (4): 654–672.
11 Yan, L., Zhang, S., Chen, P. et al. (2012). Magnetotactic bacteria, magnetosomes and their application. *Microbiol. Res.* 167 (9): 507–519.
12 Wang, X., Li, Y., Zhao, J. et al. (2020). Magnetotactic bacteria: characteristics and environmental applications. *Front. Environ. Sci. Eng.* 14 (4): 56.
13 Dieudonné, A., Pignol, D., and Prévéral, S. (2019). Magnetosomes: biogenic iron nanoparticles produced by environmental bacteria. *Appl. Microbiol. Biotechnol.* 103: 3637–3649.
14 Bazylinski, D.A. and Frankel, R.B. (2004). Magnetosome formation in prokaryotes. *Nat. Rev. Microbiol.* 2 (3): 217–230.
15 Vargas, G., Cypriano, J., Correa, T. et al. (2018). Applications of magnetotactic bacteria, magnetosomes and magnetosome crystals in biotechnology and nanotechnology: mini-review. *Molecules* 23 (10): 1–25.
16 Bazylinski, D., Lefèvre, C., and Lower, B. (2014). Magnetotactic bacteria, magnetosomes, and nanotechnology. In: *Nanomicrobiology* (ed. L.L. Barton, D.A. Bazylinski, and H. Xu), 39–74. New York, NY: Springer.
17 Zhu, X., Hitchcock, A.P., Le Nagard, L. et al. (2018). X-ray absorption spectroscopy and magnetism of synthetic greigite and greigite magnetosomes in magnetotactic bacteria. *Geomicrobiol. J.* 35 (3): 215–226. https://doi.org/10.1080/01490451.2017.1362078.

18 Farzan, F., Shojaosadati, S.A., and Tehrani, H.A. (2010). A preliminary report on the isolation and identification of magnetotactic bacteria from Iran environment. *Iran. J. Biotechnol.* 8 (2): 98–102.

19 Simmons, S.L., Bazylinski, D.A., and Edwards, K.J. (2006). South-seeking magnetotactic bacteria in the Northern Hemisphere. *Science (80-)* 311 (5759): 371–374.

20 Lefèvre, C.T. and Bazylinski, D.A. (2013). Ecology, diversity, and evolution of magnetotactic bacteria. *Microbiol. Mol. Biol. Rev.* 77 (3): 497–526.

21 Arakaki, A., Nakazawa, H., Nemoto, M. et al. (2008). Formation of magnetite by bacteria and its application. *J. R. Soc. Interface* 5 (26): 977–999.

22 Barber-Zucker, S. and Zarivach, R. (2016). A look into the biochemistry of magnetosome biosynthesis in magnetotactic bacteria. *ACS Chem. Biol.* 12 (1): 13–22.

23 Ghaisari, S., Winklhofer, M., Strauch, P. et al. (2017). Magnetosome organization in magnetotactic bacteria unraveled by ferromagnetic resonance spectroscopy. *Biophys. J.* 113 (3): 637–644.

24 Yan, L., Da, H., Zhang, S. et al. (2017). Bacterial magnetosome and its potential application. *Microbiol. Res.* 203: 19–28.

25 Bazylinski, D.A., Frankel, R.B., Heywood, B.R. et al. (1995). Controlled biomineralization of magnetite (Fe_3O_4) and greigite (Fe_3S_4) in a magnetotactic bacterium. *Appl. Environ. Microbiol.* 61 (9): 3232–3239.

26 Lin, W., Pan, Y., and Bazylinski, D.A. (2017). Diversity and ecology of and biomineralization by magnetotactic bacteria. *Environ. Microbiol. Rep.* 9 (4): 345 356.

27 Malvankar, N.S., King, G.M., and Lovley, D.R. (2015). Centimeter-long electron transport in marine sediments via conductive minerals. *ISME J.* 9 (2): 527–531.

28 Liu, F., Rotaru, A.E., Shrestha, P.M. et al. (2015). Magnetite compensates for the lack of a pilin-associated c-type cytochrome in extracellular electron exchange. *Environ. Microbiol.* 17 (3): 648–655.

29 Kiseleva, L., Briliute, J., Khilyas, I.V. et al. (2015). Magnet-facilitated selection of electrogenic bacteria from marine sediment. *Biomed Res. Int.* 2015: 582471. https://doi.org/10.1155/2015/582471.

30 Sun, J.B., Duan, J.H., Dai, S.L. et al. (2008). Preparation and anti-tumor efficiency evaluation of doxorubicin-loaded bacterial magnetosomes: magnetic nanoparticles as drug carriers isolated from *Magnetospirillum gryphiswaldense*. *Biotechnol. Bioeng.* 101 (6): 1313–1320.

31 Mickoleit, F., Borkner, C.B., Toro-Nahuelpan, M. et al. (2018). In vivo coating of bacterial magnetic nanoparticles by magnetosome expression of spider silk-inspired peptides. *Biomacromolecules* 19 (3): 962–972.

32 Cheng, L., Ke, Y., Yu, S., and Jing, J. (2016). Co-delivery of doxorubicin and recombinant plasmid pHSP70-Plk1-shRNA by bacterial magnetosomes for osteosarcoma therapy. *Int. J. Nanomed.* 11: 5277–5286.

33 Han, L., Li, S., Yang, Y. et al. (2007). Comparison of magnetite nanocrystal formed by biomineralization and chemosynthesis. *J. Magn. Magn. Mater.* 313 (1): 236–242.

34 Muscas, G., Singh, G., Glomm, W.R. et al. (2015). Tuning the size and shape of oxide nanoparticles by controlling oxygen content in the reaction environment: morphological analysis by aspect maps. *Chem. Mater.* 27 (6): 1982–1990.

35 Fischer, H., Mastrogiacomo, G., Löffler, J.F. et al. (2008). Ferromagnetic resonance and magnetic characteristics of intact magnetosome chains in *Magnetospirillum gryphiswaldense*. *Earth Planet. Sci. Lett.* 270 (3–4): 200–208.

36 Hu, L.L., Zhang, F., Wang, Z. et al. (2010). Comparison of the 1H NMR relaxation enhancement produced by bacterial magnetosomes and synthetic iron oxide nanoparticles for potential use as MR molecular probes. *IEEE Trans. Appl. Supercond.* 20 (3): 822–825.

37 Frankel, R.B., Dunin-Borkowski, R.E., Pósfai, M., and Bazylinski, D.A. (2008). Magnetic microstructure of magnetotactic bacteria. *Handb. Biominer. Biol. Asp. Struct. Form.* 1: 126–144.

38 Lins, U., Mccartney, M.R., Farina, M. et al. (2005). Habits of magnetosome crystals in coccoid magnetotactic bacteria. *Appl. Environ. Microbiol.* 71 (8): 4902–4905.

39 Liu, R.T., Liu, J., Tong, J.Q. et al. (2012). Heating effect and biocompatibility of bacterial magnetosomes as potential materials used in magnetic fluid hyperthermia. *Prog. Nat. Sci.: Mater. Int.* 22 (1): 31–39. https://doi.org/10.1016/j.pnsc.2011.12.006.

40 Matsunaga, T. and Sakaguchi, T. (2000). Molecular mechanism of magnet formation in bacteria. *J. Biosci. Bioeng.* 90 (1): 1–13.

41 Safarik, I. and Safarikova, M. Magnetic nanoparticles and biosciences. In: *Nanostructured Materials* (ed. H. Hofmann), 1–23. Springer Verlag.

42 Sestier, C., Da-Silva, M.F., Sabolovic, D. et al. (1998). Surface modification of superparamagnetic nanoparticles (ferrofluid) studied with particle electrophoresis: application to the specific targeting of cells. *Electrophoresis* 19 (7): 1220–1226.

43 Shi, Y.J., Shen, R.N., Lu, L., and Broxmeyer, H.E. (1994). Comparative analysis of retroviral-mediated gene transduction into CD34+ cord blood hematopoietic progenitors in the presence and absence of growth factors. *Blood Cells* 20 (2–3): 517–524. http://europepmc.org/abstract/MED/7538354.

44 Sato, R., Takeyama, H., Tanaka, T., and Matsunaga, T. (2001). Development of high-performance and rapid immunoassay for model food allergen lysozyme using antibody-conjugated bacterial magnetic particles and fully automated system. *Appl. Biochem. Biotechnol. - Part A Enzym. Eng. Biotechnol.* 91–93: 109–116.

45 Tanaka, T. and Matsunaga, T. (2000). Fully automated chemiluminescence immunoassay of insulin using antibody – protein A – bacterial magnetic particle complexes. *Anal. Chem.* 72 (15): 3518–3522.

46 Edouard AlphanderyStéphanie FaureImène Chebbi. 2011. Treatment of cancer or tumor induced by the release of heat generated by various chains of magnetosomes extracted from magnetotactic bacteria and submitted to an alternating magnetic field. https://patents.google.com/patent/US10252073B2/en

47 Alphandéry, E., Faure, S., Seksek, O. et al. (2011). Chains of magnetosomes extracted from AMB-1 magnetotactic bacteria for application in alternative magnetic field cancer therapy. *ACS Nano* 5 (8): 6279–6296.

48 Alphandéry, E., Faure, S., Raison, L. et al. (2011). Heat production by bacterial magnetosomes exposed to an oscillating magnetic field. *J. Phys. Chem. C* 115 (1): 18–22.

49 Xu, J., Hu, J., Liu, L. et al. (2014). Surface expression of protein A on magnetosomes and capture of pathogenic bacteria by magnetosome/antibody complexes. *Front. Microbiol.* 5 (APR): 4–11.

50 Yoshino, T., Hirabe, H., Takahashi, M. et al. (2008). Magnetic cell separation using nano-sized bacterial magnetic particles with reconstructed magnetosome membrane. *Biotechnol. Bioeng.* 101 (3): 470–477.

51 Sun, J.B., Duan, J.H., Dai, S.L. et al. (2007). In vitro and in vivo antitumor effects of doxorubicin loaded with bacterial magnetosomes (DBMs) on H22 cells: the magnetic bio-nanoparticles as drug carriers. *Cancer Lett.* 258 (1): 109–117.

52 Xiang, L., Bin, W., Huali, J. et al. (2007). Bacterial magnetic particles (BMPs)-PEI as a novel and efficient non-viral gene delivery system. *J. Gene Med.* 9 (8): 679–690.

53 Tanaka, T., Yamasaki, H., Tsujimura, N. et al. (1997). Magnetic control of bacterial magnetite-myosin conjugate movement on actin cables. *Mater. Sci. Eng., C* 5 (2 C): 121–124. https://doi.org/10.1016/S0928-4931(97)00027-1.

54 Tanaka, T., Maruyama, K., Yoda, K. et al. (2003). Development and evaluation of an automated workstation for single nucleotide polymorphism discrimination using bacterial magnetic particles. *Biosens. Bioelectron.* 19 (4): 325–330.

55 Maruyama, K., Takeyame, H., Nemoto, E. et al. (2004). Single nucleotide polymorphism detection in aldehyde dehydrogenase 2 (ALDH2) gene using bacterial magnetic particles based on dissociation curve analysis. *Biotechnol. Bioeng.* 87 (6): 687–694.

56 Ota, H., Takeyama, H., Nakayama, H. et al. (2003). SNP detection in transforming growth factor-β1 gene using bacterial magnetic particles. *Biosens. Bioelectron.* 18 (5–6): 683–687.

57 Nakayama, H., Arakaki, A., Maruyama, K. et al. (2003). Single-nucleotide polymorphism analysis using fluorescence resonance energy transfer between DNA-labeling fluorophore, fluorescein isothiocyanate, and DNA intercalator, POPO-3, on bacterial magnetic particles. *Biotechnol. Bioeng.* 84 (1): 96–102.

58 Jacob, J.J. and Suthindhiran, K. (2020). Immobilisation of lipase enzyme onto bacterial magnetosomes for stain removal. *Biotechnol. Rep.* 25: e00422.

59 Jacob, J.J. and Suthindhiran, K. (2021). Efficiency of immobilized enzymes on bacterial magnetosomes. *Appl. Biochem. Microbiol.* 57 (5): 603–610.

60 Honda, T., Tanaka, T., and Yoshino, T. (2015). Stoichiometrically controlled immobilization of multiple enzymes on magnetic nanoparticles by the magnetosome display system for efficient cellulose hydrolysis. *Biomacromolecules* 16 (12): 3863–3868.

61 Pedroso, M.M., Hine, D., Hahn, S. et al. (2021). Pesticide degradation by immobilised metalloenzymes provides an attractive avenue for bioremediation. *EFB Bioeconomy J.* 1 (April): 100015. https://doi.org/10.1016/j.bioeco.2021.100015.

62 Sannigrahi, S., Arumugasamy, S.K., Mathiyarasu, J., and Suthindhiran, K. (2020). Development of magnetosomes-based biosensor for the detection of *Listeria monocytogenes* from food sample. *IET Nanobiotechnol.* 14 (9): 839–850.

63 Taherkhani, S., Mohammadi, M., Daoud, J. et al. (2014). Covalent binding of nanoliposomes to the surface of magnetotactic bacteria for the synthesis of self-propelled therapeutic agents. *ACS Nano* 8 (5): 5049–5060.

64 Mittmann, E., Mickoleit, F., S. Maier, D. et al. (2022). A magnetosome-based platform for flow biocatalysis. *ACS Appl. Mater. Interfaces* 14 (19): 22138–22150.

65 Bahaj, A.S., James, P.A.B., and Moeschler, F.D. (1998). Wastewater treatment by bio-magnetic separation: a comparison of iron oxide and iron sulphide biomass recovery. *Water Sci. Technol.* 38 (6 pt 5): 311–317. https://doi.org/10.1016/S0273-1223(98)00591-5.

66 Serhan M, Sprowls M, Jackemeyer D, Long M, Perez ID, Maret W, et al. 2019. Total iron measurement in human serum with a smartphone. *AIChE Annu Meet Conf Proc*, 2019 November.

67 Nan, X., Lai, W., Li, D. et al. (2021). Biocompatibility of bacterial magnetosomes as mri contrast agent: a long-term in vivo follow-up study. *Nanomaterials* 11 (5): 1–16.

68 Boucher, M., Geffroy, F., Prévéral, S. et al. (2017). Genetically tailored magnetosomes used as MRI probe for molecular imaging of brain tumor. *Biomaterials* 121: 167–178.

69 Mériaux, S., Boucher, M., Marty, B. et al. (2015). Magnetosomes, biogenic magnetic nanomaterials for brain molecular imaging with 17.2 T MRI scanner. *Adv. Healthcare Mater.* 4 (7): 1076–1083.

70 Gandia, D., Gandarias, L., Rodrigo, I. et al. (2019). Unlocking the potential of magnetotactic bacteria as magnetic hyperthermia agents. *Small* 15 (41): 1–12.

71 Le Fèvre, R., Durand-Dubief, M., Chebbi, I. et al. (2017). Enhanced antitumor efficacy of biocompatible magnetosomes for the magnetic hyperthermia treatment of glioblastoma. *Theranostics* 7 (18): 4618–4631. http://www.thno.org.

72 Wu, S., Ma, F., He, J. et al. (2021). Fusion expression of nanobodies specific for the insecticide fipronil on magnetosomes in *Magnetospirillum gryphiswaldense* MSR-1. *J. Nanobiotechnol.* 19 (1): 1–9. https://doi.org/10.1186/s12951-021-00773-z.

73 He, J., Tian, J., Xu, J. et al. (2018). Strong and oriented conjugation of nanobodies onto magnetosomes for the development of a rapid immunomagnetic assay for the environmental detection of tetrabromobisphenol-A. *Anal. Bioanal.Chem.* 410 (25): 6633–6642.

10

Biofabrication of Nanoparticles in Wound-Healing Materials

Nishat Khan[1], *Isha Arora*[1], *Amrish Chandra*[2], *and Seema Garg*[1]

[1] *Department of Chemistry, Amity Institute of Applied Sciences, Amity University, Noida, UP, India*
[2] *Amity Institute of Pharmacy, Amity University, Noida, UP, India*

Abstract

The methodology for mending wounds has been significantly studied to increase and interpret the best results that achieve good and fastest recovery and reduce the effect to the minimum or least amount, ensuring feature preservation. The classic theory for wound administration is reflected in treatments, like antibacterial agents, to prevent impurities and contamination and promote an ideal wound-healing process. Nanotechnology relates to interrelated processes and submicroscopic particles having a maximum diameter of 100 nm, because they such a significant impact on promoting and expediting wound healing as well as treating and preventing bacterial infections, metal nanoparticles (NPs) like silver, gold, and zinc are more extensively used in dermatology. In addition to a wet wound area, fewer permitted dressing changes and more applications result in increased handling. This evaluation shows the recent highlights regarding the NPs used in wound healing.

Keywords *nanoparticles; nanocomposite; scaffolds; green synthesis; Arnebia nobilis; silver nanoparticles; gold nanoparticles; honey-based nanoparticles; acute chronic wound; silver mesoporous silica nanoparticles; basic fibroblast growth factor; silver nitroprusside nanoparticles*

Green Synthesis of Nanomaterials: Biological and Environmental Applications, First Edition.
Edited by Archana Chakravarty, Preeti Singh, Saiqa Ikram, and R.N. Yadava.

10.1 Introduction

Major medical issues like wound infections have a significant risk of endangering and infecting human health. According to recent studies, composite materials may help regulate endogenous/exogenous electrical conduction, direct wound migration, and accelerate wound healing. A unique silver nanoparticle (AgNP) doped with a conductive polymer based on a hydrogel system was created to speed up the healing of infected wounds. A unique AgNP was conjugated with a conductive polymer and created based on a hydrogel system. It has potent mechanical, biocompatibility, electrical, and antibacterial characteristics. The host matrix ingredients for the hydrogel binding were Ag NPs, polyvinyl alcohol (PVA), and gelatin. Phytic acid was used to cross-link the semiconducting polyaniline into the matrix gauge sheet by impregnation, and antibacterial AgNPs were preloaded [1]. Two types of wounds that often do not heal normally are chronic wounds and delayed acute wounds. To address the microenvironmental excesses in developing wound therapy solutions, a deeper understanding of the cellular mechanisms involved in the treatment of chronic wounds is required. Honey is a feasible alternative for treatment methods in a variety of circumstances due to its therapeutic benefits, particularly its antimicrobial activity. Integration with nanotechnology has created new opportunities for other medical applications in addition to wound healing. Recent developments in honey-based nanoparticles (NPs) for the healing process are discussed in this review. Additionally covered are perspectives on the difficulties and potential future directions of using honey-based NPs as well as the mechanism of action of NPs in the wound-healing process. Honey is used to treat wounds, and the underlying mechanisms are thought to involve hydrogen peroxide, high osmolality, acidity, nonperoxide components, and phenols. Due to the rising demand for combination dressings in the global market for wound dressings, which contain two or more types of chemical and physical properties to ensure optimal functionality, incorporating honey into various wound dressings has become a major trend. They are also a preferred option among researchers due to their numerous benefits, including affordability, biocompatibility, and swelling index, as well as their versatility in terms of fabrication techniques (electro-spun fibers, hydrogels, etc.).

Three biological phases are involved in wound healing: inflammation, which typically lasts up to 66 days; proliferation, which typically follows 2 weeks later; and remodeling, which can last up to 2 years. It is important to note that the phases overlap *in vivo* due to various anatomical and physiological factors, such as constant intercellular signaling, which implies real-time feedback control of pro-inflammatory and anti-inflammatory cytokine release. The inflammatory phase begins with a vascular response that ensures hemostasis via blood clot formation. Second, the distress molecules released by injured cells act as a chemoattractant for leukocytes, these have two primary functions: recognition and destruction of infectious agents and cytokine release, which stimulates cells in the proliferative phase [2]. A wound healing process can become impaired if one of these four stages is interfered with. An underlying illness (such as diabetes, aging, and ischemia) or environmental factors (such as infections) often contribute to this impaired healing. The blood flow to a wound is frequently partially or completely blocked due to these systemic preconditioning situations, which interferes with the healing process by limiting the amount of nutrition, oxygen, waste removal, and homeostasis the lesion can receive. Over the past few years, the field of research and development into wound dressing materials has advanced to a new level of standardization, which appears to have resulted in a better understanding of the etiologic of persistent wounds.

Whether or not novel approaches to promote wound healing can be developed, wound management has found solace in medical tradition. It has adopted some of the traditional treatments used in the past. One of the traditional ingredients that acts as a strong antimicrobial agent with various effects is honey. Honey has long been known to have positive health effects, and many of these effects have been investigated to learn more about how they operate. Since the early twentieth century, numerous studies into honey's makeup and useful properties have been conducted worldwide, and the findings have been astounding. The use of honey for wound healing involves several components, including sugar, acid, hydrogen peroxide, and phenolic content. These substances are all there in varying amounts.

Wound dressings come in a wide range of sizes and shapes and can be divided into different categories based on many characteristics. First, there are three different categories of wound dressings: natural (such as gauze and gauze/cotton composites), biomaterial-based (such as allografts, tissue derivatives, and xenografts), and artificial (such as film, membrane, foam, gel, composites, and spray). They can then be divided into two groups: primary dressings, which are used to cover the primary dressings, and secondary dressings, which are applied directly to the wound. Several wound dressing design solutions have been created to address particular issues with wound repair mechanisms, each with distinctive features and a variety of functions. The composition of wound dressings must comply with all applicable regulations to have the desired effect. Surgical and wound dressings are regulated by the US Food and Drug Administration under the Medical Device Regulation Act and are divided into two categories: category 1 and category 2 are generally used on mind wounds, and there are few regulatory requirements for their approval.

Granulation tissue covering the wound surface signals the start of the proliferative phase. Neo angiogenesis and the stimulation of fibroblasts, which create collagen and other extracellular matrices, are key elements in this second phase. Restoring the tissue's shape and functionality is the third phase, remodeling (Table 10.1; Figure 10.1).

Infectious agent colonization is the most frequent problem with wound healing. The bacteria that comprise the skin's microbiota help to avoid colonizing other diseases, however, if they accumulate to a dangerous level near a lesion, they can obstruct healing, especially if they form biofilms. Moreover, foreign microorganisms can directly infect the wound, impairing the healing. *Staphylococcus aureus* and methicillin-resistant *Staphylococcus aureus* (MRSA) are the most prevalent, affecting the early stages of wound healing, in contrast, *Pseudomonas aeruginosa* and *Escherichia coli* are linked to chronic wounds and infect deeper skin layers [5].

Necrotic tissue shows the variability of the healing process, which can be disturbed by internal and external variables, bacterial growth, foreign matter confinement, and underlying comorbidities (e.g. diabetes mellitus). Any of these variables could prolong each phase, leading to an outcome that is both physically and functionally undesirable. Chronic wound infections are generally developed by synergistic bacterial growth. Sudden

Table 10.1 List of some wound-healing dressing examples.

Types	Level of risk	Examples	Classification of food and drugs administration	Requirements
Fabric dressing	Low risk	Hydrophilic wound dressings, hydrogel-based wound dressings	Category 1	Only needs to be informed; approval is optional. The manufacturer is the organizer of maintaining the product's quality and safety
Advanced wound dressing	Medium risk	Honey, prisma	Category 2	510(K) approval is required

Source: Adapted from Bahari et al. [3].

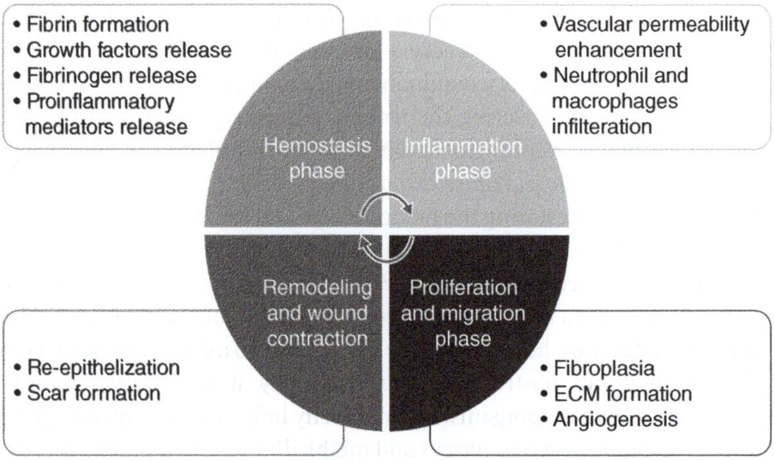

Figure 10.1 Phases of wound healing along with mediators/cells involved in wound-healing cascade. *Source:* Shah et al. [4]/with permission of Elsevier.

growth of aerobic bacteria can create a hypoxic tissue environment because they need more oxygen to maintain cell division. As a result, anaerobic bacteria that can create short-chain fatty acids to prevent host defense cells from phagocytosing them flourish. More pathogens may colonize the wound as a result.

Moreover, several microbes can exchange nutrients to support one another's growth. This explains why *P. aeruginosa* and *S. aureus* co-occur in some chronic wounds. The issue with wound healing that occurs most frequently is colonization with an infectious agent. Although the skin's microbiota helps to avoid the colonization of other infections, when it reaches a critical level at the site of a lesion, it can obstruct healing, especially if it forms biofilms. Moreover, foreign pathogens can directly infect the wound, impairing recovery. The most frequent are MRSA and *S. aureus*, obstructing the early stages of wound healing. On the other hand, *E. coli* and *P. aeruginosa* are connected to infected wounds and infect the inner layers of skin (Figure 10.2).

Understanding the creation, composition, and kinetics of molecular and atomic micro/nano NPs (with a maximum diameter of 100 nm), also known as NPs, is the basis of nanotechnology, from whom nanoproducts are created. When a particle is downscaled to the nanometric range, its floor increases exponentially while its quantity decreases, resulting in unique Physio-chemical houses that account for several clinical applications. Applying dynamic light scattering, fluorescence correlation spectroscopy,

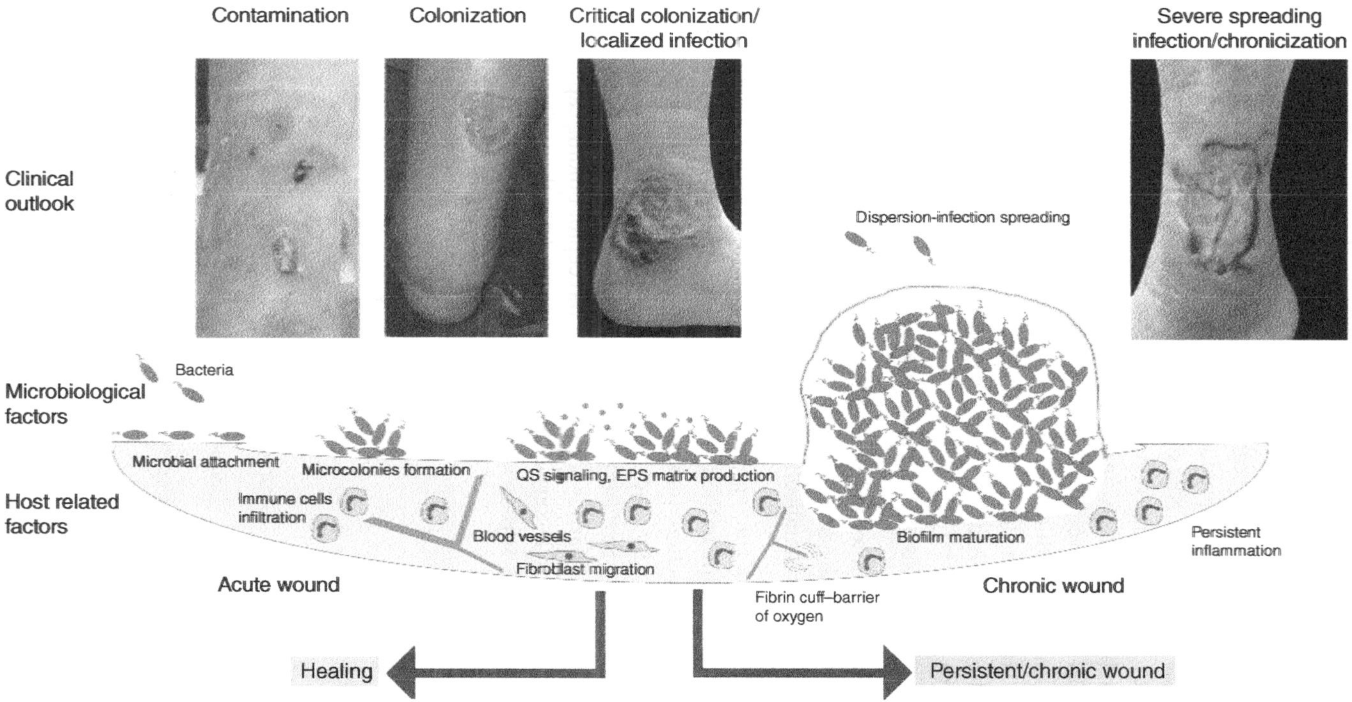

Figure 10.2 Schematic correlation diagram between the clinical outlook, microbiological, and host-related factors in acute and chronic wounds. *Source:* Mihai et al. [5]/Reproduced from MDPI/CC BY 4.0.

and Raman scattering can decide the hydrodynamic measurements distribution of NPs. In contrast, circular dichroism, infrared spectroscopy (IR), and mass spectrometry (MS) can be selected. Additionally, transmission electron microscopy and scanning electron microscopy, two types of electron microscopy, can produce resolutions of less than 1 nm [6].

The size and shape of NPs significantly impact the energetic ingredient vector for transportation either directly through the cytoplasmic membrane or by phagocytosis and cellular responses (receptor recognition). Metal NPs such as silver, gold, and zinc are the greatest options for wound dressing integration due to their antibacterial properties and low reduced toxicity. The following list illustrates the most typical types of nanomaterials used in wound care: NPs, nanocomposites, coatings, and scaffolds (Figure 10.3).

There are two types of NPs commonly used in wound healing [7]:

1) NPs have built-in wound-healing capabilities. Nonmetallic nanomaterials and conductive or metal oxide NPs are other classifications.
2) NPs are used as therapeutic agents.

Silver, gold, and zinc compounds are the most studied metallic and metal oxide NPs due to their special qualities, minimal skin penetration, and antibacterial action. Important traits such as size and shape influence its biological activity and toxicity. (The smaller particles are more biologically active, Surface potential, zeta potential, and polydispersity index.) The cellular barrier permeability and receptor binding capacity are influenced by particle surface charge, estimated by the zeta potential. Additional particle surface characteristics, including porosity, chemical

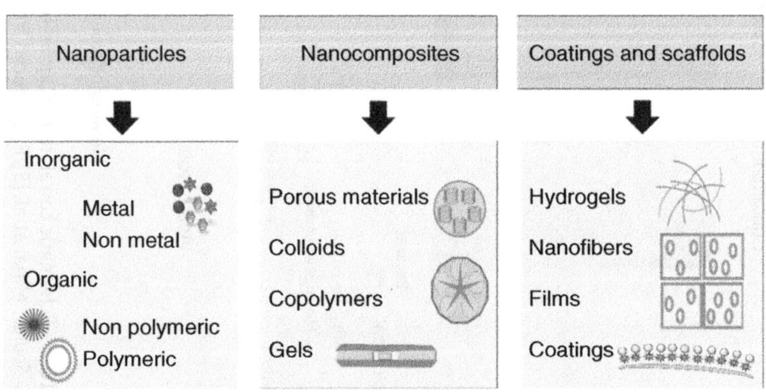

Figure 10.3 Schematic representation of classification of nanomaterials. *Source:* Adapted from Mihai et al. [5].

heterogeneity, and hydrolytic stability, impact the biological behavior of nanomaterials, including interactions with other biomolecules that may impact biodistribution.

Small polymeric matrix systems known as nanospheres comprise a fixed porous polymer that can bind to active molecules like amino acids, minerals, or chemical compounds. Active substances demonstrate improved pharmacological characteristics, increased biocompatibility, and increased stability.

10.2 Nanoparticles

10.2.1 Silver Nanoparticles

AgNPs are synthesized to deal with the drawbacks of conventional silver compounds. The high surface-to-volume ratio of AgNPs makes them more efficient even at very low concentrations, which also decreases its toxicity, which lowers their toxicity. Pure AgNPs have tuneable properties and anti-inflammatory modes of action like cytokine release, allowing for faster wound closure without increasing scarring. AgNPs work to heal the injury by stopping the differentiation of myofibroblasts from normal fibroblasts, fastening the reaction process. Furthermore, AgNPs help to produce epidermal re-epithelialization by stimulating keratinocyte proliferation and migration. Thus, the higher the concentration of AgNPs, the concentrations of AgNPs also decreased keratinocyte viability via inducing dose-dependent DNA damage and the activation of caspase (proteases implicated in programmed cell death), metabolism, migration, and differentiation [8].

Little concentrations of AgNPs can be added to antimicrobial medications to boost effectiveness while minimizing negative effects. In a recent study, AgNPs and tetracycline dramatically decreased the amount of germs in the surface and deeper tissue layers of a mouse model, hastening the healing process. These results suggest that AgNPs can be used to treat infected wounds in addition to conventional antibacterial treatments or dressings. For open wounds, cellulose functionalized with AgNPs was used as an antibacterial covering.

This nanomaterial demonstrated effective bacterial killing against Gram-negative pathogens while also promoting wound healing. Holban reported similar results when He used AgNPs in polyester-nylon dressings, which prevented bacterial colonization and biofilm creation while maintaining the integrity of the wound dressing [5].

The principal antibacterial action of AgNPs is the creation of sulfuric bonds with either the bacterial cell protein coat or the thiol groups of various enzymes in respiration, which causes death. Moreover, AgNPs can obstruct DNA synthesis during cell division, inhibiting bacterial expansion because DNA contains sulfurous and phosphorous linkages. In animal experiments, inorganic particles like silica were discovered to cling to open wounds. As a result, they created a novel substance with good antibacterial capabilities and low cellular toxicity by integrating AgNPs into mesoporous silica nanoparticles (Ag-MSNs) via disulfide bonds [8].

10.2.2 Gold Nanoparticles

Due to its chemical characteristics, optical stability, and simplicity of modification, AgNPs and gold nanoparticles (AuNPs) are useful in medicine for conditions including wound healing. Prior to use in wound healing, AuNPs must first have their surfaces modified with other biomolecules. When polysaccharide peptides are added to AuNPs, their potential, and activity to boost healing tendency are increased. Recent studies reveal that the antibacterial properties and healing properties are *in vitro* and *in vivo*, supported by histopathologic examination and detections. By tweaking the surface sites, AuNP gets more stable toward thermal stress. AuNPs can either bind to or directly target the bacterial cell wall. AuNPs also prevent DNA strands from uncoiling or unwinding, which promotes faster wound healing and improves bacteriocidal activity. Hence, multidrug-resistant bacteria like *P. aeruginosa* and *S. aureus* can be suppressed [9].

10.3 Nanocomposites or Composite Nanoparticles

The production of NPs manually is only possible to a very limited extent due to the expensive cost, high energy consumption, and additional resources needed to neutralize harmful by-products. Plant extracts have become increasingly popular as they are inexpensive, recyclable, and include alkaloids, phenols, amino acids, and proteins that are utilized to reduce and stabilize the Ag ions in AgNPs. Prosopis juliform phenolic chemicals, for example, have been employed in AgNP manufacturing to lower Ag+. Also, it measures how well commercial antibacterial treatments limit the growth of *S. aureus* and *P. aeruginosa*-like microorganisms. Gelatin-chitosan-Ag is a brand-new composite material with pores that have a diameter of 100–250 m and a high density of AgNPs that were first combined with chitosan before being cross-linked with tannic acid and cryodesiccated. The substance exhibits antibacterial and wound-healing activities while retaining minimal cytotoxicity [9, 10].

Polymeric nanomaterial uses polymeric NPs (e.g. chitosan) as wound dressings or carriers due to their characteristics for re-epithelialization and antimicrobial defense. Biocompatible polymeric networks called biopolymers can soak up a lot of liquid and moisten the wound.

The exoskeleton of crustacean arthropods is widely used to extract the hydrophilic biopolymer known as chitosan. It is low toxic and has good bioavailability. This biopolymer is chemically a linear polysaccharide containing D-glucosamine and *N*-acetyl glucosamine units. Because of its cationic nature, chitosan can form complexes with anions such as sulfate and phosphate in addition to metals, proteins, and dyes. Furthermore, its degradation products can stimulate the synthesis of extracellular matrix components [11].

Chitosan in various forms, such as hydrogels, membranes, films, sponges, and scaffolds, has been studied for wound-healing therapies. High permeability, antimicrobial, immune-modulating, and less toxic to living organisms are all characteristics of chitosan NPs. Chen et al. developed a brand-new acellular porcine dermal matrix gauge in two steps. In order to attain improved physicochemical qualities, the acellular network was first cross-linked with a naturally occurring oxidized chitosan oligosaccharide. The newly generated compound's remaining aldehyde groups were then combined with Ag ions to produce AgNPs through situ (redox reaction). The resulting scaffold had broad-spectrum antibacterial characteristics against *E. coli* and *S. aureus*, as well as faster wound healing. As a result, it was extremely biocompatible [9].

Created high-temperature hydrogel-based drug reservoirs that can release active ingredients on demand when stimulated by near-infrared light. They combined complementing antibacterial defense mechanisms by mixing ciprofloxacin-loaded polydopamine nanomaterials and glycol chitosan. First off, bacteria were drawn to and trapped on the gel's surface by the positively charged ions of the glycol chitosan. Second, polydopamine NPs can be activated by exposure to near-infrared light. This results in a photothermal reaction that produces local hyperthermia and kills bacteria. Moreover, ciprofloxacin was loaded into polydopamine NPs and released on demand after stimulation with near-infrared light, despite the hydrogel complex showing only negligible leakage under physiological settings. Cellulose, which comprises most plant cell walls and is a readily available biopolymer, is another substance employed in wound healing. It is structurally made up of cellobiose repeating units, which are made up of 1,4-linked D-glucose units. By producing numerous growth factors locally, including basic growth factors such as fibroblast growth factor and epidermal growth factor, cellulose speeds up the healing process. Because of its anti-infectious qualities and the improved mechanical characteristics of its scaffolds, nanocellulose has been used in dressings [13].

Nanocarriers for Healing Wounds Moreover, medicinal medicines can be transported via nanomaterials, regulating the release of the drugs. Angiogenesis, inflammatory pathways, cellular proliferation, extracellular matrix deposition, and remodeling are all influenced by nitric oxide.

In addition, nitric oxide exhibits wide-spectrum antibacterial properties and interferes with biofilm synthesis. As a result, many studies have tried to produce a delivery system with high loading capacity, controlled release, and reduced cytotoxicity. Nitric oxide release from polyethyleneimine NPs is extended; they are effective against MRSA and *P. aeruginosa* and speed up wound healing *in vivo* [12].

Protamine NPs and hyaluronan oligosaccharides in a pH-responsive calcium alginate hydrogel treat diabetic wounds. The idea was that the alkaline fluids would lessen as the hydrogel absorbed wound exudates, causing a pH shift that caused the calcium alginate to release active protamine NPs, killing both Gram-positive and Gram-negative bacteria. Overexpression of blood vessel growth factor increases angiogenesis, proliferation, and motility.

Curcumin is the active substituent of the nutritious spice turmeric, and it has good tendency because of its antibacterial and anti-inflammatory properties as well as its capacity to encourage the creation of granulation tissue. The MRSA and enhanced wound closure activities of curcumin were successfully fought *in vitro* in a recent study. The low solubility of curcumin was overcome by encasing it in a deionized water nanoparticle carrier. Moreover, Moradi et al. investigated the impact of pulse photobiomodulation using curcumin-loaded iron oxide NPs in a mouse model experiment. They found that it expedited wound healing and significantly decreased *S. aureus* aggregation [13].

10.4 Coatings and Scaffolds

Another application for NPs is the construction of scaffolds that imitate the properties of the extracellular matrix. Many processes, like electrospinning, self-arrangements, and phase separation, are used to create them. The most prevalent method is electrospinning, which generates polymeric composite nanofibers that may be used as hybrid scaffolds to promote fibroblast adhesion and wound growth. Some nano-polymers, like dendrimers, can build networks that contain antibacterial compounds and have anti-inflammatory effects (e.g. silver) [14].

Nanoscale networks that might facilitate stem cell migration and differentiation, boosting re-epithelialization and angiogenesis, could be useful in stem cell research. Further study is required in order to accurately predict the dynamics of stem cells during wound healing.

Nanomaterial's antibacterial characteristics, capacity to promote wound healing by promoting angiogenesis, re-epithelialization, granulation, and/or collagen synthesis and ability to prevent aberrant scarring all outweigh conventional therapies as reasons to use them in clinical practice [15].

In order to construct a three-dimensional framework that can hold liquids like liquid or wound exudates, hydrogels are made of a polymeric network containing several hydrophilic groups that interface with one another. Hydrogel-based wound dressings get beyond the limitations of conventional dressings by creating a suitable atmosphere that also permits for gas diffusion. By absorbing wound exudates, they also significantly reduce the frequency of infections that are dependent on dressings. Alginate, chitin, and chitosan are some examples of natural and synthetic polymers that have been utilized as wound dressings. Several of these materials naturally possess antibacterial qualities (i.e. chitosan-based materials).

Nanoscale system chemicals were employed to distribute growth factors like recombinant human epidermal development factor to full-thickness diabetic lesions to promote healing. Their use was constrained by the extremely proteolytic surroundings, as well as the decreased expression of related receptors for growth factor and chemical messengers in chronic lesions. On the other hand, several authors claimed that employing basic fibroblast signaling pathways produced better outcomes (bFGF) [16].

bFGF could be released from a hydrogel that used Ca^{2+} as a cross-linker between carboxyl groups (−COO). Both calcium and bFGF promoted fibroblast differentiation and proliferation. Still, the effect was more pronounced during the first stages of wound contraction induction and re-epithelialization, both in *in vitro* as well as *in vivo*. Moreover, the hydrogel displayed two bFGF release profiles: an early burst that comprised 40% of the first 12 hours and a second, protracted phase that ensured bFGF release remained constant for eight days. These dynamics suggest that the compound increases local bFGF levels and enhances their long-term stability (Table 10.2).

Table 10.2 Properties of some nanomaterials designed for wound-healing bFGF-(basic fibroblast growth factor).

Material	Antibacterial	Wound healing activity	Prevention of wound
Nanoparticles	Silver nanoparticles	Positive	Positive
	Gold nanoparticles	Positive	Unknown
	Zinc oxide nanoparticles	Positive	Unknown
	Nanocomposite	Positive	Positive
Nanocarriers	Polyethyleneimine nanoparticles, curcumin-loaded super-paramagnetic iron oxide	Positive	Positive
Coatings and scaffolds	Hydrogel preloaded with bFGF; hydrogels reinforced with AgNPs	Positive	Positive

10.5 Green Synthesis of Silver Nanoparticles

The plant extract is now commonly used to make NPs, and its popularity has also increased. Many researchers and pharmaceutical companies use plant extracts and then tune them with other materials to create NPs that can be used as wound healing material. AgNPs have a great mutation-resistant property showing good antimicrobial activity; they have been successfully utilized in antibacterial clothing, burn ointments, and as a coating for medical devices. It has previously been reported that nanoscale metals can be synthesized using plant materials [17]. Onion *Syzygium cumini* leaf extract, basil *Saraca indica* leaf extract, and *Piper nigrum* were used in the pure metallic silver and AuNPs created through the fabrication of Au and AgNPs, which can be used right away. Although chemical methods are very efficient, harmful chemicals may present significant biological and environmental concerns. Both aerobic and anaerobic bacteria are successfully combated by silver compounds, which are potent antibacterial. When compared to its ionic form, NPs appear to have lower cellular toxicity but no antibacterial efficacy in the application of silver. The formation of free radicals is clearly shown to be the cause of AgNPs' superior antibacterial properties. These bacteria, which were discovered to be antibiotic-resistant, were affected by AgNPs [13].

Additionally, it has been shown that combining antibiotics with AgNPs has synergistic effects against microorganisms. It has been shown that using dendrimers to modify silver sulfadiazine increased its antibacterial efficacy. AgNPs have been quickly adopted in medical practice thanks to their accessibility. Their applications are broadly classified as diagnostic and therapeutic. In terms of therapeutics, wound healing is one of the most well-documented and widely used applications of AgNPs [18].

Many studies have shown that AgNPs are more effective than other silver compounds in terms of healing time. It was discovered that wounds treated with AgNPs had better collagen alignment after healing when compared to controls, resulting in higher working strength; still, a complete mechanism of this process for its biological applications is yet to be determined [19].

Wounds that have more tissue damage are difficult to treat. Chronic wounds, burning injuries, and other treatments for skin allergies are the most common causes of skin damage and breakage. Burn healing is extremely difficult, for burns with a large burn surface area, where recovery remains a major challenge [18]. Due to a shortage of potential donors and the danger of spreading infectious agents, organic skin grafting cannot guarantee skin regeneration. The development of artificial bioengineering synthetic materials is required to ensure optimal skin regeneration. Numerous types of wound dressings have been reported, each with a different purpose and function. Because it is nontoxic, cotton is one of the most commonly used wound dressing materials. Materials with biodegradable and other characteristics are also used when creating scaffolds for wound closure. Some research team has already shown that silver nitroprusside nanoparticles(SNPNPs) are extremely helpful for antibacterial activity.

Even after washing it several times, the formed cotton dressing with NPs still reflects the greater antibacterial activity against Gram-positive (*P. aeruginosa* and *Bacillus subtilis*) and Gram-negative (*Klebsiella aerogenes* and *E. coli*) bacteria. They also demonstrated that it has an ideal water contact angle (113–130°) than a fine cotton gauze (60°). The biocompatibility of silver nitroprusside nanoparticles (SNPCFs) was determined using an *in vivo* membrane assay, which revealed neither a blockage in the blood coagulation vessels. Nanoscale particles made up of silver atoms having the chemical constitution $Ag_2[Fe(CN)_5NO]$ are known as silver nitroprusside and known as SNPCFs, due to its special qualities, silver nitroprusside itself has potential uses in a variety of industries, including medicine. The ability to serve as a treatment for ailments including hypertension and its antibacterial qualities. Our theory that SNPCFs can effectively treat wounds when applied topically to the skin is supported by the fact that no silver was found in the heart, liver, or kidney; all these findings show that SNPCFs tend to work as an antimicrobial and wound-healing agent [19]. Due to the complexity of the wounds and infections, many factors can interfere with their healing, with delay in wound healing, increasing patient mortality, and causing low cosmetic successions, significant discomfort, and distress. Various growth factors are released at the site of wound healing, and cytokine is also released at wound sites to heal that wound rapidly. From a deep perspective, wound healing depends on many factors, like patient age, injury or wound size, depth, infected area location, and the cause of any local disease [20].

Other factors, including nutritional and immunological tendencies, stress, smoking, diabetes, obesity, and hypertension, can affect the wound-healing process. as can the elderly population's overall increased longevity, which has increased the ulcer healing duration [20]. Microcirculation, which is impaired in aging skin, collagen, which is found in all tissues in the body, are known to play an important role in a molecule's geometrical stability, elasticity, and tensile strength. It is not an amazed that collagen is important for restoring the structural informatics of the infected tissues. Except for epidermal wounds, it has been noted that the development of scar tissue is a necessary component of wound healing in the majority of situations. This scar tissue is mostly made up of collagen. Fibroblasts release connective tissue and fibronectin to create a fresh, temporary extracellular matrix during granulation tissue and fibroplasia development. Myofibroblasts grip the wound edges and contract the wound using a mechanism similar to that of smooth muscle cells, after which epithelial cells crawl across the wound bed to cover it. So, using biological sources, our research team intended to synthesize AgNPs; as a result, collagen synthesis is a significant part of the wound healing process [21]. Collagen supplements can thus be used to supplement and speed up the healing process and improve the mechanical properties and other intrinsic characteristics of the tissue. It was demonstrated that collagen derived from a jellyfish species was effective in hastening tissue repair, making it a feasible product for use in treating major wounds. Because of their therapeutic properties and unique structure, nanofiber scaffolds/mats are primarily used in wound dressing to replace damaged ECM, the extracellular matrix (ECM) is a complex network made up of several multi-domain macromolecules arranged, based on the specific cell/tissue type. The ECM links together to create a structurally reliable composite, which influences the mechanical characteristics of tissues. Fibroblasts will attach to the wound as the scaffolds are applied, and a matrix will form that serves as a ground substrate and promotes quicker wound healing. Electrospinning is used to create uniform nanofibers to fabricate nanofibrous scaffolds [21].

10.5.1 Synthesis of Silver Nanoparticles by Aqueous Extract of *Arnebia nobilis* Roots

Arabia noblis roots were taken from the plants, and it was taken and washed correctly. *Arabia nobilis* roots were taken and appropriately washed with deionized water to remove impurities, then dried in the dark to completely remove the moisture, and then powdered in a mixer; after powdering, sieved, add a small amount of batch in an Erlenmeyer flask, add water to it and microwave for three to four minutes extract the raw material obtained the raw extract was then hot filtered to remove fibrous impurities using Whatman filter paper. The creation of AgNPs used the resulting clear extract. *A. nobilis* root extract was combined for the reduction of Ag^+ ions. The solution mixture was exposed to microwave radiation at 450 W and a fixed frequency of 2450 MHz. Aliquots of the reaction solution were periodically taken out and analyzed using UV–visible (UV–vis) spectroscopy. The hydrogel preparation process started right away with half of the colloidal solution that was thus obtained. The remaining colloidal solution was then twice re-dispersed in deionized water after being centrifuged at 8000 rpm for 10 minutes. To remove any unbound biological molecules, the resulting clear extract was twice re-dispersed in deionized water. In excision wound models, the AgNPs-*A. nobilis* hydrogel exhibits significant wound-healing activity. Despite recent advances in the field of synthetic drugs, it has been discovered that many of them have side effects, whereas plants still have their own distinct properties without causing any negative side effects, of course. In order to use plants as herbal wound-healing agents, a systematic approach should be taken to determine their effectiveness against wounds. According to healing experiments analysis of their wound healing activity, Animals treated with AgNPs had faster rates of wound contraction and epithelialization. AgNPs were topically applied to excision wounds in rats to speed up wound healing and shorten the time it took for the epithelium to form. Through the proliferation of surviving cells and a connective tissue response characterized by the formation of granulation tissue, wound healing involves the regeneration of specialized cells.

Hemostasis, re-epithelialization, and extracellular matrix remodeling are additional characteristics. Myofibroblast activity is primarily responsible for wound contraction. In contrast, the epithelialization process, which is the epithelium renewal after injury, includes the cell migration and invasion of epithelial cells toward the center of the wound. As a result, the effect of the ethanolic extract and the acetone fraction on wound contraction and epithelialization suggests that they may promote the migration and proliferation of epithelial cells as well as the development, movement, and activity of myofibroblasts. The use of plant materials (the roots of *A. nobilis*) in the production of AgNPs, hydrogel, and its ability to treat wounds. *A. nobilis* root extracts, which are eco-friendly and renewable and function as reducing and capping agents, were used in the experiments. *A. nobilis* root extract is used in the quick and easy green synthesis of AgNPs to create spherical, fcc-structured NPs.

Rectovaginal fistulas are not considered life-threatening, but in some instances, these conditions make surgery difficult. The majority of doctors focus on anorectal surgery to treat, operate, and remove the entire fistula. Patients can treat fistulas with recently developed laser treatments without any discomfort [11].

10.5.2 Honey-Based Nanoparticles in Wound-Healing Process

Due to the antimicrobial, anti-inflammatory, and antioxidant nature of honey, it has been used since ancient times to treat wounds. The reason behind its nature is endogenous hydrogen peroxide, which contributes to the antimicrobial properties of honey; other than this, honey has lower water content with high sugar levels and low pH. A strong osmotic gradient created by the honey's high sugar content and other solutes draws fluid up the subdermal tissue. Because honey has a high sugar content, it also serves as an additional source of glucose for the proliferation of cells like fibroblasts and endothelial cells. Additionally, it has been noted that the sugar in honey affects bacterial quorum sensing, and more recently, it has been discovered that osmotic pressure affects bacteria's ability to form biofilms [20, 21].

Honey's antioxidant properties have also been extensively researched as a potential medical benefit. The antioxidant properties of natural honey are attributed to a variety of phenolic compounds. A large number of phenolic compounds, which are generally secondary metabolites, are produced by plants in unfavorable conditions, either biotic or abiotic. Through nectar, these substances are transferred to honey. Flavonoids and phenolic acids are the two metabolites present in honey. Aromatic acids and flavonoids scavenge free radicals, avoiding tissue damage and managing inflammation. When mixed with antibiotics, honey has a synergistic impact and has been used in medicine to treat inflammation, burns, and surface wounds. Due to their enhanced wound-healing activity and combination of nanofiber structural advantages, honey-incorporated nanofibers (NFs) are growing in popularity. In 2014, Researchers studied the impact of honey on the electrospinning procedure. In their study, polyethylene terephthalate (PET), PET/chitosan, PET/honey, and PET/chitosan/honey solutions were electrospurned to produce fibrous mats using varying concentrations. They claimed that honey altered chitosan-containing fibers' previously beaded or ribbon-like/branched morphology. Additionally, the water content of the PET/chitosan and PET/honey fibrous mats reached equilibrium in 15 minutes, as did their capacity for absorption ranging from 280% to 430% on a dry basis which is crucial for exuding wounds. Additionally, a cytotoxicity test showed that the fibers lacked any toxic activity (Figure 10.4) [21].

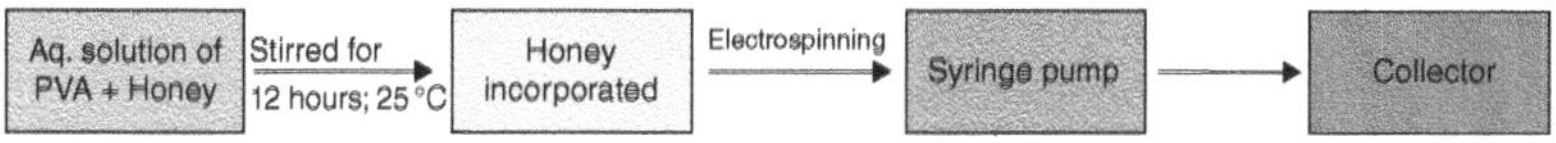

Figure 10.4 Flow diagram of the preparation of solution and electrospinning process.

10.6 Conclusion

In a nutshell, nanomaterials and nanocomposites reflect the benefits of wound healing, which was the major focus of this review. It is observed that the nanomaterials show a great extent of properties, which makes them ideal materials for wound healing. The wound healing done with nanomaterials was found to be more effective than classical and earlier wound treatment therapy, which is based on the dressing.

Inflammation, proliferation, and redesigning are three key parameters of the complex wound-healing process. Internal and external factors are the most important in infection, which can interfere, with and delay healing. Researchers and scholars have attempted to develop many products that can provide information on moist environments and antibacterial activity to achieve a good recovery with minimal scars. Due to bacteria biofilms and multidrug-resistant microorganisms, chronic wounds are difficult to treat. Because of their high surface-to-volume ratio, NPs can be used in a wide range of medical applications, including wound therapy. They have great characteristics like low toxicity and good antibacterial activity for making metal NPs like silver, gold, and zinc the ideal choice for inclusion in wound dressings. When looking for the future parameters of NPs, researchers are currently working on nano compounds that contain growth factors, genes, stem cells, and many more biological conditions that lead to a new way to heal wounds. Looking forward to its application, it is necessary to understand the potential work and risk related to nanomaterials [20] as we saw the production of AuNPs from the plant roots of *A. nobilis* hydrogel, and its ability to treat wounds. *A. nobilis* root extracts, the plant used for wound healingare eco-friendly and renewable, their function is to reduce the agents and mask the effects. *A. nobilis* root extract synthesizes AuNPs to create the spherical, face-centered structure NPs. As a result, and in future aspects, researchers are working on biocompatibility and more modulation, which should strive to create biodegradable nanomaterials capable of curing all stages of wounds.

Abbreviations

Ag-MSNs	Silver mesoporous silica nanoparticles
Ag NPs	Silver nanoparticles

Au NPs	Gold nanoparticles
bFGF	Basic fibroblast growth factor
CD	Circular dichroism
DLS	Dynamic light scattering
FCS	Fluorescence correlation spectroscopy
FDA	Food and Drug Administration
IR	Infrared spectroscopy
MS	Mass spectroscopy
MRSA	Methicillin-resistant Staphylococcus aureus
PVA	Polyvinyl alcohol
PA	Phytic acid
RS	Raman scattering
SEM	Scanning electron microscopy
SNPNPs	Silver nitroprusside nanoparticles
TEM	Transmission electron microscopy

References

1 Arafa, M.G., El-Kased, R.F., and Elmazar, M.M. (2018). Thermoresponsive gels containing gold nanoparticles as smart antibacterial and wound healing agents. *Sci. Rep.* 8 (1): 13674.

2 Moura, J., Børsheim, E., and Carvalho, E. (2014). The role of microRNAs in diabetic complications—special emphasis on wound healing. *Genes* 5: 926–956.

3 Bahari, N., Hashim, N., Md Akim, A., and Maringgal, B. (2022a). Recent advances in honey-based nanoparticles for wound dressing: a review. *Nanomaterials* 12 (15). https://doi.org/10.3390/NANO12152560.

4 Shah, S.A., Sohail, M., Khan, S. et al. (2019). Biopolymer-based biomaterials for accelerated diabetic wound healing: a critical review. *Int. J. Biol. Macromol.* 139: 975–993.

5 Mihai, M.M., Dima, M.B., Dima, B., and Holban, A.M. (2019). Nanomaterials for wound healing and infection control. *Materials* 12: 2176.

6 Lin, P.C., Lin, S., Wang, P.C., and Sridhar, R. (2014). Techniques for physicochemical characterization of nanomaterials. *Biotechnol. Adv.* 32: 711–726.

7 George, S., Lin, S., Ji, Z. et al. (2012). Surface defects on plate-shaped silver nanoparticles contribute to its hazard potential in a fish gill cell line and zebrafish embryos. *ACS Nano* 6 (5): 3745–3759.

8 Niska, K., Zielinska, E., Radomski, M.W., and Inkielewicz-Stepniak, I. (2018). Metal nanoparticles in dermatology and cosmetology: interactions with human skin cells. *Chem. Biol. Interact.* 295: 38–51.

9 Sudheesh Kumar, P.T., Lakshmanan, V.K., Anilkumar, T.V. et al. (2012). Flexible and microporous chitosan hydrogel/nano ZnO composite bandages for wound dressing: in vitro and in vivo evaluation. *ACS Appl. Mater. Interfaces* 4: 2618–2629.

10 Biranje, S.S., Madiwale, P.V., Patankar, K.C. et al. (2019). Hemostasis and anti-necrotic activity of wound-healing dressing containing chitosan nanoparticles. *Int. J. Biol. Macromol.* 121: 936–946.

11 Garg, S., Chandra, A., Mazumder, A., and Mazumder, R. (2014). Green synthesis of silver nanoparticles using *Arnebia nobilis* root extract and wound healing potential of its hydrogel. *Asian J. Pharm.* 8 (2): 95.

12 Amer, S., Attia, N., Nouh, S. et al. (2020). Fabrication of sliver nanoparticles/polyvinyl alcohol/gelatin ternary nanofiber mats for wound healing application. *J. Biomater. Appl.* 35 (2): 287–298.

13 Jamil, B., Abbasi, R., Abbasi, S. et al. (2016). Encapsulation of cardamom essential oil in chitosan nano-composites: in-vitro efficacy on antibiotic-resistant bacterial pathogens and cytotoxicity studies. *Front. Microbiol.* 7: 1580.

14 Jura, J., Szmyd, R., Goralczyk, A.G. et al. (2013). Effect of silver nanoparticles on human primary keratinocytes. *Biol. Chem.* 394 (1): 113–123.

15 Tayeb, A.H., Amini, E., Ghasemi, S., and Tajvidi, M. (2018). Cellulose nanomaterials-binding properties and applications: a review. *Molecules* 23: 2684.

16 Shao, F., Yang, A.J., Yu, D.M. et al. (2018). Bio-synthesis of *Barleria gibsoni* leaf extract mediated zinc oxide nanoparticles and their formulation gel for wound therapy in nursing care of infants and children. *J. Photochem. Photobiol. B* 189: 267–273.

17 Abdel-Sayed, P., Kaeppli, A., Siriwardena, T. et al. (2016). Anti microbial dendrimers against multidrug-resistant *P. aeruginosa* enhance the angiogenic effect of biological burn-wound bandages. *Sci. Rep.* 6: 23872.

18 Moradi, A., Kheirollahkhani, Y., Fatahi, P. et al. (2019). An improvement in acute wound healing in mice by the combined application of photobiomodulation and curcumin-loaded iron particles. *Lasers Med. Sci.* 34 (4): 779–791.

19 Rao, B.R., Kumar, R., Haque, S. et al. (2021). $Ag_2[Fe(CN)_5NO]$-fabricated hydrophobic cotton as a potential wound healing dressing: an in vivo approach. *ACS Appl. Mater. Interfaces* 13 (9): 10689–10704.

20 Pereira, R.F. and Bártolo, P.J. (2016). *Traditional therapies for skin wound healing.* 5, Adv. Wound Care: 208–229.

21 Tang, Y., Lan, X., Liang, C. et al. (2019). Honey loaded alginate/PVA nanofibrous membrane as potential bioactive wound dressing. *Carbohydr. Polym.* 219: 113–120.

11

Cellulosic Nanomaterials for Remediation of Greenhouse Effect

Athanasia Amanda Septevani, Melati Septiyanti, Annisa Rifathin, David Natanael Vicarneltor, Yulianti Sampora, Benni F. Ramadhoni, and Sudiyarmanto

National Research and Innovation Agency, Tangerang Selatan, Indonesia

Abstract

Human activities and industrialization are responsible for the increased amount of greenhouse gases (GHG), rising the global temperature and thus causing global warming. The innovation of nanocellulose as one of the most promising nature-based nanomaterials has driven the development of advanced materials in myriad fields, offering benefits for the remediation of greenhouse effects due to multifunctional characteristics and extraordinary properties. This includes the reduced usage of energy, leading to the significant reduction of GHGs and creating a new generation of membranes for capturing GHGs. Nanocellulose used in automotive parts offers a lower density, thus reducing fuel consumption and CO_2 release. It is also known as an insulating agent to reduce thermal conductivity by 30%, leading to lower energy usage. Membranes containing nanocellulose increase the permeation and selectivity for CO_2 capture at up to 65% and 80%, respectively. In this chapter, the magnificent use of cellulosic nanomaterial in GHG reduction is discussed and elaborated in detail.

Keywords *nanocellulose; greenhouse gases; remediation; superlight polymer; insulation; CO_2 capture*

Green Synthesis of Nanomaterials: Biological and Environmental Applications, First Edition.
Edited by Archana Chakravarty, Preeti Singh, Saiqa Ikram, and R.N. Yadava.

11.1 Introduction

11.1.1 Fundamentals of the Greenhouse Effect

Greenhouse gases (GHG), gas components that are present in the troposphere, dominantly consist of water vapor, methane (CH_4), nitrous oxide (N_xO), ozone (O_3), and carbon dioxide (CO_2). The phenomenon of the GHG effect occurs when about 70% of the incoming solar radiation is absorbed by the atmosphere, and then at infrared wavelengths, it is re-emitted back to the surface of the earth. When most of the radiation is caught by clouds, it is then returned to the earth's surface, causing global warming [1]. GHGs are one of the factors that drive climate change. It is estimated that the global temperature will increase by 3–5 °C in 2100 due to high CO_2 concentrations on Earth [2].

The rapid development of industrial sectors and urbanization have caused a considerable increase in GHGs. The major cause of the increased concentration is the extortion of fossil fuel to generate power, heat, and electricity for mankind, and also the exploration of land use, particularly for agriculture and urbanization. It is reported that the energy for industrial sectors is responsible for increasing GHG effects at about 75%, while about 25% is due to the agricultural sector, including deforestation. Accordingly, mankind is responsible for the major cause of the GHG effect [3]. The GHG phenomenon harms ecological and social structures, including the ocean's thermal expansion, the glacier's melting condition, the destruction of the ecosystem, and natural resources for human welfare, such as agriculture, shipping, forestry, nutrition, and health.

The reduction in fuel consumption and carbon dioxide emission as well as life cycle energy assessment, are the main factors to address these problems, wherein one of them can be done by the use of abundantly available and eco-friendly plant fibers in nature. In this chapter, we present the nanomaterial, particularly based on the cellulose structure, that can be applied in the automotive, insulation, and gas capture fields as promising pathways to reduce the effect of GHGs.

11.1.2 Cellulosic Contribution to the Remediation of Greenhouse Effect

In recent years, nanomaterials have offered a promising solution to detect or reduce GHGs, organic pollutants, chemical contaminants, and biological agents. These materials are available in many kinds of morphologies with diverse functions (e.g. adsorbents, catalysts, or membranes) [4]. The development of natural resources as advanced nanomaterials has gained great interest to the scientific, technological, and industrial areas owing to their abundant availability, as well as renewable and sustainable purposes. In this context, cellulose is one of the most plentiful natural polymers, holding enormous appealing capabilities and opportunities.

At the nanoscale, cellulose materials, known as nanocellulose (NC), have acquired a greater advantage compared to conventional cellulose fibers as consequence of higher surface area, aspect ratio, and young's modulus with unprecedented mechanical properties [4]. The fundamental characteristics of cellulose include its hydrophilicity and easily chemically tunable surface functionality for modifications [5, 6]. NC is a fascinating material that can be extracted from various biomass waste to transform into a fibrous structure of cellulose in the form of needle-like crystals known as cellulose nanocrystals (CNC) or long-entangled and flexible-like fibers, known as cellulose nanofibers (CNF). Depending on the process, acid hydrolysis is the most prevalent method to produce CNC, wherein strong acids could aggressively attack the amorphous regions, without disturbing the desired higher crystalline regions. Meanwhile, mechanical deconstruction and enzymatic processes, have been known to produce effectively CNF with a higher aspect ratio and lower crystallinity than CNC. Yet, they are tougher and more flexible than CNC [7].

Sustainable solutions to protect the environment from the GHG effect can be realized through "direct" and "indirect" pathways. Since "going green" is the current motion in various industries, massive investigations have been conducted to develop products based on natural fibers in various applications. In the role of reducing the GHG effect, NC, one of the promising natural fibers, is being developed in various applications mostly as a composite material applied in various products including automotive and insulation products to reduce energy usage and, in turn, to reduce GHGs. In the former approach, gas capture is one of the global warming mitigation technologies that can directly reduce GHGs such as CO_2 emissions in the atmosphere. Currently, NC-based membranes have been considered and tested for CO_2 capture, showing higher capture selectivity toward CO_2 compared to other commercial membranes [8]. Hence, combining these two approaches could be a promising solution to simultaneously reduce both "directly" or "indirectly" the GHGs and thus protect the environment (Figure 11.1).

11.2 Cellulosic Nanomaterials in Automotive Application

The transportation sector has shown a great transformation to replace steel alloys with aluminum alloys and is now toward the recent technology, natural fiber-reinforced polymer (NFRP) composites [9]. Compared to other

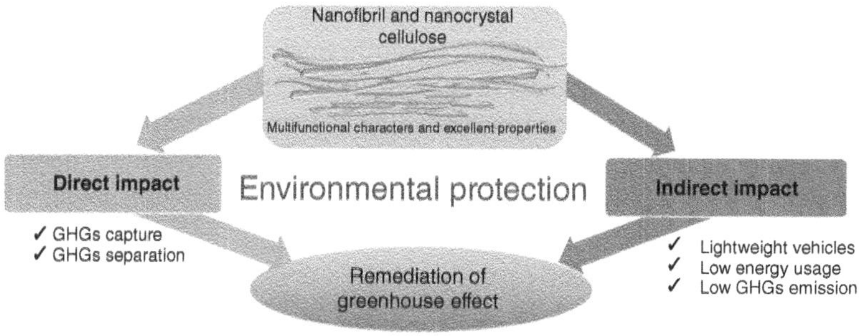

Figure 11.1 Nanocelluloses as promising materials for remediation of GHGs.

composite materials, the density of NFRP is lower, yet it still maintains excellent mechanical properties and is importantly cost-effective. It has been used in automotive products specifically for the car's internal parts, namely door panels, trim components in the dashboard, shelves, spare tire covers, and parcels. Nevertheless, a few NFRPs can be applied for exterior parts due to weather conditions, humidity, and temperature [10, 11].

The use of natural fibers in vehicle's parts is reported to reduce the vehicle's weight without compromising functional performance [12]. The reduction of vehicle weight can affect fuel consumption, thus the CO_2 emission. It is reported that a 10% reduction of total vehicle weight can improve fuel efficiency by 7%, and the weight reduction per 1 kg offers about 20 kg less CO_2 being released into the earth's atmosphere [13]. The reduction of vehicle weight corresponds to the reduction in fuel consumption, which affects CO_2 emission deflation, thus subsequently preserving the earth from global warming [2, 14].

11.2.1 Nanocellulose-Enabled Lightweight Vehicles

In current years, thermoplastic-reinforced NC fibers are utilized in automobile applications. Cellulose nanofiber (CNF) become a notable alternative for synthetic fibers like glass and carbon fibers. Mostly the researchers are focusing on the application of NC with chemical and physical fuctionalization as reinforcing fillers to improve the nanocomposite performances on the thermal and mechanical properties [15]. This improvement is caused by the high surface area of NC and its hydrophilic nature, leading to a strong inter-intramolecular hydrogen bond [16].

The most interesting merit of NC in automotive sectors is that it offers comparable tensile strength, design flexibility, and lower weight compared to aluminum [17]. These properties become the key factor for automotive industries since they can increase the stiffness, leading to the high mechanical resistance of the vehicle components [18]. Other important features of NC are thermal stability, lower thermal expansion, good optical tensile strength, and toughness in consequence of the inter-intramolecular hydrogen bond and closely packed structure of NC matrixes [19]. CNF, as a reinforcing material, can produce lower density yet maintain higher strength due to the high surface area that can be strongly functionalized with the matrix [20].

11.2.2 Processing and Performance of Nanocellulose in Automotive Parts

Processing the NC into the NFRP can be done using (i) solvent casting, (ii) melt-extrusion, (iii) injection molding, and (iv) electrospinning. The casting method is the most popular method for the production of composite materials by dissolving the NC in a suitable solvent; nonetheless, it is limitedly used for low-scale production since it is time-consuming. Strong interaction occurs when the water-soluble polymer is used because it is compatible with the hydrophilicity of NC. Water-based polymers like polyvinyl alcohol create strong interface bonding between hydroxyl groups of CNF with the polymer matrix and thus increase the tensile strength of the composite [21]. However, this technique is limitedly used for hydrophobic polymer matrixes because of the weak dispersion and the interaction between the filler and the matrix [22].

A thermomechanical method such as melt-extrusion and injection molding is one of the common methods to produce thermoplastic polymers, yet it is rarely used to prepare NC-reinforced polymer composite due to its natural incompatibility and thermal stability issues. To address this problem, the compatibilizer promotes dispersion between NC and the polymer matrix. This dispersion of NC and compatibilizer is freeze-dried before the melt compounding process in a twin-screw compounder. The compound is then extruded and formed into composite pellets [23]. This process reports an improvement in the dispersibility and thermal stability due to a strong affinity between compatibilizer and NC [24].

In the case of electrospinning, CNF and polymer matrix are first dissolved in a suitable solvent, resulting in a homogenous solution. The solution is then pumped into electrospinning apparatus with several controlled parameters such as voltage supply and needle tip distance to produce nanocomposite mats. The composite result of this process can achieve high flexural strength due to uniform dispersion, higher surface area, and load-bearing characteristics. This creates a lighter and stronger NC-based composite suitable for aeronautics and automobile application [25], yet this

Table 11.1 Nanocellulose composite performance for automotive application.

Nanocellulose sources	Polymer matrixes	Applications	Performances	Reference
Glucose biosynthesis CNF	Polyacrylic acid	Automobile, aircraft, and aerospace materials	Increase mechanical properties at a low density of ~1.35 g/cm^3 with strength up to 198 MPa/(Mg/m^3), and impact toughness up to 67 kJ/m^2/(Mg/m^3)	[26]
CNF	Polyethylene/polyvinyl alcohol	Side and back door, dashboard, under-the-hood radiator end tanks	25–40% increase in the mechanical properties (tensile strength, flexural strength and modulus of elasticity)	[27]
Rutabaga CNF	Polyvinyl alcohol	Automotive part	Increase tensile strength up to 70%	[21]
Softwood pulp CNF	Bio polyurethane	Automotive interior part	Increase compressive strength by 48.5% and increase the impact strength by 20%	[28]
Tunicin (CNC)	Epoxy resin, diglycidyl ether of bisphenol-A	Automotive part	Increase the storage modulus by 100%, increase Tg up to 13.5, and increase tensile strength up to 50%	[29]

process is only limited to the hydrophilic polymer matrix and can be challenging during large-scale processes. Table 11.1 summarizes the selected NC composites in automotive, corresponding to their application and performances.

11.3 Cellulosic Nanomaterials in the Application of Thermal Insulation

Energy for the building is currently one of the main contributors to global carbon emissions [30, 31]. Around 40% of the world's energy usage is consumed by the building sector, particularly for air conditioners and heaters.

In this context, the development of green building construction in the presence of material insulation offers a promising solution for reducing energy demand as well as carbon emissions.

Polymeric foam, such as rigid polyurethane foam, expanded polystyrene (EPS) foam, phenolic foam, and extruded polystyrene (XPS) foam, are well-known as good thermal insulators [30, 32–34]. In general, bulk polymers consist of defective morphologies and structures, for example voids, chain ends, entanglements and amorphous regions, which could suppress the heat propagation. This randomly oriented chain structures creates the poor interaction forces, mainly van der Waals, and thus result the average phonon free path becoming smaller and thus leading to lower thermal conductivity than well-ordered polymer crystals [35]. Nevertheless, the defective structures should not compromise the mechanical strength, which is also one of the key performances of insulation building.

Thermal conductivity is one of the essential properties of thermal insulation polymeric materials. There are many studies conducted to maximize the insulation properties by reducing the thermal conductivity value, such as developing a new blowing agent offering low thermal conductivity, eliminating gaseous including carbon dioxide in cell spaces, and incorporating filler which inhibits the radiation heat transfer and/or serves as a diffusion barrier. Over the last few years, many studies have emerged on the use of inorganic filler (such as boron nitride, graphene oxide, aluminum oxide, and SiO_2) and natural-based filler (such as sawdust, cellulose, and peanut shell) in various bulk polymer to improve thermal insulation properties [36, 37]. Nevertheless, the cellular structure can possibly be disrupted when such micro-scale fillers are used in the foam formulation [38]. Thus, the utilization of nanofillers as nucleating agents has been identified as a promising solution for lowering thermal conductivity.

11.3.1 Nanocellulose Reinforced Polymeric Insulation Toward Zero Energy Usage

NC has been reported to improve insulation properties by lowering thermal conductivity while increasing the mechanical strength of polymer composites [38, 39]. Adding NC to polymeric foams has been shown to reduce thermal conductivity [34, 38, 40] because NC can act as a nucleating agent in smaller cell sizes in a controlled NC alignment, as shown in Figure 11.2 [38, 39, 41]. Furthermore, NC can also provide resistance to cell expansion during the foaming process, reducing the possibility of defective cell formation, thereby increasing closed-cell content [42]. Cell size and closed-cell content are attributed to the changes in the thermal conductivity of polymeric foams.

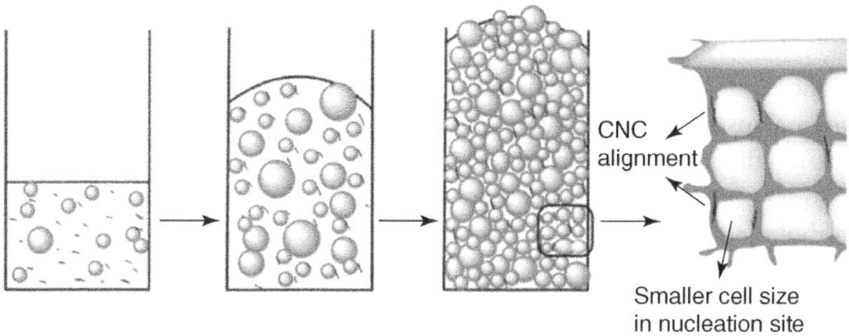

Figure 11.2 Schematic of the possible role of nanocellulose during the foaming reaction. *Source:* Septevani et al. [38] with permission of Elsevier.

The mechanism of heat transfer within polymer foam occurs via polymer matrix (solid conductivity), gas (including both gas convection and gas conductivity), and thermal radiation. Gas conductivity exerts the dominant effect on total heat transfer, which accounts for 80%, while the remaining total goes to solid conductivity, gas convection, and thermal radiation at only about 20% [40]. Gas conduction is accomplished through the collisions and diffusion of gas molecules in cells. Hence, it is highly dependent on the free paths of its molecules. The smaller the cell size, the smaller the free path of gas molecules; thus, the movement of gas molecules can be significantly limited [42, 43]. As a result, the thermal conductivity of the gas decreases and thus improves the insulation.

In addition to the enhancement of insulation properties, NC can potentially improve mechanical properties, which is one of the important factors for insulated panels [38, 44]. This improvement is due to the extraordinary intrinsic properties of NC, such as its high bending strength, elastic modulus, and high specific surface area. In addition, a high surface area provides a higher potential for chemical cross-linking and hydrogen bonding between the NC and the CO, and NH of the polyurethane matrix. Nevertheless, the concentration of NC added as reinforcement in polymeric foams must be considered for optimum enhancement [38].

11.3.2 Processing and Performance of Nanocellulose in Insulation Material

Processing of NC-based insulation material consists of two main steps: (i) mixing/compounding and (ii) foam formation. During the mixing/compounding of NC and polymeric composites, it is vital to understand the thermal behavior of NC. The compounding temperature must be kept below

200 °C during the manufacturing process to avoid degradation of NC, which decomposes at roughly 200–300 °C.

There are three main methods to produce NC-based polymeric foam for thermal insulation: (i) molding, (ii) slabstock, and (iii) aerogel method [45]. The molding method is proceeded by injecting molten polymeric foam into a closed or open mold. The slabstock method begins by spreading the foam formulation onto a conveyor from the mixing head reactor and cutting the cured foam into large slabs. Practically, the molding method is more efficient in controlling the foam density and thus the foam quality [46], but it is more costly than the slabstock method [45].

Meanwhile, NC aerogels are prepared in three steps: (i) dispersion of NC, (ii) gelation of NC, and (iii) gel drying process [47]. The gel drying process is the most important stage in preparing the NC aerogel because it is important to maintain the original structure while replacing the liquid solvent with gas to fill the porous structure. There are several effective drying methods including atmospheric pressure drying, supercritical drying, and freeze-drying, but they are mostly limited to small-scale production. Table 11.2 shows the summaries of the selected NC polymers applied as thermal insulators corresponding to their application and performances upon NC incorporation.

11.4 Cellulosic Nanomaterial for Gas Capture and Separation

Technologies for capturing and/or separating GHGs, especially carbon dioxide (CO_2) are promising solutions to "directly" reduce the GHG concentration in the atmosphere. They can be defined into three major categories, i.e. precombustion, oxy-fuel-combustion, and postcombustion carbon capture (PCCC) technique. Among the methods, PCCC is more economically viable and safe because it can be installed for the short or medium term without major changes to the existing power generation or chemical process [51–57].

PCCC is of importance in carbon capture and sequestration (CCS) projects and can be divided into three subcategories: chemical absorption, adsorption, and separation membrane [57, 58]. Although membrane techniques are relatively new in CCS, they offer the most promising benefit among others. The membrane requires simple operation procedures, smaller space, and lower energy. It can be manufactured with environmentally benign materials compared to other methods, which poses some drawbacks such as the higher production and regeneration cost and instability in high-temperature and lower chemical stability [59, 60]. A small-scale system using a commercial Polaris™ class membrane can capture CO_2 up to 91% during more than 9000

Table 11.2 Nanocellulose composite performance for insulation application.

Nanocellulose sources	Polymer matrixes	Insulation applications	Performances	Reference
CNF	Polyurethane foam	Structural insulated panels in wood construction	Lowering thermal conductivity from 0.0439 to 0.02724 W/mK	[40]
Silanized microcellulose (SiMC) and silanized NC (SiNC)	Polyurethane foam	Cryogenic insulation	1.5% SiMC and 1.5% SiNC can decrease thermal conductivity to 1.4% and 3.3%, respectively	[48]
CNC	Polyurethane foam	Building insulation	A 5% reduction in thermal conductivity by 0.4 wt.% of CNC	[38]
CNF aerogel based on pinewood	—	Building insulation	21% reduction in thermal conductivity from conventional aerogel	[47]
CNF aerogel	—	Building insulation	37% reduction in thermal conductivity from conventional bio-aerogel	[49]
CNC	Branching poly (butylene succinate)	Automobile parts, building insulations	12.5% reduction in thermal conductivity from pure branching poly (butylene succinate)	[50]

operation hours [61]. However, improvements in membrane performance still have to be carried out, especially for CO_2 permeability and selectivity.

11.4.1 Nanocellulosic Membrane for Capturing/Separating Greenhouse Gases

The membrane property, by means of membrane permeability, is inversely proportional to the selectivity. The higher the permeability value, the lower the selectivity value, and vice versa. NC fillers can play a vital role in improving the membrane performances by elevating the gas permeation and selectivity via a facilitated transport mechanism by converting targeted molecules to other high diffusion constant species; for example, CO_2 is converted to HCO^{3-} or CO_3^{2-} [8, 62]. The hydrophilicity of NC is also reported to increase the selectivity of CO_2 over CH_4 and N_2. Notably, O_2 is more hydrophilic than CH_4 and N_2 [63].

NC fillers are also reported to enhance physical performances including thermal and mechanical properties [64, 65]. Although, by nature, NC has a lower degradation temperature than polymer, interestingly, introducing NC into the polymer matrix can enhance the degradation temperature of the composite membrane [66]. The improvement in thermal properties and mechanical performances of polymer/NC composite membranes corresponded to the high compatibility and inherent hydrogen bonding between NC and polymer membranes [67, 68]. However, an excessive concentration of NC could decrease the mechanical and thermal properties due to nanofiller agglomeration [69, 70].

11.4.2 Processing and Performance of Nanocellulose Membrane for Gas Capture and Separation

CCS membrane polymers such as polyvinyl alcohol (PVA) or polyvinyl amine (PVAm) are commonly used as matrixes, while polysulfone (PSf) or poly(p-phenylene oxide) (PPO) are utilized as supporting substrates. NC fillers in the form of both CNC and CNF are used to improve membrane performance. To boost the performance, it could be done by improving the charge number of NC via further 2,2,6-tetramethylpiperidine-1oxyl radical TEMPO oxidation or phosphorylation before the polymer mixing process [71]. It is also possible to increase the degree of crosslink between polymer and NC by adding crosslink agents such as glyoxal to the mixture of polymer and NC [69, 72].

There are two common processing methods to obtain polymer/NC composite membranes for CCS applications: (i) solvent casting [73] or (ii) dip-coating [63, 71, 74, 75]. Although there are still shortcomings due to the possible defects in the thin layer surface during dip-coating [76], this method has been used more frequently, in particular, for the active polymer/NC composite membrane, compared to the self-assembly solvent casting. Further, the casting process is also time-consuming and prone to wrinkled membrane surfaces, leading to poor mechanical properties [77].

Before these two processes, mechanical stirring and/or ultrasonication were commonly applied to produce a homogenous solution of NC/polymers [63, 71, 74, 75]. Air bubbles must be removed to prevent the formation of defects in the membrane, thus increasing membrane selectivity

Table 11.3 Nanocellulose composite performance for gas capture and separation.

Nanocellulose sources	Polymer matrixes	Applications	Performances	Reference
Wood pulp CNC and CNF	PVA and PPO as supporting substrate	TFC hollow fiber membranes for improving CO_2 capture	65% and 15% increase in permeation by adding 80% CNC and CNF, respectively. 29% increase in selectivity by adding 80% CNF	[74]
CNC	PVA and PSf as supporting substrate	CO_2/CH_4 separation	250% and 82% increase in permeation and selectivity, respectively, by adding 1.5 wt.% CNC	[75]
CNC	PVA and PSf as supporting substrate	CO_2 separation from flue gas	25% and 8% increase in permeation and selectivity, respectively, by adding 4 wt.% CNC	[71]

where the bigger molecules such as N_2 or CH_4 cannot pass through the membrane via those defects [63, 78]. The prepared solution is then cast on a petri dish to obtain a self-supporting polymer/NC membrane, while the dip-coating process is done by dipping the supporting substrate onto the prepared solutions to obtain substrate-supported polymer/NC membrane followed by a drying process at 20–70 °C [63, 71, 74, 75]. Table 11.3 summarizes the selected NC/polymer composite membranes corresponding to their application and performances upon NC incorporation.

11.5 Conclusion and Future Prospective

Nanocellulose (NC) has been identified as a building block for materials suitable for various applications including for the remediation of GHG effects due to its multifunctional characteristics and extraordinary properties.

This chapter has reported the fundamental overview and challenges by exploring the use of NC in the form of CNC and CNF to reduce the GHGs: (i) as reinforcing fillers and matrixes to boost the performance of lightweight polymers in automotive and effective thermal insulating material for indirectly lowering the energy usage and gas emissions, (ii) as reinforcing fillers to improve the mechanical properties and the selectivity of membrane composites for directly capturing/reducing GHGs.

The potential application in the remediation of GHGs is achievable by the incorporation of nanocellulose-based material onto existing polymers/matrixes, which has mostly been reported in laboratory experiments. Yet, these intensive efforts should be sustained to establish reliable solutions based on green and sustainable nanomaterials for the ongoing protection of the environment.

Abbreviation

CCS	Carbon capture and sequestration
CH_4	Methane
CNC	Cellulose nanocrystal
CNF	Cellulose nanofiber
CO_2	Carbon dioxide
EPS	Expanded polystyrene foam
GHG	Greenhouse gas
N_2	Nitrogen
NC	Nanocellulose
NFRP	Natural fiber-reinforced polymer
N_xO	Nitrous Oxide
O_3	Ozone
PCCC	Post-combustion carbon capture
PPO	Poly(p-phenylene oxide)
PSf	Polysulfone
PVA	Polyvinyl alcohol
PVam	Polyvinyl amine
RPUF	Rigid polyurethane foam
SiO_2	Silicon dioxide
TEMPO	2,2,6,6-tetramethyl-1-piperidinyl oxy
TFC	Thin film composite
XPS	Extruded Polystyrene

References

1 Kweku, D.W., Bismark, O., and Maxwell, A. (2018). Greenhouse effect: greenhouse gases and their impact on global warming. *J. Sci. Res. Rep.* 17: 1–9.
2 Cassia, R., Nocioni, M., Correa-Aragunde, N., and Lamattina, L. (2018). Climate change and the impact of greenhouse gasses: CO_2 and NO, friends and foes of plant oxidative stress. *Front. Plant Sci.* 9 (March): 1–11.
3 Kumar, A., Thanki, A., Padhiyar, H. et al. (2021). Chemosphere greenhouse gases emission control in WWTS via potential operational strategies: a critical review. *Chemosphere* 273: 129694.
4 Xu, T., Du, H., Liu, H. et al. (2021). Advanced nanocellulose-based composites for flexible functional energy storage devices. *Adv. Mater.* 33 (48): e2101368.
5 Anirudhan, T.S. and Rejeena, S.R. (2012). Adsorption and hydrolytic activity of trypsin on a carboxylate-functionalized cation exchanger prepared from nanocellulose. *J. Colloid Interface Sci.* 381 (1): 125–136.
6 Lubis, R., Wirjosentono, B., Eddyanto, and Septevani, A.A. (2020). Preparation, characterization and antimicrobial activity of grafted cellulose fiber from durian rind waste. *Colloids Surf., A* 604: 125311. https://doi.org/10.1016/j.colsurfa.2020.125311.
7 Anderson, S.R., Esposito, D., Gillette, W. et al. (2014). Enzymatic preparation of nanocrystalline and microcrystalline cellulose. *Tappi J.* 13 (5): 35–42.
8 Ansaloni, L., Salas-Gay, J., Ligi, S., and Baschetti, M.G. (2017). Nanocellulose-based membranes for CO_2 capture. *J. Membr. Sci.* 522: 216–225. https://doi.org/10.1016/j.memsci.2016.09.024.
9 Chauhan, V., Kärki, T., and Varis, J. (2019). Review of natural fiber-reinforced engineering plastic composites, their applications in the transportation sector and processing techniques. *J. Thermoplast. Compos. Mater.* 35: 1–41.
10 John, M.J. and Thomas, S. (2008). Biofibres and biocomposites. *Carbohydr. Polym.* 71 (3): 343–364.
11 Koronis, G., Silva, A., and Fontul, M. (2013). Green composites: a review of adequate materials for automotive applications. *Composites, Part B* 44 (1): 120–127. https://doi.org/10.1016/j.compositesb.2012.07.004.
12 Wu, Y., Xia, C., Cai, L. et al. (2018). Development of natural fiber-reinforced composite with comparable mechanical properties and reduced energy consumption and environmental impacts for replacing automotive glass-fiber sheet molding compound. *J. Cleaner Prod.* 184: 92–100.
13 Agarwal, J., Sahoo, S., Mohanty, S., and Nayak, S.K. (2020). Progress of novel techniques for lightweight automobile applications through innovative eco-friendly composite materials: a review. *J. Thermoplast. Compos. Mater.* 33 (7): 978–1013.

14 AL-Oqla, F.M., Sapuan, S.M., Ishak, M.R., and Nuraini, A.A. (2015). Predicting the potential of agro waste fibers for sustainable automotive industry using a decision making model. *Comput. Electron. Agric.* 113: 116–127. https://doi.org/10.1016/j.compag.2015.01.011.

15 Lin, N., Huang, J., and Dufresne, A. (2012). Preparation, properties and applications of polysaccharide nanocrystals in advanced functional nanomaterials: a review. *Nanoscale* 4 (11): 3274–3294.

16 Nurazzi, N.M., Jenol, M.A., Kamarudin, S.H. et al. (2022). Nanocellulose composites in the automotive industry. In: *Industrial Applications of Nanocellulose and Its Nanocomposites* (ed. S.M. Sapuan, M.N.F. Norrahim, R.A. Ilyas, and C. Soutis), 439–467. Elsevier.

17 Trache, D., Hussin, M.H., Haafiz, M.K.M., and Thakur, V.K. (2017). Recent progress in cellulose nanocrystals: sources and production. *Nanoscale* 9 (5): 1763–1786.

18 Rusli, R. and Eichhorn, S.J. (2008). Determination of the stiffness of cellulose nanowhiskers and the fiber-matrix interface in a nanocomposite using Raman spectroscopy. *Appl. Phys. Lett.* 93 (3): 033111.

19 Nechyporchuk, O., Belgacem, M.N., and Bras, J. (2016). Production of cellulose nanofibrils: a review of recent advances. *Ind. Crops Prod.* 93: 2–25. https://doi.org/10.1016/j.indcrop.2016.02.016.

20 Xu, X., Liu, F., Jiang, L. et al. (2013). Cellulose nanocrystals vs. cellulose nanofibrils: a comparative study on their microstructures and effects as polymer reinforcing agents. *ACS Appl. Mater. Interfaces* 5 (8): 2999–3009.

21 Bhatnagar, A. and Sain, M. (2005). Processing of cellulose nanofiber-reinforced composites. *J. Reinf. Plast. Compos.* 24 (12): 1259–1268.

22 Ferreira, F.V., Lona, L.M.F., Pinheiro, I.F. et al. (2019). Polymer composites reinforced with natural fibers and nanocellulose in the automotive industry: a short review. *J. Compos. Sci.* 3 (51): 1–17.

23 Mukherjee, T., Czaka, M., Kao, N. et al. (2014). Dispersion study of nanofibrillated cellulose based poly(butylene adipate-co-terephthalate) composites. *Carbohydr. Polym.* 102 (1): 537–542. https://doi.org/10.1016/j.carbpol.2013.11.047.

24 Azouz, K.B., Ramires, E., Van Fonteyne, W., D. et al. (2011). A simple method for the melt extrusion of cellulose nano-crystal reinforced hydrophobic polymer. *ACS Macro Lett.* 1 (1): 236–240.

25 Danni, N., Sasikumar, T., and Ahamed Fazil, A. (2016). Mechanical properties of electrospun CNF/PVA nanofiber mats as reinforcement in polymer matrix composites. *Int. J. Appl. Chem.* 12 (2): 107–119.

26 Guan, Q.F., Yang, H.B., Han, Z.M. et al. (2020). Lightweight, tough, and sustainable cellulose nanofiber-derived bulk structural materials with low thermal expansion coefficient. *Sci. Adv.* 6 (18): 1–9.

27 Kiziltas, A., Erbas Kiziltas, E., Boran, S., and Gardner, D.J. (2013). Micro- and nanocellulose composites for automotive applications. In: *Society of Plastics Engineers – 13th Annual Automotive Composites Conference and Exhibition, ACCE 2013*, 402–414.

28 Faruk, O., Sain, M., Farnood, R. et al. (2014). Development of lignin and nanocellulose enhanced bio PU foams for automotive parts. *J. Polym. Environ.* 22 (3): 279–288.

29 Xu, S., Girouard, N., Schueneman, G. et al. (2013). Mechanical and thermal properties of waterborne epoxy composites containing cellulose nanocrystals. *Polymer (Guildf)* 54 (24): 6589–6598. https://doi.org/10.1016/j.polymer.2013.10.011.

30 Abu-Jdayil, B., Mourad, A.H., Hittini, W. et al. (2019). Traditional, state-of-the-art and renewable thermal building insulation materials: an overview. *Constr. Build. Mater.* 214: 709–735.

31 Smalyukh, I.I. (2021). Thermal management by engineering the alignment of nanocellulose. *Adv. Mater.* 33 (28): 2001228. https://onlinelibrary.wiley.com/doi/10.1002/adma.202001228.

32 Hasanzadeh, R., Azdast, T., Doniavi, A., and Lee, R.E. (2019). Multi-objective optimization of heat transfer mechanisms of microcellular polymeric foams from thermal-insulation point of view. *Therm. Sci. Eng. Prog.* 9: 21–29. https://linkinghub.elsevier.com/retrieve/pii/S2451904918304657.

33 Kim, B.G. and Lee, D.G. (2008). Development of microwave foaming method for phenolic insulation foams. *J. Mater. Process. Technol.* 201 (1): 716–719. https://www.sciencedirect.com/science/article/pii/S0924013607012526.

34 Wang, P., Aliheidari, N., Zhang, X., and Ameli, A. (2019). Strong ultralight foams based on nanocrystalline cellulose for high-performance insulation. *Carbohydr. Polym.* 218: 103–111. https://linkinghub.elsevier.com/retrieve/pii/S0144861719304527.

35 Uetani, K. and Hatori, K. (2017). Thermal conductivity analysis and applications of nanocellulose materials. *Sci. Technol. Adv. Mater.* 18 (1): 1–16. https://doi.org/10.1080/14686996.2017.1390692.

36 Kurańska, M., Pinto, J.A., Salach, K. et al. (2020). Synthesis of thermal insulating polyurethane foams from lignin and rapeseed based polyols: a comparative study. *Ind. Crops Prod.* 143: 111882.

37 Prociak, A., Kurańska, M., Uram, K., and Wójtowicz, M. (2021). Bio-polyurethane foams modified with a mixture of bio-polyols of different chemical structures. *Polymers (Basel)* 13 (15): 2469. https://www.mdpi.com/2073-4360/13/15/2469.

38 Septevani, A.A., Evans, D.A.C., Annamalai, P.K., and Martin, D.J. (2017). The use of cellulose nanocrystals to enhance the thermal insulation properties and sustainability of rigid polyurethane foam. *Ind. Crops Prod.* 107: 114–121. https://linkinghub.elsevier.com/retrieve/pii/S0926669017303527.

39 Khoo, R.Z., Ismail, H., and Chow, W.S. (2016). Thermal and morphological properties of poly (lactic acid)/nanocellulose nanocomposites. *Procedia Chem.* 19: 788–794. https://linkinghub.elsevier.com/retrieve/pii/S1876619616001327.

40 Leng, W. and Pan, B. (2019). Thermal insulating and mechanical properties of cellulose nanofibrils modified polyurethane foam composite as structural insulated material. *Forests* 10 (2): 200. http://www.mdpi.com/1999-4907/10/2/200.

41 Septevani, A.A., Evans, D.A.C., Hosseinmardi, A. et al. (2018). Atomic layer deposition of metal oxide on nanocellulose for enabling microscopic characterization of polymer nanocomposites. *Small* 14 (46): 1803439. https://doi.org/10.1002/smll.201803439.

42 Notario, B., Pinto, J., Solorzano, E. et al. (2015). Experimental validation of the Knudsen effect in nanocellular polymeric foams. *Polymer (Guildf)* 56: 57–67. https://linkinghub.elsevier.com/retrieve/pii/S0032386114009021.

43 Wang, P., Zhao, J., Xuan, R. et al. (2014). Flexible and monolithic zinc oxide bionanocomposite foams by a bacterial cellulose mediated approach for antibacterial applications. *Dalton Trans.* 43 (18): 6762–6768.

44 Li, Y., Ren, H., and Ragauskas, A.J. (2011). Rigid polyurethane foam/cellulose whisker nanocomposites: preparation, characterization, and properties. *J. Nanosci. Nanotechnol.* 11: 6904–6911.

45 Dutta, A.S. (2018). 2 – Polyurethane foam chemistry. In: *Recycling of Polyurethane Foams* (ed. S. Thomas, A.V. Rane, K. Kanny, et al.), 17–27. William Andrew Publishing https://www.sciencedirect.com/science/article/pii/B9780323511339000024.

46 Kaushiva, B.D. and Wilkes, G.L.1998). Influence of diethanol amine (DEOA) on structure-property relationships in molded flexible polyurethane foams. *Annu Tech Conf – ANTEC, Conf Proc.*, Vol. 2, 1850–1854.

47 Gupta, P., Singh, B., Agrawal, A.K., and Maji, P.K. (2018). Low density and high strength nanofibrillated cellulose aerogel for thermal insulation application. *Mater. Des.* 158: 224–236. https://linkinghub.elsevier.com/retrieve/pii/S0264127518306440.

48 Sture, B., Vevere, L., Kirpluks, M. et al. (2021). Polyurethane foam composites reinforced with renewable fillers for cryogenic insulation. *Polymers (Basel)* 13 (23): 4089.

49 Jiménez-Saelices, C., Seantier, B., Cathala, B., and Grohens, Y. (2017). Spray freeze-dried nanofibrillated cellulose aerogels with thermal superinsulating properties. *Carbohydr. Polym.* 157: 105–113. https://linkinghub.elsevier.com/retrieve/pii/S0144861716311249.

50 Yin, D., Mi, J., Zhou, H. et al. (2020). Fabrication of branching poly (butylene succinate)/cellulose nanocrystal foams with exceptional thermal insulation. *Carbohydr. Polym.* 247 (July): 116708.

51 Dutcher, B., Fan, M., and Russell, A.G. (2015). Amine-based CO_2 capture technology development from the beginning of 2013 – a review. *ACS Appl. Mater. Interfaces* 7 (4): 2137–2148.

52 Kheirinik, M., Ahmed, S., and Rahmanian, N. (2021). Comparative techno-economic analysis of carbon capture combustion operations. *Sustainability* 13 (13567): 1–14.

53 Kunze, C. and Spliethoff, H. (2012). Assessment of oxy-fuel, pre- and post-combustion-based carbon capture for future IGCC plants. *Appl. Energy* 94: 109–116. https://doi.org/10.1016/j.apenergy.2012.01.013.

54 Mukherjee, A., Okolie, J.A., Abdelrasoul, A. et al. (2019). Review of post-combustion carbon dioxide capture technologies using activated carbon. *J. Environ. Sci. (China)* 83: 46–63. https://doi.org/10.1016/j.jes.2019.03.014.

55 Norahim, N., Yaisanga, P., Faungnawakij, K. et al. (2018). Recent membrane developments for CO_2 separation and capture. *Chem. Eng. Technol.* 41 (2): 211–223.

56 Zhou, Q., Manuilova, A., Koiwanit, J. et al. (2014). A comparative of life cycle assessment of post-combustion, pre-combustion and oxy-fuel CO_2 capture. *Energy Procedia* 63: 7452–7458. https://doi.org/10.1016/j.egypro.2014.11.782.

57 Ho, N.A.D. and Leo, C.P. (2021). A review on the emerging applications of cellulose, cellulose derivatives and nanocellulose in carbon capture. *Environ. Res.* 197: 111100. https://doi.org/10.1016/j.envres.2021.111100.

58 Krishnamurthy, S., Lind, A., Bouzga, A. et al. (2021). Post combustion carbon capture with supported amine sorbents: from adsorbent characterization to process simulation and optimization. *Chem. Eng. J.* 406: 127121. https://doi.org/10.1016/j.cej.2020.127121.

59 Anselmi, H., Mirgaux, O., Bounaceur, R., and Patisson, F. (2019). Simulation of post-combustion CO_2 capture, a comparison among absorption, adsorption and membranes. *Chem. Eng. Technol.* 42 (4): 797–804.

60 Vakharia, V., Salim, W., Wu, D. et al. (2018). Scale-up of amine-containing thin-film composite membranes for CO_2 capture from flue gas. *J. Membr. Sci.* 555: 379–387.

61 White, L.S., Wei, X., Pande, S. et al. (2015). Extended flue gas trials with a membrane-based pilot plant at a one-ton-per-day carbon capture rate. *J. Membr. Sci.* 496: 48–57. https://doi.org/10.1016/j.memsci.2015.08.003.

62 Deng, L., Kim, T.J., and Hägg, M.B. (2009). Facilitated transport of CO_2 in novel PVAm/PVA blend membrane. *J. Membr. Sci.* 340 (1–2): 154–163.

63 Saeed, U., Jahan, Z., Niazi, M.B.K. et al. (2020). Biogas upgrading with novel cellulose nano-crystals and polyvinyl amine nanocomposite membranes. *Polym. Test.* 91 (September): 106867. https://doi.org/10.1016/j.polymertesting.2020.106867.

64 Cho, M.J. and Park, B.D. (2011). Tensile and thermal properties of nanocellulose-reinforced poly(vinyl alcohol) nanocomposites. *J. Ind. Eng. Chem.* 17 (1): 36–40. https://doi.org/10.1016/j.jiec.2010.10.006.

65 Li, W., Wu, Q., Zhao, X. et al. (2014). Enhanced thermal and mechanical properties of PVA composites formed with filamentous nanocellulose fibrils. *Carbohydr. Polym.* 113: 403–410. https://doi.org/10.1016/j.carbpol.2014.07.031.

66 Mondal, S. (2018). Review on nanocellulose polymer nanocomposites. *Polym. Plast. Technol. Eng.* 57 (13): 1377–1391.

67 Stivala, S.S. and Reich, L. (1980). Structure vs stability in polymer degradation. *Polym. Eng. Sci.* 20 (10): 654–661.

68 Mehdipour-Ataei, S. and Tabatabaei-Yazdi, Z. (2015). Heat resistant polymers. In: *Encyclopedia of Polymer Science and Technology [Internet]*, 1–31. https://doi.org/10.1002/0471440264.pst636.

69 Mandal, A. and Chakrabarty, D. (2014). Studies on the mechanical, thermal, morphological and barrier properties of nanocomposites based on poly(vinyl alcohol) and nanocellulose from sugarcane bagasse. *J. Ind. Eng. Chem.* 20 (2): 462–473. https://doi.org/10.1016/j.jiec.2013.05.003.

70 Hou, Q., Wang, X., and Ragauskas, A.J. (2019). Preparation and characterization of nanocellulose–polyvinyl alcohol multilayer film by layer-by-layer method. *Cellulose* 26 (8): 4787–4798. https://doi.org/10.1007/s10570-019-02413-0.

71 Torstensen, J., Helberg, R.M.L., Deng, L. et al. (2019). PVA/nanocellulose nanocomposite membranes for CO_2 separation from flue gas. *Int. J. Greenhouse Gas Control* 81: 93–102. https://doi.org/10.1016/j.ijggc.2018.10.007.

72 Lee, H., Hamid, S.B., and Zain, S. (2014). Conversion of lignocellulosic biomass to nanocellulose: structure and chemical process. *Sci. World J.* 2014: 1–21.

73 Duan, S., Taniguchi, I., Kai, T., and Kazama, S. (2012). Poly(amidoamine) dendrimer/poly(vinyl alcohol) hybrid membranes for CO_2 capture. *J. Membr. Sci.* 423–424: 107–112. https://doi.org/10.1016/j.memsci.2012.07.037.

74 Dai, Z., Deng, J., Yu, Q. et al. (2019). Fabrication and evaluation of bio-based nanocomposite TFC hollow fiber membranes for enhanced CO_2 capture. *ACS Appl. Mater. Interfaces* 11 (11): 10874–10882.

75 Jahan, Z., Niazi, M.B.K., Hägg, M.B., and Gregersen, Ø.W. (2018). Decoupling the effect of membrane thickness and CNC concentration in PVA based nanocomposite membranes for CO_2/CH_4 separation. *Sep. Purif. Technol.* 204: 220–225. https://doi.org/10.1016/j.seppur.2018.04.076.

76 Ramasubramanian, K. and Ho, W.S.W. (2011). Recent developments on membranes for post-combustion carbon capture. *Curr. Opin. Chem. Eng.* 1 (1): 47–54. https://doi.org/10.1016/j.coche.2011.08.002.

77 Septevani, A.A., Burhani, D., and Sampora, Y. (2022). Chapter 12 – Nanocellulose in electronics and electrical industry. In: *Micro and Nano Technologies* (ed. R. Oraon, D. Rawtani, P. Singh, and C.M. Hussain), 217–246. Elsevier https://www.sciencedirect.com/science/article/pii/B9780128239636000107.

78 Venturi, D., Grupkovic, D., Sisti, L., and Baschetti, M.G. (2018). Effect of humidity and nanocellulose content on polyvinylamine-nanocellulose hybrid membranes for CO_2 capture. *J. Membr. Sci.* 548: 263–274. https://doi.org/10.1016/j.memsci.2017.11.021.

12

Ecofriendly Nanomaterials for Wastewater Treatment

Neeru Dabas[1], *Shivani Chaudhary*[2], *Ritu Rani Chaudhary*[3], *and Gautam Jaiswar*[2]

[1] *Department of Chemistry, Amity School of Applied Science, Amity University, Gurugram, HR, India*
[2] *Department of Chemistry, Dr. Bhimrao Ambedkar University, Agra, UP, India*
[3] *Department of Chemistry, B.S.A. College, Mathura, UP, India*

Abstract

Wastewater treatment is an important area of research that needs immediate attention as the water quality is directly related to human health and the environment. Wastewater from major sectors like industries, agriculture, and domestic is diversified and needs proper treatment before its discharge into the environment. Untreated wastewater contains large amount of toxic and nontoxic dissolved solids that, if not properly treated, enter into fresh water and endanger aquatic life. Considering all the adverse impacts of untreated wastewater on our environment, it needs proper treatment. Lots of efforts are being made continuously in this area; among them, the most advanced is using nanotechnology to solve the problem associated with wastewater. The use of nanomaterials is nowadays increasing in all areas of research. Nanomaterials exhibit unique physical and chemical properties, making them more attractive in their applications. As the use of nanomaterial increases with time, the question of how safe their use is also in the researcher's mind. So, despite showing wonderful applications, all nanomaterials are unsafe, and some are potentially toxic. Herein, we focus on the synthesis of nanomaterials through green method synthesis, such as from plant parts as well as from agricultural wastes, and further present the application of these environmentally friendly nanoparticles for wastewater remediation and other sciences sectors.

Green Synthesis of Nanomaterials: Biological and Environmental Applications, First Edition.
Edited by Archana Chakravarty, Preeti Singh, Saiqa Ikram, and R.N. Yadava.

Keywords *nanomaterials; wastewater treatment; toxicity; agricultural wastes; green chemistry; ecofriendly*

12.1 Introduction

Nanotechnology is a branch of science and engineering dedicated to designing, producing, and using structures, devices, and systems by employing atoms and molecules at the nanoscale, i.e. having one or more dimensions of the order of 100 nm (100 millionths of a millimeter) or less. Those materials fall in the nano range and are measured in the nanoscale. Nanomaterials find applications in the field of [1] mechanical, thermal, and optical conductivity. They have excellent biological and chemical activities because of their high surface area-to-volume ratio and high porosity. Other than these properties, changes can be possible in them for specific requirements [2, 3]. As nanomaterials are specific, they are continuously finding applications in a wide range of sectors like healthcare, pharma, food and agriculture and daily use of personal products and various other branches of industries [4–6]. Some of the most commonly used nanomaterials include silver (Ag) nanoparticles (NPs) [7], zinc oxide (ZnO) nanoparticles, graphene oxide NPs, titanium dioxide (TiO_2) nanoparticles [8], and single/multi-walled carbon nanotubes (CNTs). These materials are tailored according to specific requirements for applications in different branches of science (Figure 12.1).

Technologies like green chemistry are masterpiece for developing ecofriendly nanomaterials, reducing the environmental impact, and making products more sustainable. There are 12 principles of green chemistry to make products ecofriendly, but it is very difficult to apply all the green chemistry principles, and it is a challenge for the scientists to achieve a maximum number of green chemistry principles in the formation of single products.

Researchers are working to develop the green production methods of different metal oxides in nanoscience and nanotechnology. Nowadays, methods for producing metal NPs are utilizing more ecofriendly chemicals, biomolecules, and metabolites by reducing harmful chemicals and solvents from the experiments [9–11]. Nanomaterials production can be done through (i) top-down approach which involves physical methods like sonication, ball milling, spray pyrolysis, etc., (ii) bottom up approach, which is a chemical method and includes photochemical, sol–gel technology, etc., (iii) some bottom-up approach includes biological methods which involve fungi, plants, yeast, etc. [12].

12.2 Application of Ecofriendly Nanomaterials

Nowadays, researchers are developing nanomaterials for various environmentally friendly applications like cleaning waste and polluted water and providing clean water for industrial purposes on a large scale for household

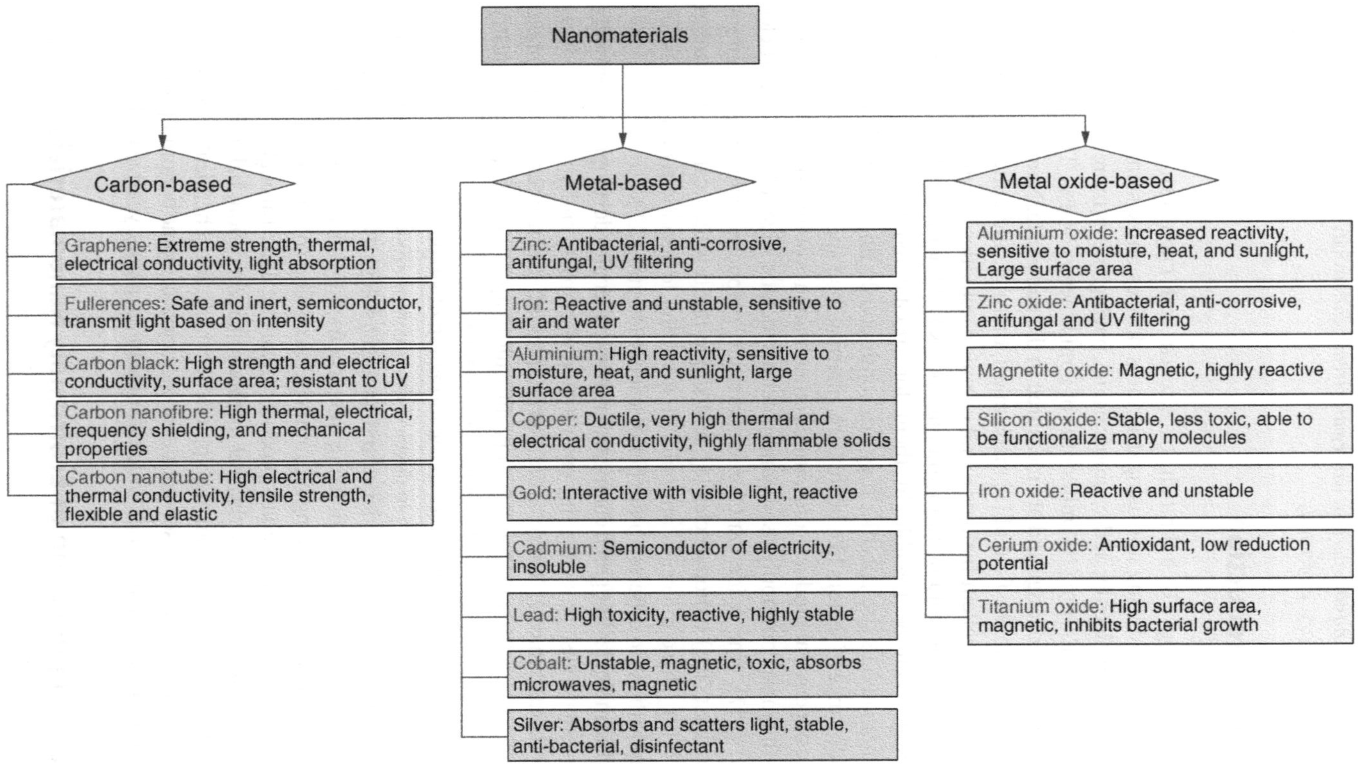

Figure 12.1 Classification of nanomaterials (carbon based, metal based and metal oxide based) on the basis of properties.

purposes and for cleaning the environment [13, 14]. Bioremediation is one of the processes through which we can clean the environment using natural materials/organisms. The conventional treatments of cleaning water are not fit now, and bioremediation has several advantages over it; it is cheaper in price, more compatible, and specific for metals and biosorbents used can be recovered and over this metal recovery chances are more in this method [15]. When [16] bioremediation technology involves various technologies like bioreactor, composting, biostimulation, and if it occurs on its own, then it is known as natural attenuation, or if it is urged to occur, then the addition of fertilizer must be done to enhance bioavailability, and known as bio stimulated remediation. The bioremediation of a contaminated site involves two methods. The first one is growing the microbes that eat up the contamination by providing suitable temperature, nutrients, and amount of oxygen. The second method involves directly using microbes of a suitable type on the site that eat up the contamination, but in both methods, microbes die when the nutrient or food finishes.

The reason behind using nanomaterials for bioremediation is that the material used is of the nanoscale range, so it has a very high surface-to-volume ratio; hence, the process occurs very smoothly, and very less material is used for large cleaning.

Nanomaterials (NMs) have less activation energy requirement because they show a quantum effect, so reactions occur at low energy, and with this, surface plasmon resonance is used for detecting different toxic materials. Sometimes, shape matters in case of NMs of metallic and nonmetallic material; we can use single, bi metallic NPs; sometimes, carbon-based NMs can be used for cleaning the environment.

NPs can be used as reductive biocatalysts for removing chlorine and Pd in Pd (0) form, and some NPs remove microbes like *Shewanella oneidensis* by depositing inside the cytoplasm or on the cell wall or by H* radical obtained from hydrogen, acetate, or formate. To remove chlorine from contaminated water, chlorinated compounds are brought into contact with *S. oneidensis* cells and charged Pd (0) deposited on them. In modern cell immobilization very small or micro-sized medium is used for fixing the surface, and ammonium oleate is used for functionalizing NPs like Fe_3O_4 on the surface of the microbe *Pseudomonas delafieldii* (Figure 12.2).

12.3 Inorganic Nanoparticles

From ancient times, inorganic NPs were practically used for the first time. From time scale in medieval times, various artistic expressions were used in religious art like in churches, and from starting, optical properties were the root of various applications of nanosized technology [17]. At the end of the

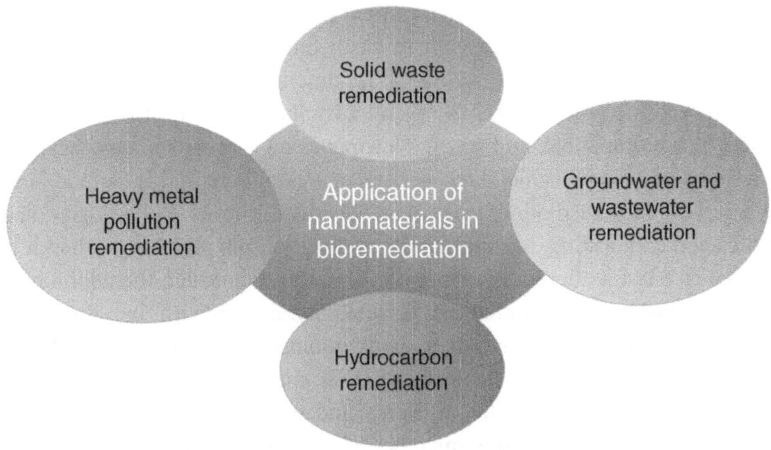

Figure 12.2 Applications of nanomaterials in bioremediation.

twentieth century, nanometric scale was used for modern technologies and became the new field for various areas of medicine, energy production, materials, etc. [18, 19]. After this industrial revolution occurred, it resulted from nanotechnology's impact and would result in a precursor for developing new technology [20].

As the world is developing, nanotechnology is also growing day by day; according to the data from the Nanotechnology Consumer Products Inventory and The United States Bureau of Labor Statistics, thousands of companies are focusing on producing nanotechnology-based materials, including metal oxide NPs like tin, iron, silver, gold, etc. The development of nanotechnology plays an excellent role in human civilization and physical, social, mental well-being of humans, and the credit goes to the lab to land development of nanotechnology [19, 20].

12.4 Synthesis of Green Nanomaterials

Green nanomaterials are highly desirable because they are environmentally friendly and nontoxic, which encourages researchers across the globe to find innovative methods/technologies for their synthesis in a facile manner. Synthesis of green nanomaterials involves natural products and reagents that are not hazardous and are termed green reagents. One of the focused areas of synthesis of green nanomaterial is using green solvents during the entire synthetic protocol, starting from the preparation to the isolation and, finally the purification of the product. The green synthetic technologies often involve the replacement of corrosive, toxic solvents with noncorrosive solvents such as water, ionic liquids, supercritical fluids, liquid polymers, deep eutectic solvents, and biosolvents for preparation of the biocompatible green nanomaterial. Nowadays, various biomolecules, microorganism, and plant products are increasingly used for the synthesis of green nanomaterial with novel physical, chemical, biological, optical, and magnetic properties. Various biomolecules derived from plants and microorganisms offer an attractive, cost-effective, sustainable source for synthesizing green nanomaterial with multifaceted applications. Plant products and microorganisms are the natural source of reduction of different precursor metal ions used to synthesize sustainable nanomaterials. Moreover, the biowaste derived directly from the decomposed plants and the agriculture industry is also explored and utilized to synthesize green nanomaterials. It is advantageous, as it reduces the burden of waste and is also an economical and nature-friendly way of waste management. Biomolecules have natural ingredients that are very active and catalyze various chemical transformations as efficiently as synthetic reagents. Biomolecules have an added advantage over synthetically driven chemicals for synthesizing nanomaterials. They reduce the precursor ions to nanosize material and stabilize them by preventing their aggregation. Various biomolecules offer an opportunity to control nanomaterials' size and shape, which is extremely important to design the smart green nanomaterials required for healthcare, energy, environmental, electronics, communications, aviation, and agricultural sector applications.

12.5 Nanocellulose Nanomaterials for Water Treatment

Water treatment by use of ecofriendly nanomaterial is the need of recent times, when the development and environment both are necessary, but one cannot be compromised at the cost of the other. Nanocellulose term is used for small-sized cellulose-containing materials derived from plants or bacteria. It includes cellulose nanocrystals and nanofibers. The use of such materials is increasing daily as it is safe to use them because of their environmentally friendly nature. Interest in nanocellulose for advanced applications also lies in the fact that they display numerous physical, chemical, and synthetic properties which are fascinating due to high surface area, high specific young modulus, high elasticity, high tensile strength, nontoxic, biodegradable, facile synthesis.

Water treatment by use of ecofriendly nanomaterial is the need of recent times, when the development and environment both are necessary, but one cannot be compromised at the cost of the other. The use of nanocellulose is increasing day by day as it is safe to use them because of their environmentally friendly nature.

Carbon dots were prepared from renewable resources; conventionally, they were synthesized from different carbon-rich precursors that originated from petroleum products. However, the synthesis method had particular disadvantage such as limited availability, greenhouse gas effect, poor cost stability, and climate change; the alternative of these resources were renewable feedstocks. The advantage of utilizing these methods for synthesis includes a comprehensive option for source selection, an abundance of raw material, and environment-friendly options.

Green biocomposites are those materials that are derived from biological or renewable origin and comprise of the reinforced and polymeric matrix. Naturally derived polymers are used as protective agents in the packaging as they have antioxidant and antibacterial activity by the addition of nano clays and nanoparticles in the polymer complex. Compared to their macrocounterparts, nanocomposites offer advantages like high surface-to-volume ratio and surface activity with improved biodegradability and biocompatibility with myriad of bioinspired materials as renewable alternatives to reduce the use of hazardous chemicals and derivatives. The synthesized materials showed high surface areas, which is very attractive for catalytic applications owing to their unique, highly porous structures with active sites, particularly those decorated by metal and metal oxide NPs. In this direction, new approaches are described in the literature including copolymerization of two types of monomers: efficient and sustainable route. Principally, extracts from leaves, seeds, roots, flowers, and fruits have been used. The extracts are composed of phytochemicals such as protein, amino acids, vitamins, polysaccharides, terpenoids, and organic acids.

12.6 Graphene-CNT Hybrid/Graphene Hybrids (GO and Biopolymer)

Graphene-based nanomaterials have numerous applications in field of pharma, sensing, energy, and electronics. Nowadays, graphene-based material has proliferated in other fields, and one of the promising applications is its use in wastewater treatment. Clean water is the foremost priority of the human; however, accessing clean water in today's time is also challenging to fulfill the needs of the growing population. There is a need to treat the water effluent coming out of various sectors such as industries, institutions, commercial, and residential complexes to solve the problem. Carbon-based nanomaterials are a promising solution to wastewater treatment as they are porous in nature and have good adsorption capacity. The wastewater often contains several dyes and pigments that imparts color to wastewater. Moreover, pH, taste, and quality of water is also adversely affected. A green assembly of graphene-based hydrogel was reported for the removal of various organic dyes (triphenylmethane and thiazine dyes) from aqueous medium. The assembly was prepared by intercalation of glutathione in between the layered assembly of graphene. Intercalation of glutathione between layers of graphene occurs by the reaction of reactive functional groups present on the surface of glutathione and graphene [21]. Such reduced graphene oxide (RGO) also has potential in wastewater treatment. Often, the synthesis of RGO involves chemical reduction, which is not environmentally friendly. Following the synthetic, environmentally benign, and economical procedure is necessary. Nowadays, RGO is synthesized using principles of green chemistry, which is an ecofriendly approach to synthesizing such materials. The synthetic methodology in green chemistry uses nontoxic materials/biomaterials/natural products in place of harmful, toxic chemicals to obtain the desired products. RGO is also obtained by reduction of GO using green plant-based products as reducing agent. RGO obtained using grape juice as a reducing agent is reported for the removal of malachite green from water [22]. Polysaccharide is one of the ecofriendly materials which is also popular nowadays in various applications. Polysaccharides can easily be mixed with graphene to form hybrid graphene with advanced structural properties and practical applications. Graphene-polysaccharide hybrids are useful materials that have relevance in water treatment. One of the very common problems in water is the presence of calcium and magnesium salts, which impart hardness to water. The water is considered very hard when the $CaCO_3$ equivalents are more than 180 ppm. Excessive soluble salts of calcium or magnesium in water have several disadvantages and are unsuitable for boilers, industries, and other daily use applications. Using chemical softeners such as resins, zeolites, and chemicals like lime soda removes hardness. However, these methods are not considered ecofriendly. Hybrids of graphene oxide and biopolymers are becoming popular water softeners because they are ecofriendly; moreover, it is often observed that such hybrids have better sorption capacity of various metal ions from aqueous solution than graphene oxide. Recently, Rocha et al. reported hybrid of graphene oxide with a biopolymer, κ-carrageenan to soften the hard water. The hybrid nanomaterial removes Ca^{2+} from the water by electrostatic interaction of Ca^{2+} with the sulfonate group on the surface of the biopolymer [23].

Graphene oxide hybrids with metal-organic framework (MOF) are also used in water treatment. MOF are suitable for water treatment applications because they are porous, and have reactive functional groups which can easily uptake small organic dye molecules or other ions present in untreated water. MOF are also able to degrade the organic dyes under the UV light. Hybrids can be made into a membrane and used for water treatment. Generally, the membrane is made up of a polymer. Using biodegradable polymer, polylactic acid, and Zeolitic imidazole framework/graphene oxide (ZIF-GO) hybrid makes a membrane that removes methylene blue dye. The removal efficiency of dye is reported 90% at very low concentrations, i.e. 0.06 mg/ml of ZIF-GO hybrid [24].

12.7 Green Nanocomposite

TiO_2-CeO_2 nanocomposite (TCN) of varying shapes and structures have shown applications in degradation of various organic dye such as methylene blue and Rhodamine B under photocatalytic conditions either using UV light or visible light. Photocatalytic activity of TiO_2 is due to its bandgap value of 3.2 eV because of which it can absorbs UV light. Doping of TiO_2 with CeO_2 greatly reduces the band gap and the hybrid material so form can also absorb the visible light. Other than the organic dyes, chlorinated phenols are also one of the pollutants present in water present in appreciable quantity. The chlorinated phenols are toxic for aquatic life too. Synthesis of TCN by lemon extract is recently reported, which exhibits greater activity in photolytic degradation of 2,4-dichlorophenol than either metal oxide, TiO_2 or CeO_2 [25]. Lemon extract also help in stabilization of nanocomposite by electrostatic repulsion between negatively charged metal oxides. The negative charge on metal oxide arises due to adsorption of citric acid on the surface of TiO_2 and CeO_2.

Nanocomposites prepared via green methods have wider applications and are preferred over the materials synthesized by conventional methods. The use of nanocomposites synthesized by greener methods is also increasing day by day for curing the problem arisen by decreasing quality of water. The extensive use of batteries and electronics releases various heavy metal ions into the environment and is also of great concern. Heavy metals such as Pb, Cr, Hg, and Cd are highly toxic even when they are present at very less concentration as low as ppb. MOF and its nanocomposites are used extensively for the removal of metal ions from the aqueous medium. Nanocomposites are derived by the addition of thin sheets or layers of other organic compounds containing various reactive functional groups into

MOF using different solvents. Use of water as solvent is preferred over organic solvents to get composites with advanced functional and structural properties. Nanocomposites synthesized by green route are exploited in water treatment. Recently, composite derived from MOF and layered double hydroxide synthesized using water as solvent has been successfully applied in the treatment of water contaminated by metal ions, Pd(II) and Cd(II) [26].

Excessive use of fertilizers and industrially important chemical compounds have increased the level of various inorganic anions such as nitrate and phosphate in freshwater. Excess amount of nitrate and phosphate leads to eutrophication, a condition that arises due to excessive growth of algae in water bodies and as a result the level of oxygen decreases severely affecting the aquatic life.

Different natural gums and their composites are used to improve water quality by removing different organic dyes, etc. Most of the gums are hydrophilic in nature and have good adsorption capacity so are used for various water treatment applications, but they have less surface area and a limited capacity to bond with organic dyes, which limits their application. In past few years green nanocomposites derived from various gums and metal oxide nanoparticles have used extensively in wastewater treatment. The metal oxides commonly used to derive the nanocomposites are SiO_2, Al_2O_3, Fe_2O_3, Fe_3O_4, TiO_2, Cu_2O, ZnO, etc. Nanocomposite derived from Moringa gum and aluminum oxide nanoparticle removes nitrate and phosphate efficiently from synthetic water sample using red LED light. Nanocomposite at very low concentration of 10 mg/l removes nitrate and phosphate ions up to 94% and 95%, respectively. The nanocomposite was also applied to domestic wastewater using various LED light sources. The maximum removal of nitrate and phosphate was observed up to 76% and 85%, respectively, using red LED [27].

12.7.1 Guar Gum-Based Nanocomposites

Guar gum (GG) is a polysaccharide that have lots of applications in food and pharma industry. Its use in field of water treatment has recently increasing due to its good water solubility and nontoxic nature. For water treatment applications such as dye removal, GG is mixed with inorganic metal oxides to get the nanocomposite which has band gap in the range suitable for the photocatalytic reactions. The mechanism of dye removal is explained by adsorption of dye onto the nanocomposite and its degradation by irradiation with sunlight. Nanocomposite helps in rapid and efficient degradation of dye which is well corroborated with the literature studies. Several modifications of irradiation experiments give different results. In one case, it is observed that simultaneous adsorption of dye onto nanocomposite and its photo-irradiation achieves better dye degradation (90%) than adsorption followed by photo-irradiation (80% dye degradation) [28]. Other than dye degradation, structurally modified guar gums-based nanocomposites are used as flocculants material in treatment of industrial wastewater [29]. Structural modification of GG is done by grafting the polymeric chain onto the surface of gum microwave or ultrasonic-assisted method. To get the nanocomposite, sol–gel method is used where metal oxide nanoparticles mixes with structurally modified GG. Nanocomposite of GG/Fe_3O_4 prepared by the coprecipitation route is a good adsorbent nanomaterial for removing various organic dyes from the water stream [30]. A study was carried out to measure the adsorption efficiency of nanocomposite for different classes of dyes. Among various dyes, the highest efficiency was obtained for Congo red. The highest absorption efficiency (97%) for the dye is due to the NH_2 group of dye, which forms a strong hydrogen bond with abundant hydroxyl group on the surface of composite.

Nanocomposite of GG with $NiWO_4$ efficiently removes xanthene dye and crystal violet. The absorption efficiency for phloxine B, a xanthene dye, and crystal violet dyes are 220.21 and 170.42 mg/g, respectively. The optimum time for removal of phloxine B and crystal violet with high efficiency is reported 40 and 50 minutes, respectively [31].

A few nanocomposites of gum are reported to have antibacterial properties and can inhibit bacteria growth. *E. coli* and *S. aureus* are the most common bacteria that are found in wastewater. Removing these bacteria and organic dyes is equally important to get pathogen-free clean water. In one example, a cost-effective nanocomposite derived from xanthan gum and nanoclay, montmorillonite of variable concentrations removes malachite green dye and pathogenic bacteria found in wastewater. The maximum removal of malachite green was achieved upto 99.99% in the basic medium for 90 minutes. In another example, nanocomposite of xanthan gum was derived with SiO_2 nanoparticle by ultrasonication and also exhibits similar properties and applications. Herein, the maximum removal of dye was achieved as 99.5% at neutral pH within 6 hours [32].

A green nanocomposite of xanthan gum with natural clay is reported and applied to remove Pb(II)ion from different wastewater samples. The nanocomposite exhibits good recycling ability in the presence of HCl and can be regenerated up to maximum of five cycles with appreciable absorption capacity, with the maximum adsorption reported 92% at the first cycle [33].

Another nanocomposite derived from xanthan gum and Methionine-bentonite nanocomposite removes anionic, Congo red dye from aqueous media with a removal efficiency of 85%. The nanocomposite also works well and is reused successfully by treatment with 0.1 M NaOH for up to five cycles without an appreciable loss in activity [34].

12.8 Ecofriendly Nanomaterials from Agricultural Wastes

India generates about 350 million tons of agricultural waste every year. Its management without causing pollution is necessary. Nanotechnology-based technology is the latest approach to this. Recent research on silica has focused on nanoparticles. Silica nanoparticles can be prepared from either organic chemicals or biomass. To prepare silica nanoparticles, most investigations use organic precursors of alkoxysilane such as tetraethylorthosilicate (TEOS) and tetramethylorthosilicate. Nonetheless, silica from natural resources is used in biomedical and materials fields due to its low cost, eco-friendliness, and availability. Silica can be successfully extracted from biomass such as sugarcane bagasse, rice husk, corn cob, coffee husk, and wheat husk. Rice husk is an abundant silicon source and contains about 75–90% of cellulose, hemicellulose and lignin, totaling 17–20% of ash content. The ash contains more than 90% of silica and few metallic impurities. Sugarcane bagasse ash, a major by-product of the sugarcane industry, contains 40–50% of silica; acid pretreatment allows to increase amount of silica up to 80%. Corn cobs from maize corn contain more than 60% of silica. Wheat husk ash contains up to 90% silica, which is close to that of dry silica sand, of 99.4%. Bamboo leaf ash also has a large silica content of 75.90–82.86 wt.%. Raw teff straw contains about 52% of silica, which can be increased to 97% after thermal treatment. Examples of silica-rich biomass are shown in Figure 12.3.

Rice husk has been studied for energy and for the production of silica. For instance, rice husk ash is a precursor for silica gel synthesis by the sol–gel method. For optimal production of nano silica from rice husk, the effect of the acid leaching, concentration of sodium silicate solution, reaction temperature, and time of aging and gelation, pH should be maintained. As a consequence, rich husk ash shows promising applications for construction materials and technical ceramics due to the high reactivity of the porous structure. Highly pure silica nanoparticles with high specific area and an average size of 25 nm were prepared by alkali extraction, followed by acid precipitation. Research has also focused on "waste in valuable product manufacture" using silica, revealing good performance and simple industrial implementation as antisticking agents, filter in rubber products, and paper.

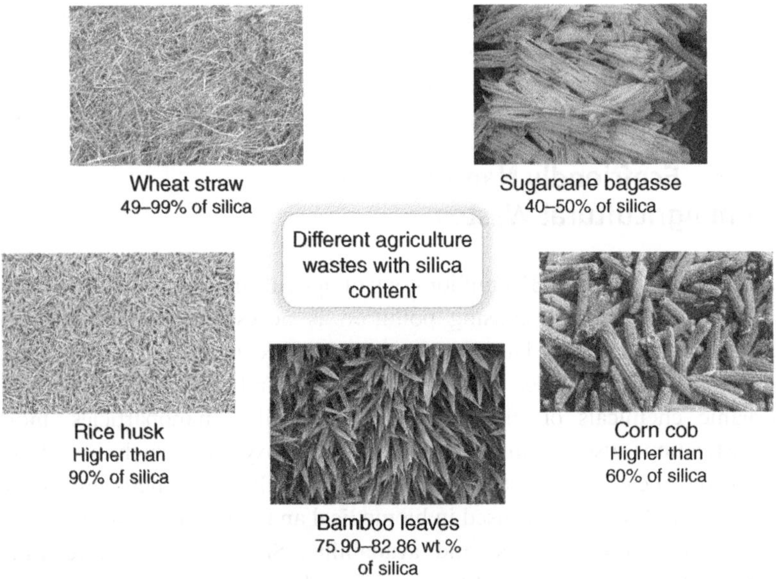

Figure 12.3 Showing different agriculture wastes with silica content.

Silica nanocomposite can be prepared with rice husk ash, which is used as a nanofiller in epoxy-silica nanocomposites; products display good tensile strength, uniform distribution, and no agglomeration.

Other than rice husk sugarcane baggage ash is also rich source of silica nanoparticle. Mesoporous silica nanoparticles prepared from sugarcane bagasse are applied in biomedical and industrial fields. Sugarcane bagasse is considered as a better option than wood fibers in producing textiles, paper, pressed wood materials, and other products.

Silica nanoparticles are used in many fields such as biomedical, electrical, textile, and rubber sectors, yet nowadays research is moving toward the biomedical field. Some applications are shown in Figure 12.4.

12.8.1 Ecofriendly Nanomaterials for Clean Water

Clean water is the basic necessity of human life without which there is no life on our planet. However, due to rapid urbanization and industrialization, it is very challenging to get clean drinking water in today's scenario. The variety of industrial liquid waste of hazardous nature and municipal solid waste is creating havoc and raised an alarming concern for us [35]. A number of solutions have been suggested and tried by scientific communities to tackle these kinds of wastes, and among these, the use of ecofriendly

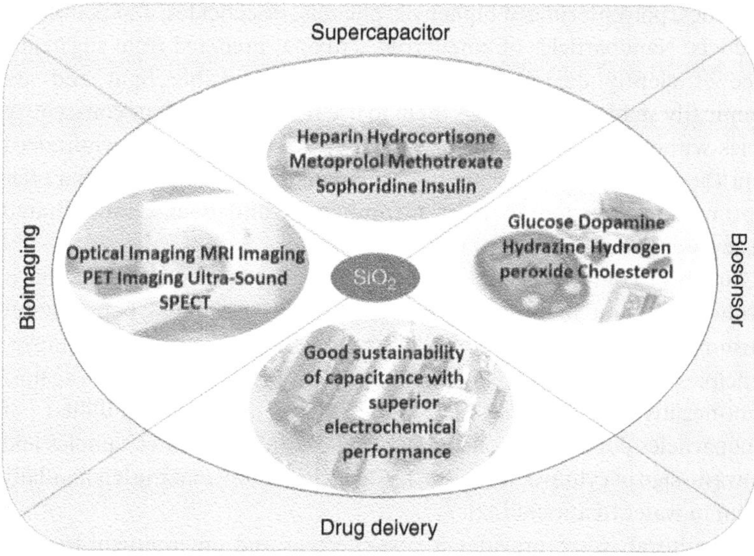

Figure 12.4 Showing different applications of silica nanoparticles derived from various agro wastes.

material is very popular as they do not cause any harm to our environment. Nanotechnology has paved an unprecedented way to solve the variety of global problems we face. From healthcare to energy to electronics to artificial intelligence to environment, nanotechnology has uses in every segment of our life. Nanotechnology is also a boon for us which not only solves the crisis of water pollution but also provides an opportunity to manage waste [36]. There is increasing use of nanomaterials to access clean water, which caters to the population's need daily. Metallic and nonmetallic nanomaterials of varying shape, size, and surface are used in bioremediation of toxic water waste faster and more efficiently. Nanoscale zerovalent iron is extensively used to treat arsenic-contaminated water pollution. The mechanism of arsenic removal by nanoscale zerovalent iron is based primarily on the adsorption technique wherein As(V) adsorbs the iron. Adsorption of arsenic onto iron-based nanomaterials is pH-dependent. At high pH, Fe_3O_4 is the dominant species of iron, which has strong adsorption affinity for As(V) and is supported by electrostatic interactions [37].

A recyclable chitosan-based magnetic nanocomposite material removes lead and cadmium from aqueous medium and could be used to treat metal-contaminated drinking water supply [38].

Photoactive nanomaterial is potentially useful material for treating polluted water contaminated by different organic pollutants such as dyes, pesticides, polychlorinated biphenyls, phenols, insecticides, and petroleum products. Nanoparticles of zinc ferrite, $ZnFe_2O_4$ prepared from sugarcane juice efficiently degrades organic dyes under visible light and are chemically stable which allows them to use repeatedly for four consecutive cycles without any loss of the activity [39]. Ag nanoparticles synthesized from *Duranta erecta* leaves' are used to prepare nanocomposites that offer a promising solution to treat nitrophenols and dyes contaminated wastewater [40].

Growth of coliform bacterial colonies in water is very common and is another pollutant of prime concern that needs treatment before water consumption [41]. CuO nanoparticles synthesized using green chemistry principles by algae are safe to use and have antibacterial properties against gram-negative *Escherichia coli.* Fifty micrograms per milliliter of nanoparticles effectively treats *Escherichia coli*-infected water samples and shows no sign of cytotoxicity below 2.5% concentration, making it potentially useful in water treatment [42].

Agricultural waste provides a cost-effective and environment-friendly way to synthesize multifunctional nanomaterial. For example, nanomaterials of SiO_2, TiO_2, graphene, carbon quantum dots, and cellulose are prepared from agricultural and plant wastes. Various kinds of popular biowastes are barley grain waste, rice husk, wheat husk, green tea waste, fruit peel agro waste, fruit pulp waste, sugarcane bagasse, leaf extracts, wheat straw, and bamboo [43]. Ecofriendly nanomaterials display superior adsorption [44] and catalytic properties and remove heavy metal ions, dissolved solids, organic pollutants, and coliforms from water. Green carbon quantum dots derived from guava leaf extracts show excellent reduction of Congo red and bromophenol blue [45]. Waste bamboo-derived nanofiber composite membrane shows excellent underwater oil adhesion capability [46], superb stability after 10 separation cycles, and high mechanical strength [47].

Use of recycled plastic waste to create nanomaterials useful for water treatment is becoming popular nowadays [48]. It is a sustainable approach to managing waste and is the need of the hour, as plastic waste is a huge burden in our society [49]. Functionalized amidoxime membrane made up of recyclable plastic with abundant nano-holes and nanostructured particles removes 100% uranium from the [50] Uranium, a radioactive waste tailings water. It provides an opportunity to treat plastic waste as well as radioactive waste-contaminated water through a unified approach [51].

So, wastewater treatment is an important area of research that needs immediate attention as the water quality is directly related to human health and environment. Wastewater from major sectors like industries, agriculture, and domestic is diversified and needs proper treatment before its discharge into the environment. Untreated wastewater contains large amount of toxic and nontoxic dissolved solids that, if not properly treated, enter into fresh water and endanger aquatic life. Considering all adverse impacts of untreated wastewater on our environment, it needs proper treatment. Much effort is being made continuously in this area; among them, the most advanced is the use of nanotechnology to solve the problem associated with wastewater. Use of nanomaterials is nowadays increasing in all areas of research. Nanomaterials exhibit unique physical and chemical properties, making them more attractive in their applications. As the use of nanomaterial increases with time, the question of how safe are their use also arises in mind of the researcher. So, despite showing wonderful applications, all nanomaterial are unsafe, and some are potentially toxic. Herein we focus on the applications of eco-friendly nanomaterials for wastewater treatment.

12.8.2 Clay-Based Material are Also Used for Wastewater Treatment

Large amount of suspended organic solids present in sewage water is a big problem in metropolitan and rural areas, which clogs the filtering unit of the sewage treatment plant. Such suspended solids have good stability and remain suspended in water due to the repulsive forces between similar charges on the suspended solids. Most of the time, the coagulant or flocculants are added into such water, reducing the repulsive forces between suspended particles and helping in their faster aggregation. Aggregated solids settle at the bottom, which does not allow the harmful organic solids to pass on further to the filtering unit. Nanocomposites derived from clay and polymers are green material that improves water quality by removing these suspended organic solids. They also reduce water's turbidity and impart aesthetic properties to water, improving the taste and quality. Nanocomposites of clay are made by dispersing it into the polymeric matrix. Use of natural polymers such as starch, cellulose, *M. olifera*, chitosan, are popular as they are biodegradable and not hazardous; however, their shelf-life is less. Mixing natural polymer with robust clay material improves their strength, life, and the adsorption capacity. Nanocomposite of clay and cellulose can be made by carrying out their reaction in aqueous medium in the presence of NaOH/Urea and cross-linking the two components. The result is the formation of a material that has excellent adsorption properties for methylene blue. It was observed that when nanocomposite concentration is 100 mg/l, 98% removal of methylene blue occurs. A novel cost-effective water disinfectant clay nanocomposite was prepared by mixing with papaya seeds and dopants (salts of zinc and copper) by solvothermal method, which, upon photo-irradiation, kills *E. coli* and also inhibits any further growth [52]. The nanocomposite carries a positive charge on the surface and electrostatically attracts the negatively charged *E. coli* stain under the given reaction conditions. The disinfection process continues further as metal ions are cytotoxic for the bacteria. Also, during photo-irradiation, reactive singlet oxygen species are produced, inhibiting the bacteria's growth.

12.9 Conclusion

The nanoparticles obtained from plant materials nowadays find versatile applications in the field of sciences because of the ecofriendly materials used during the process of the nanomaterials. Nanomaterials of silver, titanium dioxide, zinc oxide, etc., can be easily obtained from the green method of synthesis through plant materials. There are nanomaterial that are obtained from agricultural wastes, such as silicon oxide, which find many applications in biosensors and medicinal chemistry. Nanomaterials such as zinc oxide and titanium dioxide, which absorb pollutant present in the water, are used for water treatment. Nanoparticles derived from agricultural waste, cellulose-based composites, and clay-containing polymer nanocomposites are some of the highly demanding products that have shown great help in the bioremediation of the environment by helping us to provide clean water for mankind. Progress is being seen in modifying the naturally obtained plant materials for the formation of nanocomposites so that it can be made available for clean water technology.

Financial Support

This work is supported by the UP Govt Research and Development Project through Letter no 47/2021/606/Seventy-4-2021-4 (56)/2020, Dated: (30 March 2021).

Abbreviations

CNT	Carbon nanotubes
GG	Guar gum
Go	Graphene oxide
MOF	Metal organic framework
NMs	Nanomaterials

NPs Nanoparticles
RGO Reduced graphene oxide
TCN Titanium dioxide cerium oxide nanocomposites
ZIF Zeolitic imidazole framework

References

1 Mofijur, M., Siddiki, S.Y., Shuvho, M.B. et al. (2021). Effect of nanocatalysts on the transesterification reaction of first, second and third generation biodiesel sources – a mini-review. *Chemosphere* 270: 128642.

2 Portela, C.M., Vidyasagar, A., Krödel, S. et al. (2020). Extreme mechanical resilience of self-assembled nanolabyrinthine materials. *Proc. Natl Acad. Sci.* 117 (11): 5686–5693.

3 Sadri, R., Hosseini, M., Kazi, S.N. et al. (2018). A facile, bio-based, novel approach for synthesis of covalently functionalized graphene nanoplatelet nano-coolants toward improved thermo-physical and heat transfer properties. *J. Colloid Interface Sci.* 509: 140–152.

4 Dimkpa, C.O. and Bindraban, P.S. (2017). Nanofertilizers: new products for the industry? *J. Agric. Food. Chem.* 66 (26): 6462–6473.

5 Pelaz, B., Alexiou, C., Alvarez-Puebla, R.A. et al. (2017). Diverse applications of nanomedicine. *ACS Nano* 11 (3): 2313–2381.

6 Sarno, M. (2020). Nanotechnology in energy storage: the supercapacitors. In: *Studies in Surface Science and Catalysis*, vol. 179, 431–458. Elsevier.

7 Deshmukh, S.P., Patil, S.M., Mullani, S.B., and Delekar, S.D. (2019). Silver nanoparticles as an effective disinfectant: a review. *Mater. Sci. Eng., C* 97: 954–965.

8 Ziental, D., Czarczynska-Goslinska, B., Mlynarczyk, D.T. et al. (2020). Titanium dioxide nanoparticles: prospects and applications in medicine. *Nanomaterials* 10 (2): 387.

9 Das, M. and Chatterjee, S. (2019). Green synthesis of metal/metal oxide nanoparticles toward biomedical applications: boon or bane. In: *Green Synthesis, Characterization and Applications of Nanoparticles* (ed. A.K. Shukla and S. Iravani), 265–301. Elsevier.

10 Patete, J.M., Peng, X., Koenigsmann, C. et al. (2011). Viable methodologies for the synthesis of high-quality nanostructures. *Green Chem.* 13 (3): 482–519.

11 Varma, R.S. (1999). Solvent-free organic syntheses using supported reagents and microwave irradiation. *Green Chem.* 1 (1): 43–55.

12 Marslin, G., Siram, K., Maqbool, Q. et al. (2018). Secondary metabolites in the green synthesis of metallic nanoparticles. *Materials* 11 (6): 940.

13 Liu, M., Wang, Z., Zong, S. et al. (2014). SERS detection and removal of mercury (II)/silver (I) using oligonucleotide-functionalized core/shell magnetic silica sphere@ Au nanoparticles. *ACS Appl. Mater. Interfaces* 6 (10): 7371–7379.

14 Schrick, B., Hydutsky, B.W., Blough, J.L., and Mallouk, T.E. (2004). Delivery vehicles for zerovalent metal nanoparticles in soil and groundwater. *Chem. Mater.* 16 (11): 2187–2193.

15 Kratochvil, D. and Volesky, B. (1998). Advances in the biosorption of heavy metals. *Trends Biotechnol.* 16 (7): 291–300.

16 Li, Y.Y. and Li, B. (2011). Study on fungi-bacteria consortium bioremediation of petroleum contaminated mangrove sediments amended with mixed biosurfactants. *Adv. Mater. Res.* 183: 1163–1167.

17 Horikoshi, S.A. and Serpone, N.I. (2013). Introduction to nanoparticles. In: *Microwaves in Nanoparticle Synthesis: Fundamentals and Applications* (ed. S.A. Horikoshi and N.I. Serpone), 1–24. Wiley.

18 Ghosh, B. and Giri, T.K. (2018). Recent advances of chitosan nanoparticles as a carrier for delivery of antimicrobial drugs. In: *Polysaccharide-Based Nano-Biocarrier in Drug Delivery* (ed. T.K. Giri and B. Ghosh), 63–79. CRC Press.

19 Sohail, M.I., Waris, A.A., Ayub, M.A. et al. (2019). Environmental application of nanomaterials: a promise to sustainable future. In: *Comprehensive Analytical Chemistry*, vol. 87, 1–54. Elsevier.

20 Halada, G.P. and Orlov, A. (2018). Environmental degradation of engineered nanomaterials: Impact on materials design and use. In: *Handbook of Environmental Degradation of Materials* (ed. M. Kutz), 225–239. William Andrew Publishing.

21 Shi, Y.C., Wang, A.J., Wu, X.L. et al. (2016). Green-assembly of three-dimensional porous graphene hydrogels for efficient removal of organic dyes. *J. Colloid Interface Sci.* 484: 254–262.

22 Upadhyay, R.K., Soin, N., Bhattacharya, G. et al. (2015). Grape extract assisted green synthesis of reduced graphene oxide for water treatment application. *Mater. Lett.* 160: 355–358.

23 Rocha, L.S., Nogueira, J., Daniel-da-Silva, A.L. et al. (2021). Water softening using graphene oxide/biopolymer hybrid nanomaterials. *J. Environ. Chem. Eng.* 9 (1): 105045.

24 Dai, X., Li, X., Zhang, M. et al. (2018). Zeolitic imidazole framework/graphene oxide hybrid functionalized poly (lactic acid) electrospun membranes: a promising environmentally friendly water treatment material. *ACS Omega* 3 (6): 6860–6866.

25 Gnanasekaran, L., Rajendran, S., Priya, A.K. et al. (2021). Photocatalytic degradation of 2, 4-dichlorophenol using bio-green assisted TiO_2–CeO_2 nanocomposite system. *Environ. Res.* 195: 110852.

26 Soltani, R., Pelalak, R., Pishnamazi, M. et al. (2021). A novel and facile green synthesis method to prepare LDH/MOF nanocomposite for removal of Cd(II) and Pb(II). *Sci. Rep.* 11 (1): 1609.

27 Velu, M., Balasubramanian, B., Velmurugan, P. et al. (2021). Fabrication of nanocomposites mediated from aluminium nanoparticles/*Moringa oleifera* gum activated carbon for effective photocatalytic removal of nitrate and phosphate in aqueous solution. *J. Cleaner Prod.* 281: 124553.

28 Pathania, D., Katwal, R., Sharma, G. et al. (2016). Novel guar gum/Al_2O_3 nanocomposite as an effective photocatalyst for the degradation of malachite green dye. *Int. J. Biol. Macromol.* 87: 366–374.

29 Pal, S., Patra, A.S., Ghorai, S. et al. (2015). Modified guar gum/SiO_2: development and application of a novel hybrid nanocomposite as a flocculant for the treatment of wastewater. *Environ. Sci. Water Res. Technol.* 1 (1): 84–95.

30 Sahoo, J.K., Aniket, K., Juhi, R. et al. (2017). Guar gum-coated iron oxide nanocomposite as an efficient adsorbent for Congo red dye. *Desalin. Water Treat.* 95: 342–354.

31 Hussain, D., Khan, S.A., and Khan, T.A. (2021). Fabrication and characterization of mesoporous guar gum/$NiWO_4$ nanocomposite for efficient adsorption of phloxine B and crystal violet from aqueous solution and evaluation of its antioxidant activity. *Colloid Interface Sci. Commun.* 44: 100488.

32 Elella, M.H., Goda, E.S., Gamal, H. et al. (2021). Green antimicrobial adsorbent containing grafted xanthan gum/SiO_2 nanocomposites for malachite green dye. *Int. J. Biol. Macromol.* 191: 385–395.

33 Mirza, A. and Ahmad, R. (2018). Novel recyclable (xanthan gum/ montmorillonite) bionanocomposite for the removal of Pb (II) from synthetic and industrial wastewater. *Environ. Technol. Innovation* 11: 241–252.

34 Ahmad, R. and Mirza, A. (2017). Green synthesis of xanthan gum/ methionine-bentonite nanocomposite for sequestering toxic anionic dye. *Surf. Interfaces* 8: 65–72.

35 Bloem, E., Albihn, A., Elving, J. et al. (2017). Contamination of organic nutrient sources with potentially toxic elements, antibiotics and pathogen microorganisms in relation to P fertilizer potential and treatment options for the production of sustainable fertilizers: a review. *Sci. Total Environ.* 607: 225–242.

36 Deshpande, B.D., Agrawal, P.S., Yenkie, M.K., and Dhoble, S.J. (2020). Prospective of nanotechnology in degradation of waste water: a new challenges. *Nano-Struct. Nano-Objects* 22: 100442.

37 Wu, C., Tu, J., Liu, W. et al. (2017). The double influence mechanism of pH on arsenic removal by nano zero valent iron: electrostatic interactions and the corrosion of Fe 0. *Environ. Sci.: Nano* 4 (7): 1544–1552.

38 Li, B., Zhou, F., Huang, K. et al. (2017). Environmentally friendly chitosan/PEI-grafted magnetic gelatin for the highly effective removal of heavy metals from drinking water. *Sci. Rep.* 7 (1): 1–9.

39 Patil, S.B., Naik, H.B., Nagaraju, G. et al. (2018). Sugarcane juice mediated eco-friendly synthesis of visible light active zinc ferrite nanoparticles: application to degradation of mixed dyes and antibacterial activities. *Mater. Chem. Phys.* 212: 351–362.

40 Albukhari, S.M., Ismail, M., Akhtar, K., and Danish, E.Y. (2019). Catalytic reduction of nitrophenols and dyes using silver nanoparticles@ cellulose polymer paper for the resolution of waste water treatment challenges. *Colloids Surf., A* 577: 548–561.

41 Fung FM, Su M, Feng HT, Li SF. Extraction, separation and characterization of endotoxins in water samples using solid phase extraction and capillary electrophoresis-laser induced fluorescence. *Sci. Rep.* 2017 7(1):1-0.

42 Bhattacharya, P., Swarnakar, S., Ghosh, S. et al. (2019). Disinfection of drinking water via algae mediated green synthesized copper oxide nanoparticles and its toxicity evaluation. *J. Environ. Chem. Eng.* 7 (1): 102867.

43 Singh, S., Ramakrishna, S., and Gupta, M.K. (2017). Towards zero waste manufacturing: a multidisciplinary review. *J. Cleaner Prod.* 168: 1230–1243.

44 Bhatia, S.K., Mehariya, S., Bhatia, R.K. et al. (2021). Wastewater based microalgal biorefinery for bioenergy production: progress and challenges. *Sci. Total Environ.* 751: 141599.

45 Velmurugan, P., Kumar, R.V., Sivakumar, S., and Ravi, A.V. (2022). Fabrication of blue fluorescent carbon quantum dots using green carbon precursor *Psidium guajava* leaf extract and its application in water treatment. *Carbon Lett.* 32 (1): 119–129.

46 Nasrollahzadeh, M., Sajjadi, M., Iravani, S., and Varma, R.S. (2021). Green-synthesized nanocatalysts and nanomaterials for water treatment: current challenges and future perspectives. *J. Hazard. Mater.* 401: 123401.

47 Xiang, B., Shi, G., Mu, P., and Li, J. (2022). Eco-friendly WBF/PAN nanofiber composite membrane for efficient separation various surfactant-stabilized oil-in-water emulsions. *Colloids Surf., A* 645: 128917.

48 Jovanov, D., Vujić, B., and Vujić, G. (2018). Optimization of the monitoring of landfill gas and leachate in closed methanogenic landfills. *J. Environ. Manage.* 216: 32–40.

49 Jodar-Abellan, A., López-Ortiz, M.I., and Melgarejo-Moreno, J. (2019). Wastewater treatment and water reuse in Spain. Current situation and perspectives. *Water* 11 (8): 1551.

50 Lu, F. and Astruc, D. (2020). Nanocatalysts and other nanomaterials for water remediation from organic pollutants. *Coord. Chem. Rev.* 408: 213180.

51 Fu, M., Ao, J., Ma, L. et al. (2022). Uranium removal from waste water of the tailings with functional recycled plastic membrane. *Sep. Purif. Technol.* 287: 120572.

52 Ugwuja, C.G., Adelowo, O.O., Ogunlaja, A. et al. (2019). Visible-light-mediated photodynamic water disinfection@ bimetallic-doped hybrid clay nanocomposites. *ACS Appl. Mater. Interfaces* 11 (28): 25483–25494.

13

Bio-nanomaterials from Agricultural Waste and Its Applications

Shaily, Adnan Shahzaib, Fahmina Zafar, and Nahid Nishat**

Inorganic Materials Research Lab, Department of Chemistry, Jamia Millia Islamia, New Delhi, DL, India

Abstract

Agricultural bio-waste has prompted a lot of study into finding methods to utilize the resources as substitutes for synthetic materials due to its high strength, availability, renewability, low cost, nontoxicity, and biodegradable nature. Bio-nanomaterials receive a lot of interest due to their size, shape, and surface, and they have better physical and mechanical characteristics, which allow them to be used in a variety of fields such as aeronautical engineering, pharmacology, biosensors, and material science. With increased similar properties, seed oils, lignin, lipids, protein, starch, polysaccharide, polyesters, and cashew nut shell liquid (CNSL) vegetable/fruit waste are bio-based raw materials that are suitable for 3-D cross-linked thermally stable structures. Many possible uses for bio-nanomaterials have been investigated, including packaging, adhesives, coatings, automotive, and many more. The objective of this chapter is to provide a brief overview of the development of bio-nanomaterials from agricultural wastes and their various applications in adhesives and coating industries.

Keywords *CNSL; Nano-composite; Bio-waste; Coating; Lignin; Green route*

*Corresponding Authors: fahmzafar@gmail.com; nishat_nchem08@yahoo.com

Green Synthesis of Nanomaterials: Biological and Environmental Applications, First Edition.
Edited by Archana Chakravarty, Preeti Singh, Saiqa Ikram, and R.N. Yadava.

13.1 Introduction

Due to the hazardous environmental effects, petro-products need to be replaced by agricultural products. Rapid urbanization created a strong demand for agricultural products, as well as new opportunities for agro-products in the sector in response to disposal challenges. Agricultural waste-based biomass is the most prevalent type of biomass in nature. The term "agricultural biomass" refers to all organic waste products created as a result of agricultural production. Every year, especially in developing nations, enormous volumes of biomass are produced as byproducts of collecting and processing agricultural materials. Due to its high strength, availability, and sustainability, agricultural waste has a major potential in composites. A lot of study has been done to figure out how to use agricultural waste as a replacement for synthetic/petro materials because of its potential. The substitution of petro-derived materials for bio-based agricultural products is an adequate option for generating alternative eco-friendly materials due to their renewability, plentiful availability, low cost, nontoxic, and biodegradable nature. They are widely used in both traditional and high-tech industries, including adhesives, inks, lubricants, packaging materials, paints, and films/coatings [1, 2]. A falling availability of raw materials is a reason to worry, and in this light, biomaterials are viewed as a good alternative resource to make value-added goods such as biocomposites. Agriculture waste may be derived from fruits/vegetable peels, some raw nut waste like cashew nut shell, lipid, and lignin [3]. As a feasible alternative for the synthesis, the exceptionally rapid development of nanomaterials from agricultural waste and the usage of bio-nanomaterials has garnered a lot of attention.

Nanomaterials received a lot of interest due to their size, shape, and surface, and they have better physical and mechanical characteristics. Several greener approaches produced polymeric nano ranges particles to lower the cost-effectiveness, biodegradability, and plentiful availability in nanomaterials with low toxicity. Vegetable waste, lignin, and cashew nut shell liquid (CNSL) are the ideal precursors for synthesizing polymeric nanomaterials. The employment of such green-synthesized agricultural waste nanoparticles in environmental remediation applications, as well as the usage of recyclable nanoparticles, encourages to achieve environmentally beneficial results. Many attempts to produce bio-nanomaterials have shown that the potential of nanoparticles for filler/reinforcement in polymer composites is considered as extremely promising because bio nanoparticles have splendid and complicated structures that are vital in understanding their chemical uses. It is regarded as a novel technology that has the potential to considerably enhance, if not completely change, a wide range of fields and industries.

The chapter includes a brief review of the development and implementation of new, bio-nanomaterials from agricultural waste resources appropriate for three-dimensional cross-linked thermally stable structures in association with their usage.

13.2 Lignin

Organic compounds derived from plants or animals included fatty acids and their derivatives, such as tri- and di-monoglycerides/phospholipids, are called lignin. Lignins are generally soluble in methanol or chloroform but insoluble in water [4]. Hardwoods, softwoods, bagasse, and sugarcanes all produce lignin (Figure 13.1). Lignins produce a lot of energy with their chemical makeup of mostly carbon, oxygen, and hydrogen. p-coumaryl, coniferyl, and sinapyl, alcohol are the propyl phenol units that are used to make lignin, along with a small amount of other propyl phenols (Figure 13.2a) [5, 6].

Many efforts have been made to create products using biodegradable polymers including poly(lactide), poly(-caprolactone), and poly[(R)-3-hydroxybutyrate] [8, 9]. Considered to have the potential to displace fossil

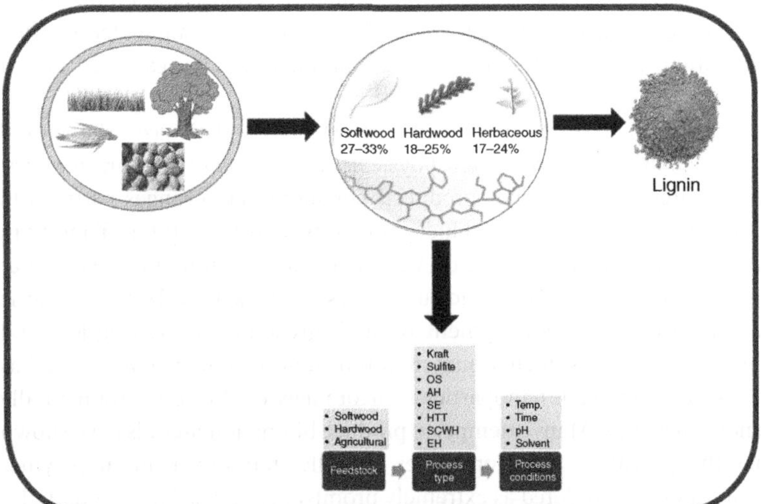

Figure 13.1 Representation of extracting lignin.

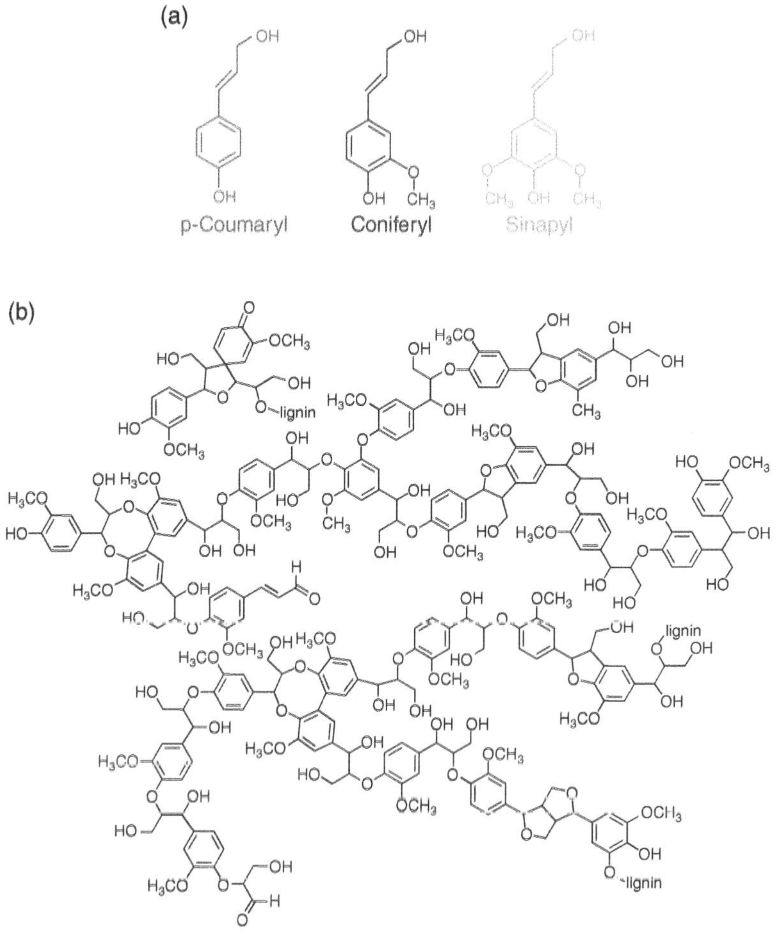

Figure 13.2 Representation of (a) lignin precursors and (b) softwood lignin structure. *Source:* Reprinted with permission from Khan *et al.* [7]. Copyright (2019), Elsevier.

fuels, lignin is a naturally plentiful and renewable resource biomass. It is the second most abundant biomass on earth, right after cellulose, heterogeneous, and amorphous, which makes up a significant component of the cell walls of vascular plants [10, 11]. To increase or improve the material characteristics, raw lignin has been added directly into the matrix without any chemical alteration. Raw lignin can be used as fillers, UV blockers, flame retardants, antioxidants, or reinforcing agents in plastics [6]. For the linkages of lignins, research has shown that oligomers or monomers frequently create lignin linkages. The process of polymerization produced around half of the bonds. There are some other significant connections which are more harder to degrade and frequently become progressively less susceptible to degradation throughout conventional manufacturing procedures as a result of significant radical and carbon-carbon cross-linking. Some of the common connections in lignin are shown in Figure 13.2b [7].

13.2.1 Lignin Nanocomposites (NCs)

Lignin polymers with nanoparticles improve polymer's physical and mechanical characteristics, known as lignin-based polymer NCs. Several studies have described various lignin-based NC synthesis in various applications. For instance, the incorporation of inorganic nano-scale clay particles improved the flame-retardant qualities of organic polymers like nylon-6 and raised the modulus. This is important for applications like vehicles. Evenly distributing the nano-clay is necessary for optimal performance (foliation) [12]. Many studies have been done on the synthesis of lignin-based NCs. Enrico *et al.* created robust bio-based polyurethane (PU) foams, and bio-based lignin NCs that were polymerized were used as a polyol [13]. Zang *et al.* synthesized lignin NC films with high performance, good water vapor barrier, and UV protection [14].

13.2.2 Lignin-Based Catalysts and Photocatalyst

Lignin has garnered a lot of interest as catalyst and photocatalyst by using bio-nanomaterials derived from green, inexpensive lignocellulose biomasses [15, 16]. Quin *et al.* synthesized graphitic carbon-encapsulated iron nanoparticles on the basis of lignin, serve as an efficient catalyst of lower olefins [17]. Srisawimon *et al.* reported photocatalysts using TiO_2/lignin-based carbon composites for improved photocatalytic conversion of lignin to high-value chemicals [18]. In order to create a variety of biomass empty fruit bunch (EFB) supported zero-valent metal nanoparticles based on Cu, Co, Ni, and Ag NPs, Akhtar *et al.* used a very straightforward wet chemical method based on EFB feedstock waste as a very resilient and comprehensive support material [19] (Figure 13.3a and b). Research on the reduction/degradation of 4-nitrophenol (4-NP), anionic, and cationic dyes from aqueous solutions using $NaBH_4$ proved that among the as-synthesized MNPs/EFB catalysts, the catalytic degradation efficiency of catalyst approached a balance (99.5%, 95.1%, 82.8%, and 96.8%) after 4 (Figure 13.3c), 7, 14, 5, and 10 minutes respectively [19].

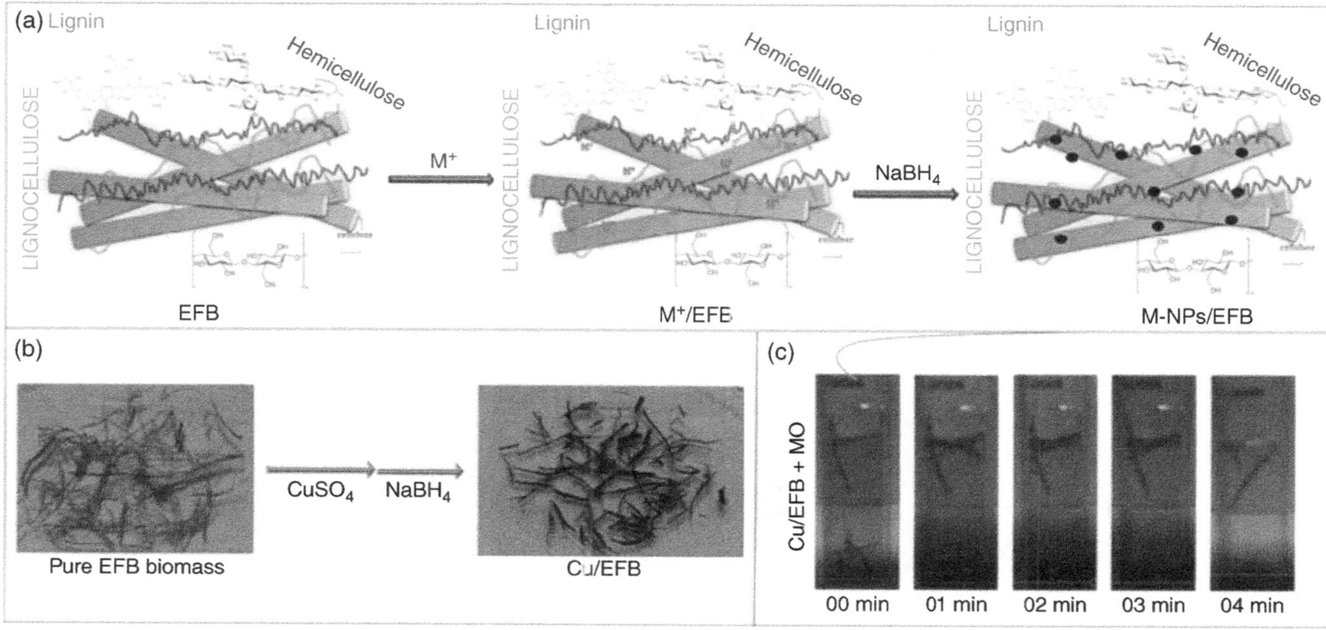

Figure 13.3 (a) Metal nanoparticles growth over the EFB bio- matrix (b) images of pure Cu/EFB and (c) reduction in the presence of Cu/EFB. *Source:* Sangon *et al.* [15]/Reproduced with permission from Elsevier.

13.2.3 Lignin-Based NC Coatings

Lignin offers enormous potential to create NC coatings with distinctive properties. Lignin-based films and coatings have drawn a lot of interest since they are elevated NC films with no hazardous environmental effects. Alcantara *et al.* successfully synthesized $CoFe_2O_4$ nanoparticles and polyelectrolyte solutions to produce ultra-thin NC films for electrochemical sensors. By immersing a solid substrate in a dispersion of nanoparticles on alternate occasions, the amount of $CoFe_2O_4$ was increased. Lü *et al.* produced well-sized dispersed lignin-poly (*N*-methylaniline) NCs (lignin-PNMA) using the chemical oxidative polymerization method [20, 21]. The enzyme hydrolyzed lignin and PMMA alone was less effective in adsorbing Ag^+ than the lignin-PNMA NCs. Silver nanoparticles were eventually created via a redox reaction and ion exchange between lignin-PNMA NCs and Ag^+, as illustrated in Figure 13.4, and the NCs had a saturation capacity of up to 1556.8 mg/g.

13.3 Cashew Nut Shell Liquid (CNSL)

The cashew plant *Anacardium occidentale* L. produces CNSL, an aromatic resource that is sustainable and renewable. It is generated in quantities of around 450,000 metric tons yearly. CNSL is regarded as a crucial precursor for commercial polymer products. Its nontoxic behavior, biodegradability, sustainability, cheap cost, plentiful availability worldwide, chemical reactivity, and other properties make it a significant precursor for a number of reasons [22]. Based on processing technique, there are basically two types of CNSL: (i) natural and (ii) technological. Due to the presence of phenolic chemicals, CNSL consists of the following four elements: cardol, 2-methyl cardol, anacardic acid, and cardanol [23] (Figure 13.5). Technical CNSL is made up mostly of cardol, 2-methyl cardol, and cardanol (62–63%) as opposed to natural CNSL, which contains a majority of Anacardic acid (64–65%). Cardanol has a synthetically versatile character since it contains hydroxyl groups (phenolic and lipidic). Additionally, the existence of a hydrophobic alkyl chain at the meta position with cis double bonds that are not isoprenoid qualifies it as functional. Because of these characteristics, cardanol is one of the best precursors for the creation of different polymeric coatings among all ingredients, and cardanol-based coatings are flexible and durable [1, 24]. The existence of an olefinic side chain and a functional hydroxyl group combined enhance the use of 3D cross-linked coatings in numerous applications. Auto oxidation with the assistance of air is possible when side chain double bonds are present.

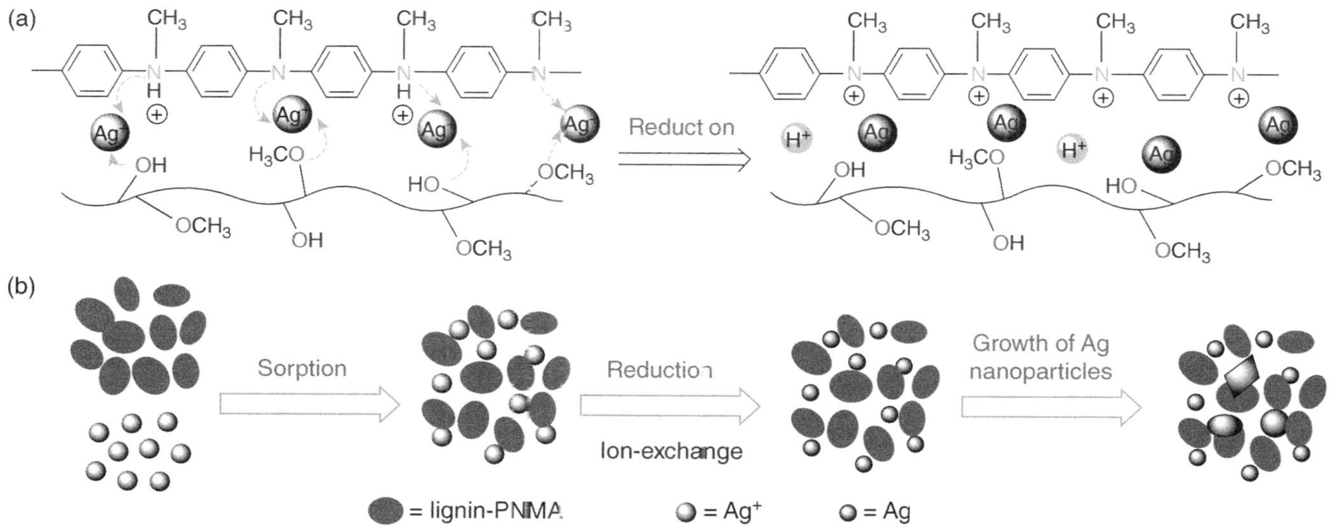

Figure 13.4 Ion-exchange mechanism showing (a) reduction and (b) growth process of nanoparticles. *Source:* Reproduce with permission from Lü *et al.* [21] Copyright 2014, American Chemical Society.

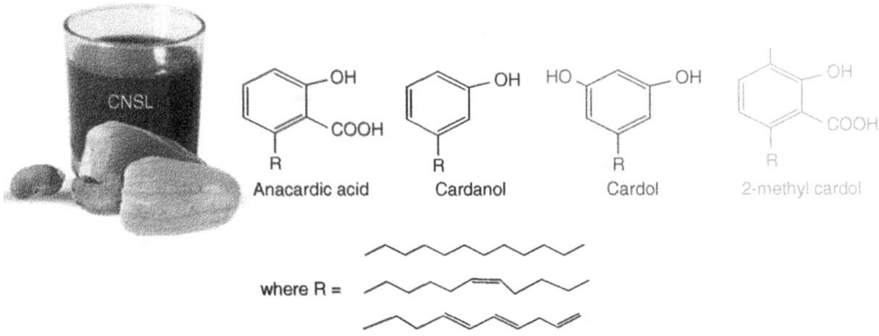

Figure 13.5 Constituents of CNSL.

13.3.1 CNSL NC-Based Surfactants

Surfactants are typically amphiphilic organic molecules that have both hydrophobic and hydrophilic groups. A surfactant, thus, consists of both a water-insoluble (or oil-soluble) component and a water-soluble component. Due to their ability to reduce a liquid's surface tension, the interfacial tension between two liquids, or the tension between a liquid and a solid, surfactants can function as detergents, wetting agents, emulsifiers, foaming agents, and dispersants. Due to a lengthy hydrophobic hydrocarbon chain and mildly polar phenolic ring, chemically unaltered cardanol naturally possesses surfactant-like characteristics. Cardanol-based surfactants are environmentally friendly with their structural characteristics for use as industrial and home surfactants compared to linear alkyl benzenes (LAB) surfactants made from petroleum fuels. Similar to the well-known liposome/membrane/lamellar stabilizer cholesterol [25]. The phenol ring of cardanol, which has been alkyl-substituted ($C_n = 15$), allows for substitution, creating a variety of amphiphilic cardanol derivatives that resemble synthetic counterparts. Tyman *et al.* have created cardanol-based polyethoxylate surfactants with remarkable biodegradability [26]. Cardanol's sulfonated derivative was created, and its surfactant characteristics were studied by Dantas *et al.* under ambient conditions, and 2,4-sodium disulfonate-5-n-pentadecylphenol spontaneously self-assembled into stable micellar aggregates [27].

Cardanol-based sulfate surfactant products not only enhance foam stability and boost the solubilization capacity of inorganic salt but also do not preserve alkylphenol ethers' permeability, high detergency, good compatibility, quick dissolving, and mild skin irritation [28]. As a result, the substance is widely employed in a variety of industries, including the textile, petroleum, and industrial cleaning industries. However, because these alkylphenol polyethenoxy ether sulfates might take a very long time to decompose in the environment, they will pollute the ecosystem. Hence, cardanol, a naturally occurring alkylphenol that may decay through biodegradation, is a suitable raw material. To replace alkylphenol polyethenoxy ether sulfates derived from petrochemical feedstocks, a novel solution was developed involving the synthesis of ammonium salt derivatives of cardanol polyethenoxy ether sulfates. The schematic depiction of mono-sulfonated cardanol surfactants is presented in Figure 13.6.

13.3.2 CNSL-Based NC Films/Coatings

CNSL-based eco-friendly green films/coatings have several industrial uses in resole [30], novolacs [31], epoxies [31], phenalkamine [32], benzoxazines [33], PUs [34], and in other polymer resins. Due to an aromatic ring, a long hydrophobic side chain, and a phenolic functional group, cardanol reacts similarly to phenol. Through various reactions, the combination of these functional groups gives it access to additional reaction sites and enhances its performance. Solvent solubility, inherent hydrophobicity, weather ability, chemical (acid and basic) resistance, film-forming capacity, and high adhesion are some of these properties. This meets CNSL requirements for coating performance.

Epoxy-group-containing thermoset polymers have improved properties including adhesion, heat resistance, and corrosion resistance (chemical and mechanical). The enhancement of these attributes is achieved primarily through coating treatments, and further optimization is accomplished by introducing additional substances, including monomers and polymers, into epoxy resins, resulting in the formation of a cross-linked curing structure. This is exemplified in various setups where epoxy curing agents like phenolic compounds, polyamides, and amines are utilized, representing just a subset of the available epoxy curing agents. To prevent steel corrosion, Ayman *et al.* created solvent-free epoxy based on CNSL as shown in Figure 13.7 [35]. Epoxidation produces epoxy resin from the hydroxyl group and the double bond of the long carbon chain. Optimized were the chemical, mechanical, salt spray (Figure 13.8) resistance and anticorrosive attributes.

13.3.3 CNSL-Based PU Coatings

CNSL can be used to prepare agro-PU coatings as an alternative approach to petro-PUs. There were various studies reported on CNSL-based PU coatings and their applications. Balgude *et al.* reported for a variety of end

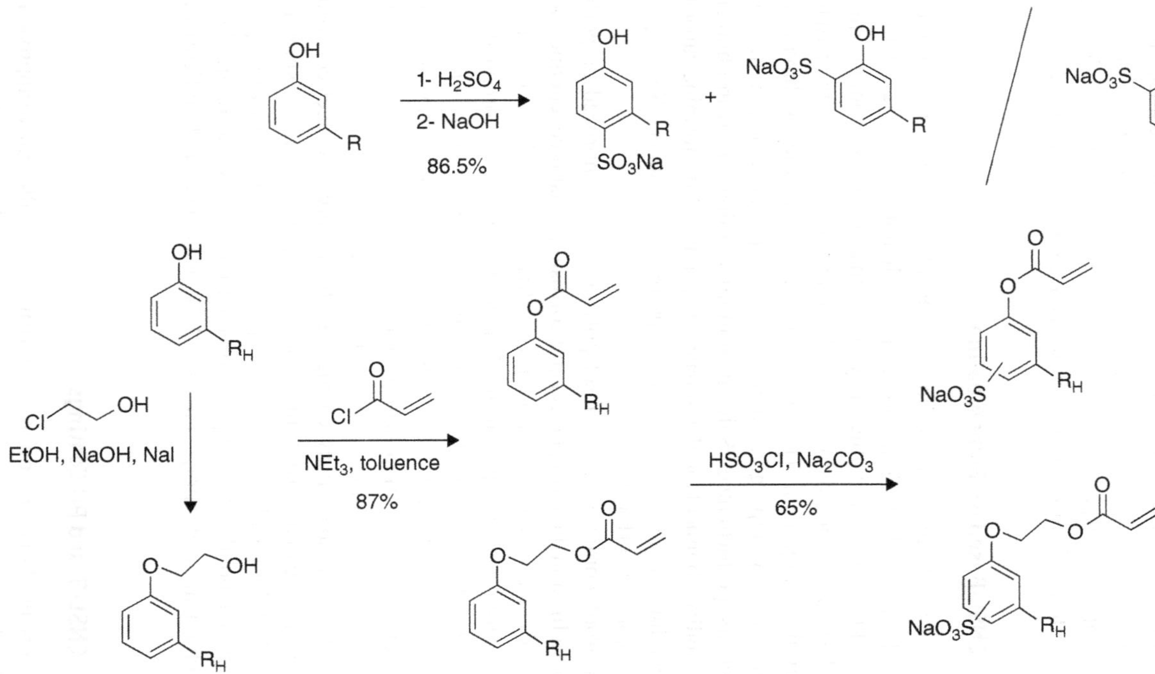

Figure 13.6 Surfactants based on monosulfonated cardanol. *Source:* Roy *et al.* [29]/MDPI/CC by 4.0.

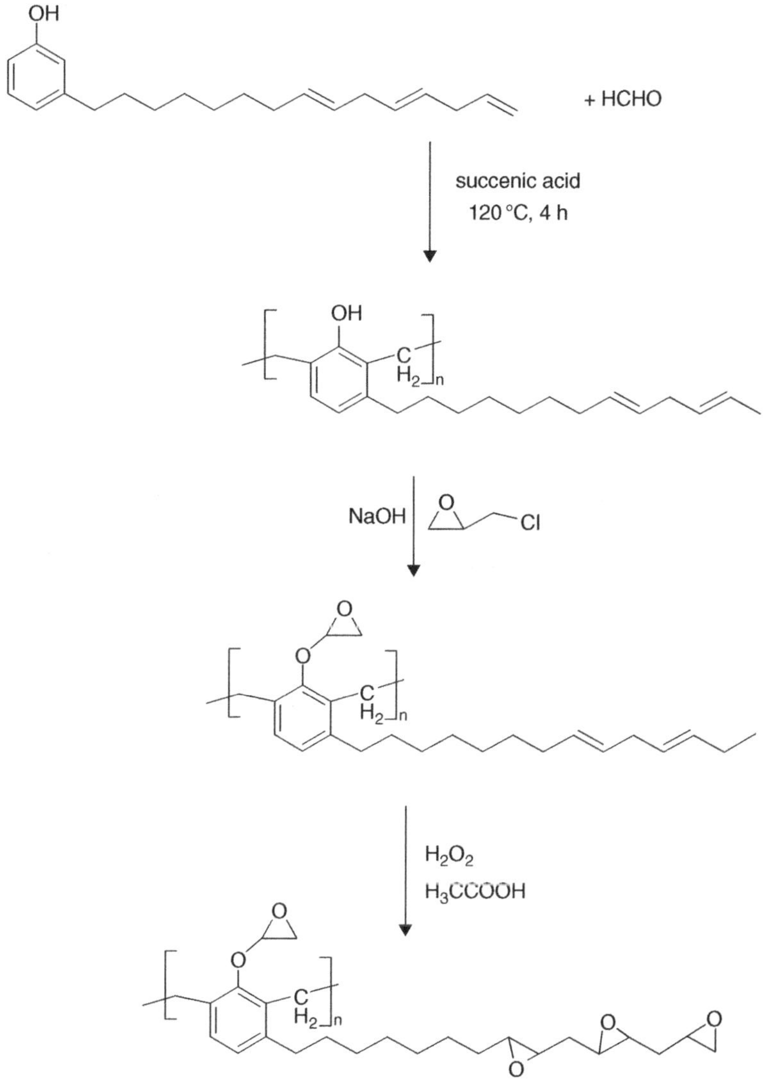

Figure 13.7 Synthesis of CNSL-based epoxy coatings. *Source:* Reproduce with permission from Atta *et al.* [35] Copyright 2017, Elsevier.

applications, including coatings, adhesives, and paints, commercially available aqueous 2K-PU coatings are created by reacting water-based polyols with water-based polyisocyanate in a stoichiometric quantity [36]. These petroleum-based raw materials are the primary source of these water-based polyols owing to rising worries about the environment and health

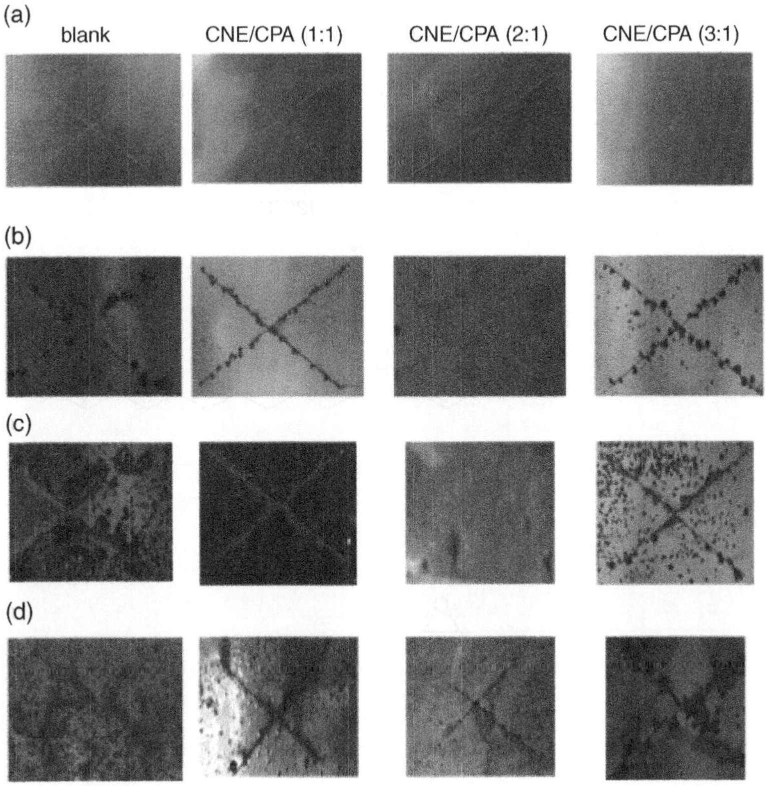

Figure 13.8 Salt spray on cured epoxy coatings with different ratios at different times (a) before exposure, (b) 21, (c) 90, (d) 180 days. *Source:* Atta *et al.* [35]/ Reproduced with permission from Elsevier.

caused by the usage of petroleum-based feedstock. Khan et al. *in situ* synthesized metal coordinated PU coatings [22]. They presented the sustainable, affordable, and environmentally friendly *in situ* synthesis of cardanol-based coordination PUs (CPUs) using cardanol formaldehyde, aliphatic diisocyanate, and transition metal ions [Mn(II), Co(II), Ni(II), and Zn(II)]. Figure 13.9 shows the scheme of the ambient cured coating. These superhydrophobic coatings (Figure 13.10) shows better antibacterial properties. Plates with ZOI values for various compositions have been evaluated for use as an antibacterial agent. It was examined and contrasted how an aliphatic diisocyanate and the usage of various transition metal ions affected the structure and ultimately characteristics. The prepared films or coatings shows advanced chemical and mechanical stability along with superhydrophobicity.

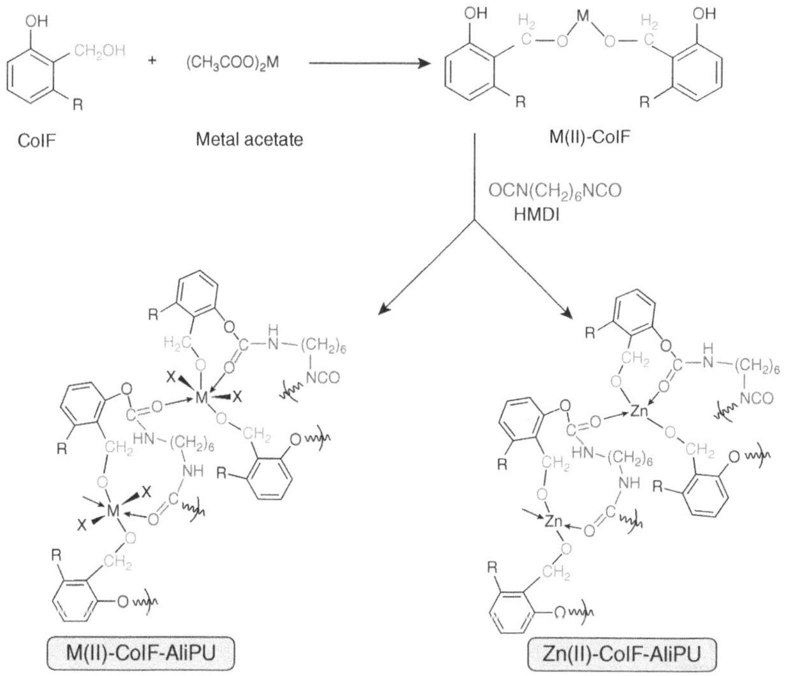

Figure 13.9 Scheme represents the metal cardanol formaldehyde-based coating. *Source:* Khan *et al.* [22]/with permission of Elsevier.

Figure 13.10 Contact angles of the coatings showing superhydrophobicity with contact angle of (a) 151° (b) 152° and (c) 153°. *Source:* Khan *et al.* [22]/Reproduced with permission from Elsevier.

Besides this, water born PUs are new class of PUs. Many researches have been published on cardanol-based water born PUs. Mestry *et al.* reported in the synthesis of waterborne PUs dispersions, a difunctional reactive dispersing agent generated from cardanol can be employed in place of dimethylol propionic acid (DMPA) [37]. After the cardanol was sulfonated with oil, the interaction between the hydroxyl group of phenol and the epichlorohydrin took place (ECH). In order to add double functionalities into the PU backbone by a chemical reaction with a diisocyanate, the resulting product was hydrolyzed after it had been produced [37].

13.4 Vegetable/Fruit Waste

Vegetable/fruit waste consider as a good precursor for bio-nanomaterials. Source reduction and recycling have shown encouraging outcomes to utilize this waste in various applications, having economics and environmental effects to be prioritized. Finding a viable strategy for efficiently exploiting fruit and vegetable waste to produce value-added goods that are more environmentally friendly, affordable, and sustainable. Table 13.1 shows waste conversion to energy with their challenges.

Table 13.1 Vegetable/fruit waste conversion to energy: Matrix and their challenges.

Waste-to-energy treatment method	Characteristics	Challenges
Landfills	Cheapest and most accessible forms of disposals	Emission of bad odor
	Has been considered more suitable for economically stable countries	Maintenance of ambient temperature profile during the process
	Considered as source of energy, when the methane produce used for energy production	
Composting	Water holding capacity and soil aggregate stability can be increased as waste consists of high organic content	Creates air, water, and soil pollution due to the emission of CO_2, and infiltration of the leachate may cause groundwater contamination
	Increase in the microbial diversity of the soil, by which they have an ability to suppress the activity of the pathogen	They produce bad odor and visual pollution
	The compost helps in the resilience of the plan-soil system	

Table 13.1 (Continued)

Waste-to-energy treatment method	Characteristics	Challenges
Anaerobic digestion	Production of biogas, which can be used for multiple applications	Presence of the heavy metal may result in digestate contamination
	Reduces the volume of waste, as compared to other treatment methods	Construction of the digestor is expensive
	Digestate can be used as the manure/fertilizer for agricultural uses	More power requirement for aeration and mixing
Enzymatic treatments	Helps in improving the overall efficiency of pH and temperatures should be controlled. the final product	pH and temperatures should be controlled
	Acts as a catalyst, aiding in the feasibility of the degradation process	Economic feasibility of enzymes should be considered

Source: Adapted from Ganesh *et al.* [38].

13.4.1 Vegetable/Fruit Waste-Induced Nanomaterials

Metallic NPs have special physical, chemical, and thermal characteristics that are used in electrocatalysis, sensing, and a variety of medical applications due to their surface area and tiny size (nano-level). Which result in increased features such as high reactivity, porosity, and cellular barriers. Both physical and chemical processes can be used to create NPs [38]. Metal NPs have been synthesized using a number of different processes, including microwave irradiation, ultrasound, laser ablation, gas phase condensation, chemical precipitation, spray pyrolysis, and solvo thermal breakdown [39]. As a result, due to their natural availability and commercial viability, fruit and vegetable waste or their composites have drawn more attention. The interplay of fruit and vegetable waste matrices for environmental restoration and pollutant removal from water. Figure 13.11 a and b shows the parameters to consider for fruit/vegetable waste as adsorbent [38].

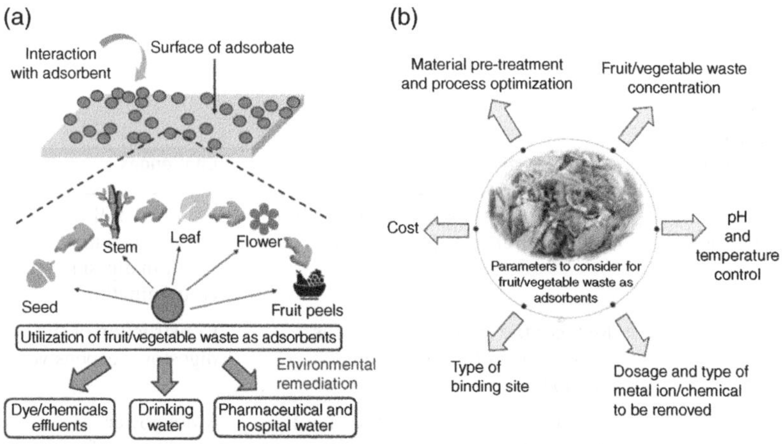

Figure 13.11 Vegetable/fruit waste matrices for environmental restoration and pollutant removal from water (a) and parameters to consider it as adsorbent (b). *Source:* Ganesh *et al.* [38]/Reproduced with permission from Elsevier.

13.4.2 Medicinal Activities of Vegetable/Fruit Waste

The extraction of NPs from various wastes has been the subject of several research to assess their antibacterial and antidiabetic capabilities. For instance, compared to bottle gourd peel, *Murraya koenigii* fibers, *Moringa oleifera* fibers, and *Luffa acutangula* waste peels, mandarin waste peels demonstrated the highest yield of 2.5 g/l [40]. The extraction from waste, which was used in the manufacture, was made by reducing Ag^+ to AgO by adding ammonium nitrate and used for medicinal purposes [41]. There are various methods for preparing vegetable/fruit waste NPs and their possible applications (Figure 13.12).

13.5 Conclusion

Bio-based agricultural waste materials, such as lignin, CNSL, vegetable/fruit waste, etc., are an easily accessible industrial product that is nonhazardous, biodegradable, and green. They are excellent precursors for the formation of 3D highly cross-linked structures. Similar to petro-based products, they have the same levels of microbiological resistance, mechanical resistance, anticorrosive properties, and many more. These properties make them suitable candidates for many applications such as

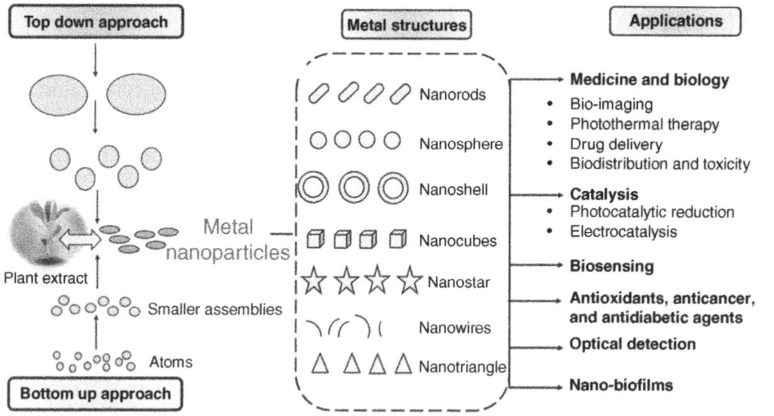

Figure 13.12 Various methods for NPs synthesis and their possible applications. *Source:* Ganesh et al. [38]/Reproduced with permission from Elsevier.

adhesives, catalyst, photocatalyst, sensors, binders, films/coatings, and food packaging and in many other applications. The materials' nano size enhances the material's properties due to its large surface area. Hence, bio-nanomaterials from agricultural waste have great potential as petro-based material alternatives with their advanced applications.

Acknowledgments

Shaily would like to acknowledge University Grants Commission, New Delhi, India, for a non-Net Fellowship. Dr. F. Zafar acknowledges, Women Scientist Scheme (WOS) for Research in Basic/Applied Science with reference DST/WOS-A/CS-83/2021 from DST, New Delhi, India. The author would also like to extend their gratitude to the Head, Dept. Of Chemistry, Jamia Millia Islamia, for providing essential facilities to carry out the work.

Abbreviation

CNSL	Cashew nutshell liquid,
LABs	Linear alkyl benzenes
NC	Nano composite
PUs	Polyurethanes

References

1 Khan, S., Masood, S., Siddiqui, K. et al. (2018). Utilization of renewable waste material for the sustainable development of thermally stable and biologically active aliphatic amine modified Cardanol (phenolic lipid) – formaldehyde free standing films. *J. Cleaner Prod.* 196: 1644–1656.

2 Zafar, F., Khan, S., Ghosal, A. et al. (2019). Clean synthesis and characterization of green nanostructured polymeric thin films from endogenous Mg(II) ions coordinated methylolated-cashew nutshell liquid. *J. Cleaner Prod.* 238: 117716.

3 Dungani, R., Abdul Khalil, H.P.S., NAS, A. et al. (2017). Bionanomaterial from agricultural waste and its application. In: *Cellulose-Reinforced Nanofibre Composites* (ed. M. Jawaid, S. Boufi, and H.P.S. Abdul Khalil), 45–88. Elsevier https://linkinghub.elsevier.com/retrieve/pii/B9780081009574000036.

4 Bell, A. and Wright, G.F. (1950). The extraction of birch lignins with acetic acid 1. *JACS* 72 (4): 1495–1499. https://pubs.acs.org/doi/abs/10.1021/ja01160a020.

5 Sannigrahi, P., Pu, Y., and Ragauskas, A. (2010). Cellulosic biorefineries—unleashing lignin opportunities. *Curr. Opin. Environ. Sustainability* 2 (5–6): 383–393. https://linkinghub.elsevier.com/retrieve/pii/S1877343510000953.

6 Kai, D., Tan, M.J., Chee, P.L. et al. (2016). Towards lignin-based functional materials in a sustainable world. *Green Chem.* 18 (5): 1175–1200. http://xlink.rsc.org/?DOI=C5GC02616D.

7 Khan, T.A., Lee, J.-H., and Kim, H.-J. (2019). Lignin-based adhesives and coatings. In: *Lignocellulose for Future Bioeconomy* (ed. H. Ariffin, S.M. Sapuan, and M.A. Hassan), 153–206. Elsevier https://linkinghub.elsevier.com/retrieve/pii/B9780128163542000098.

8 Zhu, X. and Jun Loh, X. (2015). Layer-by-layer assemblies for antibacterial applications. *Biomater. Sci.* 3 (12): 1505–1518. http://xlink.rsc.org/?DOI=C5BM00307E.

9 Zhang, D., Li, G., and Yu, J.C. (2010). Inorganic materials for photocatalytic water disinfection. *J. Mater. Chem.* 20 (22): 4529. http://xlink.rsc.org/?DOI=b925342d.

10 Norgren, M. and Edlund, H. (2014). Lignin: recent advances and emerging applications. *Curr. Opin. Colloid Interface Sci.* 19 (5): 409–416. https://linkinghub.elsevier.com/retrieve/pii/S1359029414000867.

11 Saito, T., Brown, R.H., Hunt, M.A. et al. (2012). Turning renewable resources into value-added polymer: development of lignin-based thermoplastic. *Green Chem.* 14 (12): 3295. http://xlink.rsc.org/?DOI=c2gc35933b.

12 Grossman, A. and Vermerris, W. (2019). Lignin-based polymers and nanomaterials. *Curr. Opin. Biotechnol.* 56: 112–120. https://linkinghub.elsevier.com/retrieve/pii/S0958166918300612.

13 Quinsaat, J.E.Q., Feghali, E., van de Pas, D.J. et al. (2021). Preparation of mechanically robust bio-based polyurethane foams using depolymerized native lignin. *ACS Appl. Polym. Mater.* 3 (11): 5845–5856. https://pubs.acs.org/doi/10.1021/acsapm.1c01081.

14 Zhang, X., Liu, W., Liu, W., and Qiu, X. (2020). High performance PVA/lignin nanocomposite films with excellent water vapor barrier and UV-shielding properties. *Int. J. Biol. Macromol.* 142: 551–558. https://linkinghub.elsevier.com/retrieve/pii/S014181301936310X.

15 Sangon, S., Hunt, A.J., Attard, T.M. et al. (2018). Valorisation of waste rice straw for the production of highly effective carbon based adsorbents for dyes removal. *J. Cleaner Prod.* 172: 1128–1139. https://linkinghub.elsevier.com/retrieve/pii/S0959652617325192.

16 Sajjadi, M., Ahmadpoor, F., Nasrollahzadeh, M., and Ghafuri, H. (2021). Lignin-derived (nano)materials for environmental pollution remediation: current challenges and future perspectives. *Int. J. Biol. Macromol.* 178: 394–423. https://linkinghub.elsevier.com/retrieve/pii/S0141813021004645.

17 Qin, H., Wang, B., Zhang, C. et al. (2017). Lignin based synthesis of graphitic carbon-encapsulated iron nanoparticles as effective catalyst for forming lower olefins via Fischer-Tropsch synthesis. *Catal. Commun.* 96: 28–31. https://linkinghub.elsevier.com/retrieve/pii/S1566736717301085.

18 Srisasiwimon, N., Chuangchote, S., Laosiripojana, N., and Sagawa, T. (2018). TiO_2/lignin-based carbon composited photocatalysts for enhanced photocatalytic conversion of lignin to high value chemicals. *ACS Sustainable Chem. Eng.* 6 (11): 13968–13976. https://pubs.acs.org/doi/10.1021/acssuschemeng.8b02353.

19 Akhtar, K., Ali, F., Sohni, S. et al. (2020). Lignocellulosic biomass supported metal nanoparticles for the catalytic reduction of organic pollutants. *Environ. Sci. Pollut. Res.* 27 (1): 823–836. http://link.springer.com/10.1007/s11356-019-06908-y.

20 Alcantara, G.B., Paterno, L.G., Fonseca, F.J. et al. (2013). Dielectric properties of cobalt ferrite nanoparticles in ultrathin nanocomposite films. *Phys. Chem. Chem. Phys.* 15 (45): 19853. http://xlink.rsc.org/?DOI=c3cp53602e.

21 Lü, Q.-F., Luo, J.-J., Lin, T.-T., and Zhang, Y.-Z. (2014). Novel lignin–poly(N-methylaniline) composite sorbent for silver ion removal and recovery. *ACS Sustainable Chem. Eng.* 2 (3): 465–471. https://pubs.acs.org/doi/10.1021/sc400475r.

22 Khan, S., Shaily, A.M., Ghosal, A. et al. (2022). Superhydrophobic coordination polyurethane films based on methylolated-cardanol and hexamethylene diisocyanate: synthesis, characterization and antibacterial evaluation. *Prog. Org. Coat.* 168: 106886. https://www.sciencedirect.com/science/article/pii/S0300944022001837.

23 Masood, S., Ghosal, A., Gupta, A. et al. (2022). Comparative studies on coating materials of urotropine modified furfurylolated-tCNSL and methylolated-tCNSL thermoset for anticorrosive application: switching towards a cleaner approach. *J. Cleaner Prod.* 345: 130933. https://linkinghub.elsevier.com/retrieve/pii/S0959652622005716.

24 Zafar, F., Azam, M., Sharmin, E. et al. (2016). Nanostructured coordination complexes/polymers derived from cardanol: "one-pot, two-step" solventless synthesis and characterization. *RSC Adv.* 6 (8): 6607–6622.

25 De Maria, P., Filippone, P., Fontana, A. et al. (2005). Cardanol as a replacement for cholesterol into the lipid bilayer of POPC liposomes. *Colloids Surf., B* 40 (1): 11–18. https://linkinghub.elsevier.com/retrieve/pii/S092777650400284X.

26 Schmidt, R., Griesbaum, K., Behr, A. et al. (2014). Hydrocarbons. In: *Ullmann's Encyclopedia of Industrial Chemistry* (ed. M. Bohnet), 1–74. Weinheim, Germany: Wiley-VCH Verlag GmbH & Co. KGaA https://onlinelibrary.wiley.com/doi/10.1002/14356007.a13_227.pub3.

27 Balachandran, V.S., Jadhav, S.R., Vemula, P.K., and John, G. (2013). Recent advances in cardanol chemistry in a nutshell: from a nut to nanomaterials. *Chem. Soc. Rev.* 42 (2): 427–438. http://xlink.rsc.org/?DOI=C2CS35344J.

28 Yang, X.H., Wang, Z.M., Jing, F. et al. (2013). A brief review of cardanol-based surfactants. *Appl. Mech. Mater.* 483: 83–87. https://www.scientific.net/AMM.483.83.

29 Roy, A., Fajardie, P., Lepoittevin, B. et al. (2022). CNSL, a promising building blocks for sustainable molecular design of surfactants: a critical review. *Molecules* 27 (4): 1–51.

30 Shukla, S.K., Srivastava, K., and Srivastava, D. (2015). Studies on the thermal, mechanical and chemical resistance properties of natural resource derived polymers. *Mater. Res.* 18 (6): 1217–1223.

31 Sultania, M., Rai, J.S.P., and Srivastava, D. (2010). Studies on the synthesis and curing of epoxidized novolac vinyl ester resin from renewable resource material. *Eur. Polym. J.* 46 (10): 2019–2032.

32 Wazarkar, K., Kathalewar, M., and Sabnis, A. (2017). High performance polyurea coatings based on cardanol. *Prog. Org. Coat.* 106: 96–110.

33 Arumugam, H., Krishnan, S., Chavali, M., and Muthukaruppan, A. (2018). Cardanol based benzoxazine blends and bio-silica reinforced composites: thermal and dielectric properties. *New J. Chem.* 42 (6): 4067–4080.

34 Mestry, S., Kakatkar, R., and Mhaske, S.T. (2019). Cardanol derived P and Si based precursors to develop flame retardant PU coating. *Prog. Org. Coat.* 129: 59–68.

35 Atta, A.M., Al-Hodan, H.A., Hameed, R.S.A., and Ezzat, A.O. (2017). Preparation of green cardanol-based epoxy and hardener as primer coatings for petroleum and gas steel in marine environment. *Prog. Org. Coat.* 111: 283–293.

36 Balgude, D., Sabnis, A., and Ghosh, S.K. (2016). Synthesis and characterization of cardanol based aqueous 2K polyurethane coatings. *Eur. Polym. J.* 85: 620–634.

37 Mestry, S.U., Khuntia, S.P., and Mhaske, S.T. (2021). Development of waterborne polyurethane dispersions (WPUDs) from novel cardanol-based reactive dispersing agent. *Polym. Bull.* 78 (12): 6819–6834. https://doi.org/10.1007/s00289-020-03450-7.

38 Ganesh, K.S., Sridhar, A., and Vishali, S. (2022). Utilization of fruit and vegetable waste to produce value-added products: conventional utilization and emerging opportunities – a review. *Chemosphere* 287: 132221. https://linkinghub.elsevier.com/retrieve/pii/S004565352102693X.

39 Usman, A.I., Abdul Aziz, A., and Abu, N.O. (2018). Application of green synthesis of gold nanoparticles: a review. *J. Teknol.* 81 (1): 171–182. https://journals.utm.my/index.php/jurnalteknologi/article/view/11409.

40 Omran, B.A., Aboelazayem, O., Nassar, H.N. et al. (2021). Biovalorization of mandarin waste peels into silver nanoparticles and activated carbon. *Int. J. Environ. Sci. Technol.* 18 (5): 1119–1134. https://link.springer.com/10.1007/s13762-020-02873-z.

41 Jaybhaye, S.V. (2015). Antimicrobial activity of silver nanoparticles synthesized from waste vegetable fibers. *Mater. Today Proc.* 2 (9): 4323–4327. https://linkinghub.elsevier.com/retrieve/pii/S2214785315008391.

14

Peptide-Assisted Synthesis of Nanoparticles and Their Applications

Vikas Kumar

Department of Chemistry, Government College Khimlasa, Sagar, MP, India

Abstract

Developing green technology for the synthesis of metal nanoparticles have attracted huge scientific interest. Peptides possess unique capability to provide template for the fabrication of metal nanoparticles and their encapsulation that can be used as career for targeted drug delivery. In the field of the material science, these peptide-metal nanoparticles can be used for applications as nano-catalysts, nano-photonics, theranostic agents, nutraceuticals and for fabricating nano electronic devices. This chapter includes the detailed discussion of chemical and peptide-assisted synthesis, spectroscopic and microscopic characterization, and application of peptide-MNP hybrids in the field of material and medicinal chemistry.

Keywords *peptides; nanoparticles; green synthesis; spectroscopy; microscopy*

14.1 Introduction

Green synthesis of metal nanoparticles has exponentially attracted scientific interest due to their applications in the field of nano-catalysis [1], nanoelectronics [2], nano-photonics [3], theranostics [4–7], nutraceuticals [8], development of antimicrobial nano-scaffolds for tissue regeneration [9], wastewater treatment [10], and photoinduced eradication of pollutants [10]. The applications of nanoparticles further extend for targeted drug delivery [11], chemotherapy for cancer [12–14], biofilm eradication [15–17], and treatment

Green Synthesis of Nanomaterials: Biological and Environmental Applications, First Edition. Edited by Archana Chakravarty, Preeti Singh, Saiqa Ikram, and R.N. Yadava.

of tuberculosis [18–22]. Applicability of nanoparticles depends on their size, shape, and composition as well as their arrangements on template molecules and organization of ligand molecules that are used to modify the surface of nanoparticles. This exciting field offers various challenges to scientists to obtain nanoparticles with definite size, shape, stability, and bioavailability for medicinal applications. Chemical synthesis of nanoparticles has its own drawbacks due to the use of non-eco-friendly solvents.

Bio-assisted green synthesis of nanoparticles by using plant or microorganism extracts for catalytic and antimicrobial applications has been reported; however, the nanoparticles with definite shape, size, and desired stability, expected catalytic and bioactivity could not be achieved [23, 24]. Although biological extracts are a mixture of various bioactive natural products such as flavonoids, alkaloids, and especially denatured enzymes. Enzymes consist of protein and prosthetic groups as their functional units, therefore, denatured enzymes produce short peptide derivatives in biological extract, which possibly provide template for nanoparticle fabrication. A Dutch chemist Johan Mulder coined the term "Protein" that comes from a Greek word "Proteios" that means "primary," or "which holds the first place." As per the theory of "Chemical Evolution of Life" [25], proteins were the first biologically active molecules that originated on earth, which consists of peptides as constructive units. Peptides contain amino acids that are also known as building blocks of life. they belong to a class of unique molecules with programmable structures that accommodate specific functional moieties, which possess capability for sustainable production, biocompatibility, biodegradability, and accuracy at nanoscale level [26]. They exhibit property of tailorable nanoscale self-assembly due to which they form programmed self-assembled nanostructure (Figure 14.1) that can be triggered for morphological transformation under external stimuli such as temperature, pH, and presence of organic and inorganic moieties [27], and hybridization with nanoparticles [28]. Such hybrid nanostructures of peptides offer a range of applications in the field of bio-nanofabrication, nano-catalysis, antimicrobial nano-scaffolds for tissue regeneration, wound healing, food additive and preservatives, and as new-generation therapeutics [29].

Several structural variants have been discovered for peptide-based materials that include linear and cyclic peptides, peptide amphiphiles, peptides containing α- and β-amino acids and other unnatural amino acids, peptide conjugates of natural products such as vitamins, hormones, fatty acids, and various drugs [27]. The presence of specific amino acids and certain functional groups on peptide chain and secondary structural motifs such as α-helix, β-turns, and β-sheets generate various self-assembled morphologies of peptides in

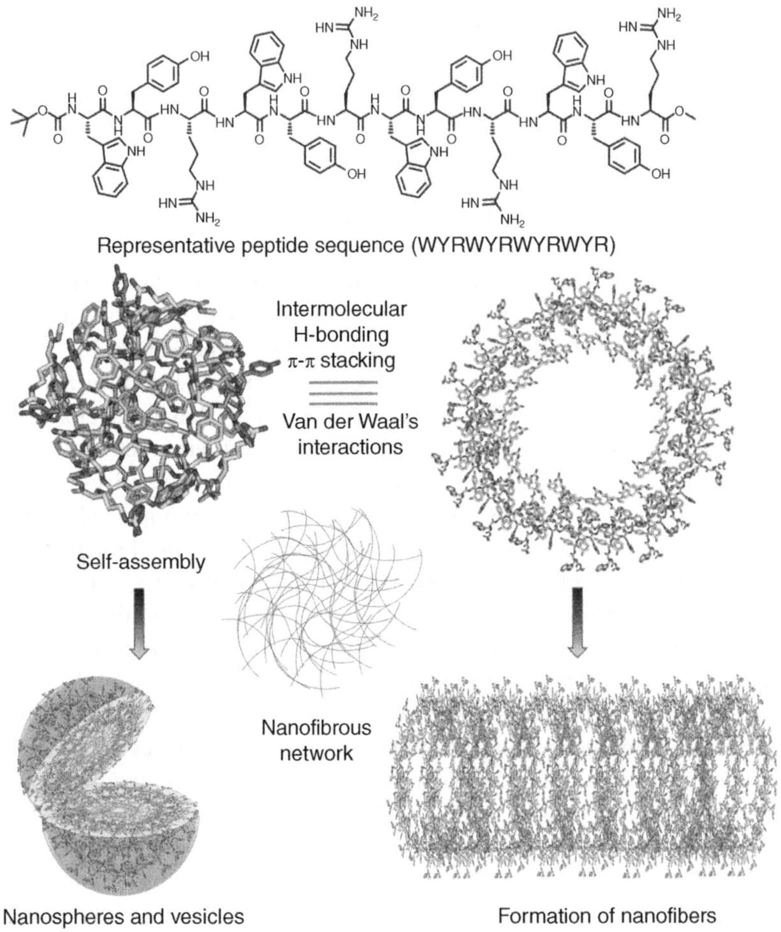

Figure 14.1 Depicts the formation of self-assembled nanostructures of peptide-based materials that can be possible templates for the fabrication of nanoparticles. Noncovalent interactions and Van der Waal's forces (aromatic, dipole–dipole, electrostatic attractions) and intermolecular hydrogen bonding leads to the formation of secondary structures of peptides.

aqueous-organic solvents [29]. These self-assembled nano-scaffolds provide template for green synthesis of metal nanoparticles. The capability of peptides to coordinate with metal ions and formation of an exciplex that further leads to the formation of metal nanoparticles in the presence of radiation offer new directions for material and medical scientists to fabricate biodegradable nano-scaffolds for advanced material and biomedical applications [29].

This chapter includes methodologies for the peptide-assisted synthesis of different metal nanoparticles and brief description of their applications in the fields of material chemistry and bio-nanotechnology. Here, we will discuss about different methods reported for peptide-assisted synthesis of nanoparticles, their spectroscopic and microscopic characterization, and their biological and environmental applications.

14.2 Synthesis of Metal Nanoparticles by Using Peptides as Template

The synthesis of metal nanoparticles on peptide scaffolds can be performed by using two different methods that are mentioned below:

14.2.1 In the Presence of Reducing Agents

In this method, peptides are mixed with solution of metal ions and certain reducing agents such as sodium citrate [30], sodium borohydride ($NaBH_4$) [31], or bis(p-sulfonatophenyl)phenylphosphine (BSPP) [32, 33], and many more are added, which help to reduce metal ions into neutral metal atoms. Nucleation of metal nanoparticle starts by following the arrangement of atoms at characteristic crystallographic positions depending on metal being used.

14.2.2 In the Absence of Reducing Agent

Here peptide itself acts as reducing agent, and nanoparticle synthesis is performed either in the presence of UV–Visible radiation [31, 34–36], NIR radiation [37], ultrasonication [38–40], or under microwaves [41, 42]. In photochemical synthesis, peptide–metal ion mixture is irradiated by radiation of certain wavelength for three to four minutes, and color change of the mixture indicates the formation of nanoparticles depending on type of nanoparticles that are being synthesized (Figure 14.2). The presence of certain amino acids facilitates photoinduced electron transfer from peptide to metal ions and these fully or partially neutralized metal ions form nanoparticles.

14.3 Characterization of Peptide-MNP Hybrids

The synthesized peptide–metal nanoparticles are characterized by using different analytical techniques such as spectroscopy and microscopy that will be briefly discussed here.

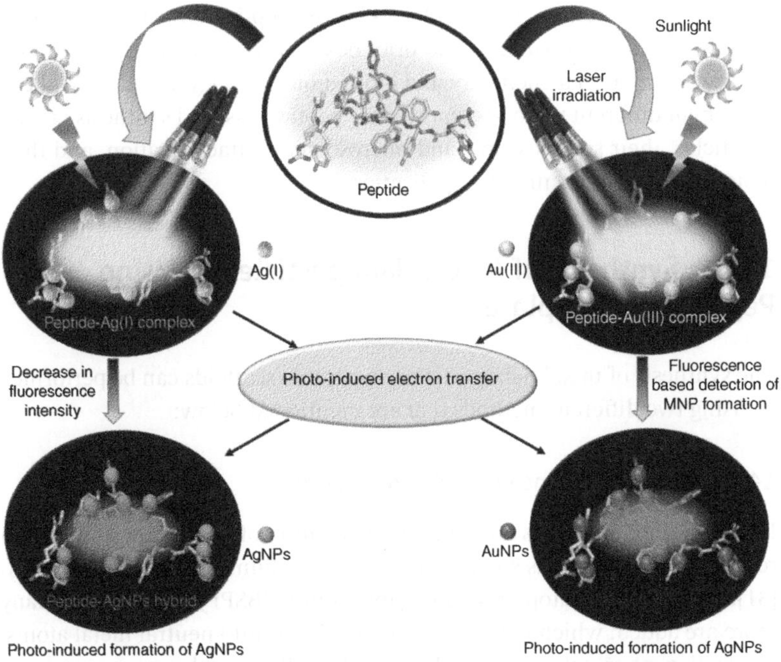

Figure 14.2 Demonstration of peptide-assisted photoinduced synthesis of metal nanoparticle especially noble metals such as silver and gold NPs. Either sunlight or pulsed/continuous wave laser of different wavelengths can be used for irradiation of peptide–metal ion mixture, which possibly leads to photoinduced electron transfer from peptide to metal ions.

14.3.1 UV–Visible Spectroscopy/Surface Plasmon Resonance (SPR) Spectroscopy

Depending on the type of metal, all nanoparticles exhibit characteristic plasmonic properties and produce signal in UV–Visible spectra. For example, silver nanoparticles show characteristic SPR band starting from 400 to 480 nm [34] that depends on their size and shape. Similarly, gold nanoparticles exhibit SPR band at 500–590 nm [31, 35, 36] that also depends on their size and shape. AgNPs of size range 2–5 nm exhibit SPR band at 395–405 nm but AuNPs of similar size range show SPR band at 510–515 nm [31, 35, 36]. Nanorods for different metals possess two SPR bands, Longitudinal SPR (LSPR) at higher wavelength and Transverse SPR (TSPR) at shorter wavelength. For copper nanoparticles, SPR bands are observed between 620 and 710 nm [43, 44]. Gelatin-coated magnetite nanoparticles exhibit SPR band between 350 and 420 nm [45]. The position of SPR bands may vary from metal to capping agents that are used for the synthesis of nanoparticles.

In various methods, capping agents are used to stabilize nano-colloid. Peptides can act as appropriate capping agents to stabilize nanoparticles and can also increase their biological relevance by enhancing their cell penetration capability, bioavailability, and biocompatibility [43, 44].

14.3.2 Fluorescence Spectroscopy

Fluorescence spectroscopy provides crucial information about the interaction of metal ions with peptides [31, 34–36]. But the technique is limited to the study of peptides consisting aromatic amino acids such as tyrosine and tryptophan. Electrostatic attractions and dipole–dipole interaction between peptide and metal ions lead to the formation of peptide–metal complex, which causes the conformational changes in peptide molecules and affects the signal intensity. Decrease in fluorescence signal intensity or hyperchromic/hypochromic or red/blue shifts are generally observed, suggesting the formation of peptide–metal noncovalent complex. Charges on metal ions and charge density affect the aromatic stacking between peptide molecules, which destabilize the intermolecular dipole–dipole attractions.

14.3.3 Circular Dichroism

Peptide offers a range of secondary structural motifs such as α-helix, β-turns, and β-sheets that can be studied by circular dichroism analysis [31, 34–36]. For example, α-helical are observed as negative maxima at 222 nm, 208 nm and positive maxima at 193–195 nm. Antiparallel β-sheets display negative bands at 218 nm and positive bands at 195 nm. The position of absorption maxima may vary with 1 or 2 nm depending on solvent system and peptide that is being analyzed. Hydrogen bonded β-structures are also observed in CD spectra as absorption maxima at 214–215 nm. Addition of metal ions with peptides interferes in the formation of secondary structures because metal ions disturb noncovalent forces between peptide molecules, hence, the formation of peptide–metal complex can be confirmed by CD analysis [31, 34–36].

14.3.4 Ultrafiltration and Centrifugation

To remove unhybridized/uncapped peptides or unreacted peptides during peptide-assisted nanoparticle synthesis, two different techniques can be used. Nano-colloid can be purified by using ultrafiltration by which unreacted peptide can be filtered and remaining nano-colloid can be used for further analysis [46]. Centrifugation at specific rpm/g the nanoparticles can be collected in the form of pallet and supernatant can be removed [31]. Further spectroscopic analysis can confirm the presence of unreacted peptide in nano-colloid, and nanoparticle pallet can be used for further microscopic analysis.

14.3.5 Zeta Potential Study

Zeta potential is electrokinetic potential at the surface of colloidal particles. It is the potential difference generated between the dispersion medium/solvent and the stationary layer of fluid/solvent/mother liquor attached to the dispersed particle. To determine the stability of nano-colloid, zeta potential study is an important analysis preferred by scientists [47]. Magnitude of zeta potential decided the stability of nanoparticles in solution such as if its magnitude is found between 10 and 30, the nano-colloid possesses incipient instability, its value between 40 and 60 suggests moderate to excellent stability of peptide-nanoparticle colloid.

14.3.6 Dynamic Light Scattering (DLS)

Dynamic light scattering is used to analyze the particle size of protein, nanoparticles, carbohydrates, peptides, and other colloids. The technique provides size distribution of nanoparticles. A monochromatic light beam is incident on nano-colloid. If particles possess size less than the wavelength of incident light (250 nm), the light gets scattered in all possible directions (Rayleigh Scattering). Scattered light beams interfere either constructively or destructively. Scattering depends on the size of the nanoparticles, concentration of nano-colloid and their motion in colloid (Brownian motion). The scattered light signals are then analyzed by different algorithms to collect the information of the size of nanoparticles [47].

14.3.7 Transmission Electron Microscopy (TEM)

TEM is the most efficient technique that is used for the analysis of nanomaterials, which provides detailed information of nanoparticles [31, 34–36, 48]. TEM gives a clear image of their morphology, detailed picture if they are embedded inside peptide nanostructures, their crystallinity or distorted crystal structure. Nanoparticles can stop high-intensity electron beam, which further falls on photographic sheet to create images of sample. Few microliters of sample are spread on carbon-coated 200 or 300 mess grids and they are/are not negatively stained with uranyl acetate to obtain good contrast images. Depending on type of metal nanoparticles, their crystal structures may vary such as AgNPs and AuNPs possess face centered cubic (FCC) crystal lattice. High-resolution transmission micrographs (HRTEM) provide clear view of lattice planes of nanoparticles, and electron beam can also be diffracted by nanoparticles to provide diffraction pattern. Selective area diffraction pattern (SADP) provides "concentric circles," which represent the different planes of nanoparticles' crystal [31, 34–36]. A sensor for detection of elements and their percentage analysis can also be coupled with TEM by which the presence of certain elements can be confirmed whether there is Ag, Au, Hg, or Fe. The sensor works on the principle of Energy Dispersive X-ray Spectroscopy (EDS or EDAX). In which high energy electron beam is used to excite inner cell electrons of atoms for electronic excitation, which is recorded by the detector. EDS sensor can also be coupled with scanning electron microscope (SEM) to collect the information of nanoparticles. Secondary electrons emitted from the samples are collected on a detector and the images represent nanoparticles as bright dots. Depending on type of metal, brightness of nanoparticles may vary hence EDS becomes an important to analyze whether the bright particles that are visible in SEM are of desired metal or not.

14.3.8 X-ray Diffraction Analysis

X-ray diffraction analysis is well known to study crystals. In case of availability of more quantity of nanoparticle sample, they can be analyzed by XRD analysis first to avoid expensive TEM analysis. XRD pattern provides crucial information about crystallinity of nanoparticles [31] and produces signals accordingly. Nanoparticles behave as nanoscale single crystals and produce diffraction patterns accordingly, which suggests about their crystallinity. Selective area diffraction pattern analysis is more advanced version of XRD in which electron beam is diffracted through the bunch of nanoparticles present of TEM grid in selective area.

14.3.9 Matrix Laboratory (MATLAB) Analysis

Sometimes it feels that nanoparticles are free of peptide nanostructures when they are analyzed by TEM. Due to poor capability of peptide soft structure to stop high energy electron beam, it seems that nanoparticles are not on the surface of peptide and independent of peptide structure. To confirm the presence of peptide structure there, sample can be negatively stained by uranyl acetate in TEM microscopy so that their capability to stop electron beam be enhanced and high contract images with clear vision of peptide with nanoparticles can be obtained. But MATLAB image analysis software can also provide crucial information about the presence of peptide nanostructure. To obtain information from MATLAB image analysis software, the digital signal for background of the image is subtracted from total digital signal, to further confirm the presence of peptide, digital signal for nanoparticles also subtracted by which signal for peptide nanostructure remain visible [34]. This analysis is crucial where peptide cannot be seen in TEM images and if one wants to confirm the presence whether the nanoparticles are embedded on peptide structures, or it is just the random arrangement of nanoparticles.

14.3.10 ImageJ Analysis

ImageJ is an interesting tool to analyze the size distribution of nanostructures by using either SEM or TEM image in case of nonavailability of DLS instrument. Based on scale bar, parameters for image analysis are selected. Interconversion of image pixels into area covered by particles is performed by suggested steps on ImageJ analysis software [49]. Area covered by single particles is selected for calculation of area covered by other nanoparticles by which diameter of all particles is calculated. Detailed analytical steps can be learned from the user manual of ImageJ analytical tool for image analysis. A histogram can be plotted in Excel to get an idea of particle size distribution (PSD) of nanoparticles. Sometimes a print of image is taken and by calculating length of scale bar, diameters of nanoparticles are calculated manually though it is the least common practice for PSD analysis but sometimes scientists use it when number of particles is found very less per image.

14.3.11 Atomic Force Microscopy (AFM)

The technique provides 3-D characterization of nanoparticles with subnanometer resolution. Once the presence of nanoparticle is confirmed in sample by other spectroscopic and microscopic analysis, they can be subjected to topological analysis by AFM [31, 34–36]. Topological study and analysis of the height of nanoparticles can be performed by AFM which can also provide information of PSD of nanoparticles. However, it cannot be said that the structure that are visible in AFM images are exactly of nanoparticles or of the other dust particles because it only records the 'shapes of particles' that are spread on AFM sample holders. It means, it cannot be confirmed whether they are silver nanoparticles or gold nanoparticles therefore confirmation of the material of nanoparticles should be done by other techniques (such as EDS analysis). The advantage of this technique is that it provides 3-D visualization of nanoparticles and provides analytical data for their diameter and heights. 3-D images of nanoparticles in software-generated different color themes can be obtained for clear picturization of nanoparticles.

The above discussion about purification of nano-colloid and its characterization contains only brief information. Details of sample preparation, specific physical conditions drying, and maintenance of sample depend on type of peptide and nanoparticles that are being used.

14.4 Biological and Environmental Applications of Peptide Nanoparticles

In Introduction part, we discussed that peptide nanoparticles can have a range of applications in the field of biomedicines where they can be used to fabricate antimicrobial nano-scaffolds for tissue regeneration, wound healing, bioimaging, biosensors, antifungal, and antibacterial activity, antimalarial and antileishmanial activity and for anticancer therapy as theranostics (Figure 14.3).

For environmental applications, they can be used for the detection of certain toxic heavy metals such Pb, Pd, Cd, and Hg. Their sensitivity for nanomolar concentration of these toxic metal ions can be used for water

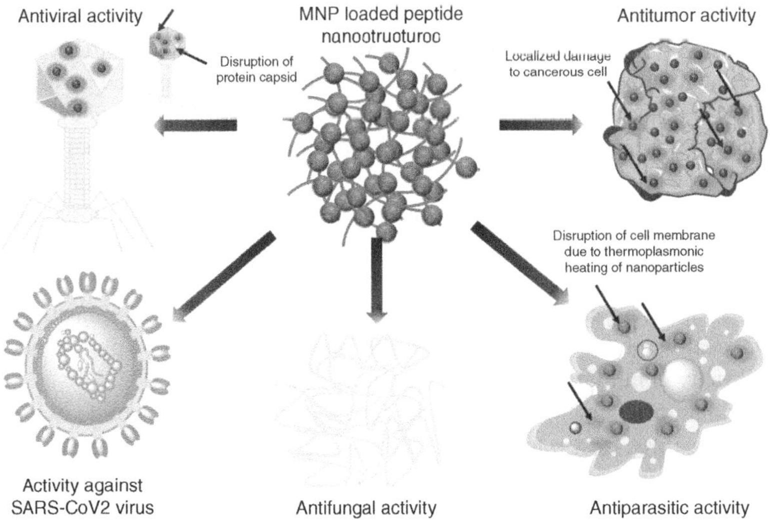

Figure 14.3 Depicts biomedical applications of peptide nanoparticles for antimicrobial, antiviral, antifungal, and anticancer activity.

filtration and wastewater treatment. Silver and gold nanoparticles have been studied for their capability to reduce *p*-nitrophenol, to detect nanomolar concentration of Hg^{+2} ions as well other trace metal ions such as (Zn^{2+}, Na^{+}, Cr^{3+}, Ni^{2+}, Ca^{2+}, Fe^{2+}, K^{+}, Hg^{2+}, Mg^{2+}, Cu^{2+}, Mn^{2+}, Al^{3+}, and Fe^{3+}). Here, the synthesis of different nanoparticles on peptide scaffolds and their biological and environmental applications will be discussed in brief.

14.4.1 Peptide-Assisted Synthesis of Silver Nanoparticles and Their Applications

Fabrication of silver nanoparticles and their applications in various fields of material chemistry and synthetic chemistry always attracted huge scientific attention. For the first time, Lea produced citrate-stabilized silver nanocolloid in 1889 [50]. Other methods for the synthesis of AgNPs were also reported. Even though chemically synthesized silver nanoparticles were very popular in material chemistry for catalytic and piezoelectric applications, however, they could not be used for medical applications due to cell toxicity, less water solubility, poor capability to cross the cell membrane and less bioavailability. Bio-assisted synthesis of AgNPs mediated by microbial and plant extracts, were reported for their enhanced biocompatibility and stability under certain physiological circumstances. Natural "extracts" contain a range of secondary metabolites and denatured enzymes, decorated with specific functional groups that help them to coordinate with metal ions and external stimuli such as radiation, thermal energy, or sonication provide support for the reduction of metal ions into nanoparticles without even using reducing agents. Peptides offer class of unique bioinspired materials that can be used for green synthesis of silver nanoparticles for their applications as antimicrobial agents against multidrug-resistant microbes. They have been widely used for their activity against Gram-positive and Gram-negative bacteria, malaria parasite, leishmaniasis, antituberculosis, and anticandidal activity. Coating of silver nanoparticles with peptides helps to reduce chemical synthesized AgNPs and enhance their bioavailability and cell penetration capability. For example, Gurunathan et al demonstrated that coating of silver nanoparticles with humanin peptide can protect neurotoxicity in human neuroblastoma cancer cells SH-SY5Y [51].

Humanin is a cytoprotective and neuroprotective small peptide, which contains three turn α-helix with no symmetry. Depending on the place of production it may either contain 21 amino acids if produced in mitochondria or 24 amino acids if produced in cytosol. They utilized delphinidin for bio-assisted synthesis of silver nanoparticles, and SPR (surface plasmon resonance) band at 420 nm in UV–vis spectra suggests the size of AgNPs about 18 nm, which was further confirmed by dynamic light scattering experiment for particle size distribution. Recently, Retout et al reported peptide-induced morphological transformation of silver nanoparticles into fractal nanostructures [33]. They synthesized silver nanoparticles by using bis(p-sulfonatophenyl)phenylphosphine (BSPP) by following seed growth method. BSPP was selected due to its capability to coordinate with silver, its negative charge and aromaticity. The morphological change of silver nanoparticles was implemented for the colorimetric detection of SARS-Cov2 by using its main protease (MPro) for enzymatic modulation of self-assembly of AgNPs, which causes color change of the colloid. They also developed a multichannel sensor array to differentiate SARS-CoV2 and influenza-infected external breath condensate (EBC) or saliva samples by using peptide-coated silver nanoparticles.

Seferji et al. synthesized silver nanoparticles by using a nonaromatic peptide Ile-Val-Phe-Lys [17]. They described that the peptide was capable to form nanoparticle when mixed with the silver nitrate and irradiated with UV radiation. The peptide-AgNPs hybrid was demonstrated for their antibacterial/antibiofilm activity with minimal cell toxicity. Pal et al. reported the inhibitory activity of antimicrobial peptide-coated silver nanoparticles against a group of multidrug-resistant pathogens named as "ESKAPE," which refers to *Enterococcus faecium*, *Staphylococcus aureus*, *Klebsiella pneumoniae*, *Acinetobacter baumannii*, *Pseudomonas aeruginosa*, and *Enterobacter species* [52]. They utilized chemically synthesized peptides Andersonin-Y1 (AY1) [FLPKLFAKITKKNMAHIR], CAY1 [CFLPKLFAKITKKNMAHIR], and AY1C[FLPKLFAKITKKNMAHIRC] and hybridized them with citrate capped silver nanoparticles for the purpose of antimicrobial screening. *In situ* NMR revealed that coating of silver nanoparticles with above-mentioned peptides increases their activity against selected pathogens by several folds as compared to peptides alone.

Ramírez-Acosta et al. reported the synthesis of magnetite–silver core-shell nanoparticles for the *in-vitro* delivery of plasmids [53]. Chemically synthesized magnetite nanoparticles were suspended with solution of silver nitrate in honey, which acts as reducing agent. These bio-chemically synthesized magnetite–silver core-shell nanoparticles were functionalized with (poly (2-dimethylamino) ethyl methacrylate) methyl chloride (pDMAEMA), which were subsequently conjugated with poly ether amine (PEA). Then (magnetite-Ag)NPs-(pDMAEMA)-PEA were finally linked with Buforin II (BUF-II) peptide through amide linkage. Buforins are a group of peptides that were first isolated from the tissues of stomach of Asian Toad *Bufo bufo gagarizans*. Buforin I possesses 39 amino acids and demonstrates broad-spectrum antimicrobial activity against range of drug-resistant pathogens. Buforin II consists of 21 amino acids, which is derived from buforin I, and possesses higher antimicrobial activity than its parent peptide Buforin I. It was observed that conjugation of magnetite–silver nanoparticles with (pDMAEMA)-PEA-Buforin II increases their biocompatibility and reduces their cell toxicity and hemolytic activity compared to free nanoparticles or the nanoparticles conjugated only with pDMAEMA.

Zharkova et al. utilized antimicrobial peptides porcine protegrin-1 (AMP: PG-1), bovine indolicidin (AMP: Ind), hen egg-white lysozyme (AP: Lyz), salmon protamine (AP: Prot), or calf thymus histones (AP: His) for preparing AgNP conjugates [54]. The peptide-AgNPs conjugates were then investigated for their antibacterial activity against Gram-positive and -negative bacteria. They observed that conjugation of AgNPs with AMPs can increase their antibacterial activity by two- to fourfolds. They also investigated the cytotoxic effects of these AMP-AgNPs hybrids that were observed to be relatively less cytotoxic compared to AgNPs alone or gelatin-coated AgNPs, however, the exact reasons for cytotoxicity were not explained in detail but it was suggested that cell membrane denaturation, decomposition of membranes of various cell organelles can be the reason for cell damage.

Peptides offer a unique character of being self-assembled into various nanostructures such as vesicles, spheres, nanofibers, nanobelts, nanosheets, and many more interesting assemblies [31, 34–36]. Various noncovalent forces such as dipole–dipole, aromatic interaction, intermolecular hydrogen bonding, and Van der Waal's attractions regulate the programmed arrangement of isolated peptide building blocks into self-assembled nanostructures. These nanostructures of peptides undergo morphological modulation when they are conjugated with silver nanoparticles. The presence of tryptophan and its unique photochemical properties provide photo-sensitive template for photoinduced fabrication of AgNPs upon peptide nano-scaffolds. Several peptides and peptide amphiphiles have been implicated for the synthesis of silver nanoparticles. Mishra et al demonstrated photoinduced synthesis of AgNPs by using Palmitoyl-Trp-Trp-Gly and Palmitoyl-Trp-Trp-β-Ala peptide amphiphiles [34]. Irradiation by sunlight for few minutes led to the synthesis of silver nanoparticles that were decorated over peptide amphiphile fibers. Prolonged irradiation led to the morphological transformation of AgNPs-loaded PA fibers into nanowreath kind of self-assembly due to the thermoplasmonic effect of silver nanoparticles. When AgNPs absorb radiation of appropriate wavelength, electronic transitions take place on surface of nanoparticles that are known as surface plasmons. When electrons come back to ground state, they lose their energy into form of heat. This effect is known as thermoplasmic effect and the heat that comes out, known as thermoplasmonic heating. Localized heating of silver nanoparticles leads to the

melting of peptide fibers at certain places, which causes the formation of nanowreath. Biotinylated di-L-tryptophan molecule was utilized for the fabrication and encapsulation of silver nanoparticles. The morphology of peptide-AgNPs was modulated by changing solvent composition, and it was observed that the morphology of peptide-AgNPs hybrids changed from vesicles to fibrous when polarity of solvent system was changed from lower to higher. Ethanol-water binary solvent system was used for the fabrication of peptide-AgNPs where percentage of ethanol from 0% to 100% was chosen to modulate the self-assembly of peptide-AgNPs hybrids [55]. Further, peptide-AgNP nanostructures were fabricated on the surface of living diatoms [56]. Diatoms are unicellular brown algae that produce their food by photosynthesis and produce fatty acids as secondary metabolites. Scientific approaches are being developed to enhance the oil production in diatoms and clean harvesting by various means in which silver nanoparticles and peptide-based supplements can help to enhance diatom oil production which high value-added substance that can be used as an additive to natural oil to fulfill high requirement of fossil fuel.

The adverse effect of silver nanoparticles on various microorganisms is well reported. Their coating and conjugation with biological molecules may alter the activity of silver nanoparticles toward soil microbiota and aquatic organisms. Hence, instead of using chemically synthesized silver nanoparticles, the use of bio-synthesized silver nanoparticles must be encouraged. Though biological extracts provide silver nanoparticles due to their nonuniformity, and no morphological precision, more appropriate template for the fabrication of AgNPs is needed. Peptides offer superior candidacy for the synthesis of nanoparticles than biological extract and provide more stable nano-colloid and functionalize the surface of nanoparticles to be more biocompatible and enhance their bioavailability. The application of silver nanoparticles is further extended to inhibition of harmful microbiota in soil to stop the growth of undesired microorganism and dose-dependent inhibition of seed germination of rice seeds [57]. Peptide coating of silver nanoparticles can also reduce their harmful effect of aquatic animals as compared to chemically synthesized AgNPs [57].

14.4.2 Peptide-Assisted Synthesis of Gold Nanoparticles and Their Applications

Peptide-conjugated gold nanoparticles show good performance for biomedical applications such as theranostics for anticancer activity, wound healing, and antimicrobial activity [58]. The sequence and type of amino acids present in peptide chain can regulate the shape, size, and morphology of gold

nanoparticles [59]. It has been investigated that the chemical, physical, and biological properties of AuNPs are the function of their size, shape, and morphology, which depends on the peptides that are used for their fabrication. Short peptide conjugates and peptide amphiphiles help to reduce Au(III) ions in the presence of radiation and form nanoparticles even in the absence of any reducing agent. They provide a template for the nucleation of gold nanoparticles and further direct the growth of gold nanostructures under constrained and confined environment. The presence of specific functional groups and side chain residues of certain amino acids support the formation of an exciplex (excited state complex) between peptide and Au(III) where electron transfer takes from peptide to metal ions possibly via Forster Resonance Energy Transfer (FRET) or photoinduced electron transfer (PET) [31].

The aromatic amino acids such as histidine and tryptophan, have shown excellent capability to interact with gold ions which was confirmed by fluorescence spectroscopy analysis, where intensity of fluorescence signal decreased upon addition of increasing concentration of gold ions [31, 34–36]. Circular dichroism analysis confirmed the change of conformation when peptides are mixed with metal ions, which could be the main reason for the change in the intensity of fluorescence signal. Intrinsic fluorescence of tryptophan provides uniqueness to the peptide, which can be used for laser induced synthesis of gold nanoparticles. Formation of gold nanoparticles may also lead to the change in self-assembly of peptides and can modulate their morphology such as Mishra et al reported that formation of gold nanoparticles can transform peptide fibers into vesicular morphology which further leads to the encapsulation of gold nanoparticles inside peptide nanostructures [31, 34–36].

Zhu et al reported that high-density energy carrier near infrared radiation may support photoinduced reduction of Au(III) ions into AuNPs or AuNCs (gold nanoclusters) on peptides surface that contain different amino acids or same amino acids but different sequences [58]. They utilized these peptide-AuNPs hybrids for stepwise killing of cancer cells and inhibit bacterial infection. Localized electronic transitions possibly take place on the surface of gold nanoparticles (thermoplamonic phenomenon), which may cause thermoplasmonic heating of gold nanoclusters that led to killing of cancer cells. Ding et al demonstrated the synthesis of ultrasmall gold nanoparticles by using peptide building block of trans-activator of transcription protein (Tat-R-EK) [6]. The peptide-coated gold nanoparticles were then implicated for their anticancer activity. Radiotherapy for cancer treatment is becoming popular and gold nanostructures whether they are nanoparticles or nanorods, play an important role to achieve this goal of developing new-generation theranostics for cancer treatment [6]. Jabir et al utilized a

pentapeptide consisting amino acid sequence CALNN conjugated with linalool-loaded gold nanoparticles for anticancer treatment of breast cancer by using MCF-7 cell line [60]. Several other reports have mentioned the applications of gold nanoparticles for drug delivery, antimicrobial applications, and their frequent use for developing anticancer medicines. Due to the unique optoelectronic properties of gold nanoparticles, they are used for molecular recognition, chemical sensing, bioimaging, trace and toxic metal ion sensing and for catalytic activity [60]. Li et al reported the applications of gold nanoclusters for the detection of Hg^{+2} ions. Here the gold nanoclusters were synthesized by using a peptide having amino acid sequence CCYRRRRRRHHHH [61]. In addition to Hg^{+2} ions, they also utilized the peptide-gold nanocluster for the detection of Cr^{3+}, Ag^{+}, Ca^{2+}, Mn^{2+}, Al^{3+}, Co^{2+}, Fe^{3+}, Cu^{2+}, Pb^{2+}, Ni^{2+}, Mg^{2+}, Cd^{2+}, Li^{+}, K^{+}, Zn^{2+}, Ba^{2+} ions as well. They observed that up to 30 nM concentration of Hg^{+2} ions can be detected in intracellular environment [61]. Gao et al. synthesized gold nanoclusters by using pristine aprotinin peptide, which possesses amino acid sequence pristine aprotinin sequence as RPDFCLEPPYTGPCKARIIRYFYNAKAGL CQTFVYGGCRAKRNNFKSAEDCMR TCGGA [62]. These peptide-gold nanoclusters were synthesized by mixing solution of peptide with Au(III) as $HAuCl_4$, at pH 7.0 by adding 1M NaOH solution. It was observed that thiol moieties of cysteine amino acids form complex with Au(III) ions and further formation of phenolic negative ions also helps to form gold nanoclusters. The synthesized peptide-gold nanoclusters were then utilized for the detection of heavy metal ions (Zn^{2+}, Na^{+}, Cr^{3+}, Ni^{2+}, Ca^{2+}, Fe^{2+}, K^{+}, Hg^{2+}, Mg^{2+}, Cu^{2+}, Mn^{2+}, Al^{3+}, and Fe^{3+}) and trypsin, which is the enzyme that helps to digest protein in small intestine. It breaks protein into small peptides that further digested to form amino acids by other enzymes [62].

Apart from the applications of gold nanoparticles in the field of medicinal and material chemistry for molecular recognition and metal ion detection, they can be implicated for wastewater treatment and detection of certain pollutants. The plasmonic effect of gold nanoparticles and nanoclusters can be useful for photoinduced degradation of certain dyes and chemical pollutants that are released in environment by degradation of plastic-based materials. Subair et al. demonstrated the application of gold nanoparticles for photoinduced reduction of *p*-nitrophenol and degradation of dyes such as Congo Red and methylene blue. Abbas et al synthesized tetrapeptide IVFK-coated gold nanoparticles for their catalytic role in reduction of *p*-nitrophenol. As per the U.S. Environmental Protection Agency list, *p*-nitrophenol is one the major environmental pollutant that may cause inflammation in eyes, nose, and skin [63]. Due to its delayed interaction with blood, it forms methemoglobin, which causes methemoglobinemia. It may result in cyanosis, confusion, and unconsciousness, therefore, its remedy is

important and gold nanocluster can play an important role in developing technologies for the photoinduced reduction of *p*-nitrophenol [63].

14.4.3 Synthesis of Core-Shell Bimetallic Nanoparticles and Their Catalytic Application of Metal Nanoparticles

Bimetallic nanoparticles or alloyed nanoparticles are gaining popularity due to their unique optoelectronic, chemical, physical, and catalytic properties. Wu et al. synthesized peanut-shaped $Au_{33}Pt_{67}$ nanoparticles on Z1 peptide template (peptide sequence-KHKHWHW) [64]. They observed the catalytic activity of alloyed AuPt nanoparticles toward oxygen reduction reaction (ORR) and hydrogen evolution reaction (HER). Core-shell nanoparticles show two absorbance bands in their UV–Vis spectra located at ~210 and ~260 nm. Wagner et al. demonstrated the synthesis of palladium nanoparticles on gold nanorods. They synthesized high aspect ratio gold nanorods that were used as template for the fabrication of palladium nanoparticles [65]. SPR study depicts the LSPR band at NIR region from 980 to 1200 nm depending on length of nanorods. Unique optoelectronic properties of gold nanomaterials provide appropriate template for nucleation of other metal nanostructures on their surface and lead to the formation of alloyed nanoparticles. Alloyed nanoparticles are prepared via chemical synthesis by using reducing agents such as $NaBH_4$ and metal salts are mixed with peptides in the presence of gold or silver nanoparticles as supporting nano-templates. Peptide named as "CPd_4" with amino acid sequence CTSNAVHPTLRHL was used for the synthesis of nanoparticles. Several other peptides with different amino acid sequences have been reported for the synthesis of bimetallic nanoparticles where the presence of various amino acids plays an important role in regulating morphology, catalytic and biological response of nanoparticles. Apart from the application of peptides and derivatives for bimetallic nanoparticles, they have also been reported for the synthesis of metal oxide and metal sulfide nanoparticles as well such as Fe_2O_3, TiO_2, ZnO, Gd_2O_3, CdS, and CuS nanoparticles have been prepared by using amyloid-β peptide sequence as template in the presence of reducing agents [65].

14.5 Conclusion

Here, we discussed peptide-assisted chemical and photoinduced synthesis of novel metal nanoparticles including metal oxide, metal sulfide, and bimetallic nanoparticles that can be used for various biomedical, material, and catalytic applications. Plasmonic nanoparticles such as AgNPs and

AuNPs can be used as theranostics for remedy of cancer or as combinatorial therapy for cancer treatment. Antimicrobial activity of silver and gold nanoparticles against various multidrug-resistant microbes, and Gram (+/−) bacteria is well known hence they can be loaded on antimicrobial peptides as formulation to obtain enhanced antimicrobial activity. Several peptides in combination with bimetallic nanoparticles and metal oxide and metal sulfide nanoparticles can be used as environmental indicators that can reduce ortho- and para-nitrophenol into less harmful aniline derivative. Several organic reactions are catalyzed by Pt or Pd/Cd nanoparticles; hence, peptide-MNPs can be used as catalysts in several reactions to obtain increased yield. The high biocompatibility and capability to encapsulate various metal nanoparticles on self-assembled nanostructures of peptides make them appropriate candidate for targeted delivery of desired metal nanoparticles inside the body. The optoelectronic properties of nanoparticles help in their *in-vivo* detection and monitoring of their locomotion inside the cell. Peptide-MNPs are becoming promising candidates to be used in medicinal and material chemistry. Several nanoparticles have been reported for their capability for photocatalytic degradation of organic dyes and pollutants. Peptides can trap several toxic chemical compounds and metal ions from water therefore, they can be used for wastewater treatment that is highly desired and demanding area of environmental science. The huge demand of drinking water is a big challenge therefore, scientists are developing advanced technologies by using peptide-nanoparticles based filters for wastewater treatment. The chronicles of peptide–metal nanoparticles for various material and biomedical applications always attract huge scientific interest and inspire us to continuous exploration of the possibilities of their uses in new materialistic applications.

Abbreviations

AFM	Atomic force microscopy
AgNPs	Silver nanoparticles
AMP	Antimicrobial peptide
AuNPs	Gold nanoparticles
BSPP	bis(p-sulfonatophenyl)phenylphosphine
BUF	Buforin
CD	Circular dichroism
DLS	Dynamic light scattering
EBC	External breath condensate

EDS/EDAX	Energy dispersive X-ray spectroscopy
FCC	Face centred cubic crystal
FRET	Forster resonance energy transfer
HER	Hydrogen evolution reaction
HRTEM	High resolution transmission micrographs
Ind	Indolicidin
Lyz	Lysozyme
LSPR	Longitudinal surface plasmon resonance
MATLAB	Matrix Laboratory
MNP	Metal nanoparticles
NIR	Near infrared
ORR	Oxygen Reduction Reaction
pDMAEMA	(poly (2-dimethylamino) ethyl methacrylate) methyl chloride
PEA	Poly ether amine
PET	Photoinduced electron transfer
PG	Porcine protegrin
PSD	Particle size distribution
SARS-Cov2	Severe acute respiratory syndrome corona virus (Strain-II)
SPR	Surface plasmon resonance
SEM	Scanning electron microscopy
TEM	Transverse electron microscopy
TSPR	Transverse surface plasmon resonance
Trp	Tryptophan
XRD	X-ray diffraction

References

1 Gawande, M.B., Shelke, S.N., Zboril, R., and *Varma, R.S. (2014).* Microwave-assisted chemistry: synthetic applications for rapid assembly of nanomaterials and organics. *Acc. Chem. Res.* 47 (4): 1338–1348. *https://pubs.acs.org/doi/10.1021/ar400309b.*

2 Shnoudeh, A.J., Hamad, I., Abdo, R.W. et al. (2019). Synthesis, characterization, and applications of metal nanoparticles. In: *Biomaterials and Bionanotechnology; Advances in Pharmaceutical Product Development and Research* (ed. R.K. Tekade), 527–612. Elsevier https://linkinghub.elsevier.com/retrieve/pii/B9780128144275000159.

3 Gogurla, N., Sinha, A.K., Naskar, D. et al. (2016). Metal nanoparticles triggered persistent negative photoconductivity in silk protein hydrogels. *Nanoscale* 8 (14): 7695–7703. http://xlink.rsc.org/?DOI=C6NR01494A.

4 Cao, M., Xing, R., Chang, R. et al. (2019). Peptide-coordination self-assembly for the precise design of theranostic nanodrugs. *Coord. Chem. Rev.* 397: 14–27. https://linkinghub.elsevier.com/retrieve/pii/S0010854519303200.

5 Akrami, M., Samimi, S., Alipour, M. et al. (2021). Potential anticancer activity of a new pro-apoptotic peptide–thioctic acid gold nanoparticle platform. *Nanotechnology* 32 (14): 14510. https://iopscience.iop.org/article/10.1088/1361-6528/abd3cb.

6 Ding, Y., Sun, Z., Tong, Z. et al. (2020). Tumor microenvironment-responsive multifunctional peptide coated ultrasmall gold nanoparticles and their application in cancer radiotherapy. *Theranostics* 10 (12): 5195–5208. http://www.thno.org/v10p5195.htm.

7 Khlebtsov, N., Bogatyrev, V., Dykman, L. et al. (2013). Analytical and theranostic applications of gold nanoparticles and multifunctional nanocomposites. *Theranostics* 3 (3): 167–180. http://www.thno.org/v03p0167.htm.

8 Spicer, C.D., Jumeaux, C., Gupta, B., and Stevens, M.M. (2018). Peptide and protein nanoparticle conjugates: versatile platforms for biomedical applications. *Chem. Soc. Rev.* 47 (10): 3574–3620. http://xlink.rsc.org/?DOI=C7CS00877E.

9 Franco-Ulloa, S., Guarnieri, D., Riccardi, L. et al. (2021). Association mechanism of peptide-coated metal nanoparticles with model membranes: a coarse-grained study. *J. Chem. Theory Comput.* 17 (7): 4512–4523. https://pubs.acs.org/doi/10.1021/acs.jctc.1c00127.

10 Sadegh, H., Ali, G.A.M., Gupta, V.K. et al. (2017). The role of nanomaterials as effective adsorbents and their applications in wastewater treatment. *J. Nanostructure Chem.* 7 (1): 1–14. http://link.springer.com/10.1007/s40097-017-0219-4.

11 Pan, L., He, Q., Liu, J. et al. (2012). Nuclear-targeted drug delivery of TAT peptide-conjugated monodisperse mesoporous silica nanoparticles. *J. Am. Chem. Soc.* 134 (13): 5722–5725. https://pubs.acs.org/doi/10.1021/ja211035w.

12 Valodkar, M., Jadeja, R.N., Thounaojam, M.C. et al. (2011). Biocompatible synthesis of peptide capped copper nanoparticles and their biological effect on tumor cells. *Mater. Chem. Phys.* 128 (1–2): 83–89. https://linkinghub.elsevier.com/retrieve/pii/S0254058411001507.

13 Satriano, C., Munzone, A., Cucci, L.M. et al. (2018). Angiogenin-mimetic peptide functionalised gold nanoparticles for cancer therapy applications. *Microchem. J.* 136: 157–163. https://linkinghub.elsevier.com/retrieve/pii/S0026265X16304179.

14 Tran, N.T.T., Wang, T.-H., Lin, C.-Y. et al. (2011). Direct synthesis of rev peptide-conjugated gold nanoparticles and their application in cancer therapeutics. *Bioconjug. Chem.* 22 (7): 1394–1340. https://pubs.acs.org/doi/10.1021/bc2001215.

15 Rajchakit, U. and Sarojini, V. (2017). Recent developments in antimicrobial-peptide-conjugated gold nanoparticles. *Bioconjug. Chem.* 28 (11): 2673–2686. https://pubs.acs.org/doi/10.1021/acs.bioconjchem.7b00368.

16 Chen, K.-J. and Lee, C.-K. (2018). Twofold enhanced dispersin B activity by N-terminal fusion to silver-binding peptide for biofilm eradication. *Int. J. Biol. Macromol.* 118: 419–426. https://linkinghub.elsevier.com/retrieve/pii/S0141813018322037.

17 Seferji, K.A., Susapto, H.H., Khan, B.K. et al. (2021). Green synthesis of silver-peptide nanoparticles generated by the photoionization process for anti-biofilm application. *ACS Appl. Bio Mater.* 4 (12): 8522–8535. https://pubs.acs.org/doi/10.1021/acsabm.1c01013.

18 Simões, M.F., Ottoni, C.A., and Antunes, A. (2020). Mycogenic metal nanoparticles for the treatment of mycobacterioses. *Antibiotics* 9 (9): 569. https://www.mdpi.com/2079-6382/9/9/569.

19 Teengam, P., Siangproh, W., Tuantranont, A. et al. (2017). Multiplex paper-based colorimetric DNA sensor using pyrrolidinyl peptide nucleic acid-induced AgNPs aggregation for detecting MERS-CoV, MTB, and HPV oligonucleotides. *Anal. Chem.* 89 (10): 5428–5435. https://pubs.acs.org/doi/10.1021/acs.analchem.7b00255.

20 Sharma, R., Raghav, R., Priyanka, K. et al. (2019). Exploiting chitosan and gold nanoparticles for antimycobacterial activity of in silico identified antimicrobial motif of human neutrophil peptide-1. *Sci. Rep.* 9 (1): 7866. https://www.nature.com/articles/s41598-019-44256-6.

21 Geng, H., Vilms Pedersen, S., Ma, Y. et al. (2022). Noble metal nanoparticle biosensors: from fundamental studies toward point-of-care diagnostics. *Acc. Chem. Res.* 55 (5): 593–604. https://pubs.acs.org/doi/10.1021/acs.accounts.1c00598.

22 Di Natale, C., De Benedictis, I., De Benedictis, A., and Marasco, D. (2020). Metal–peptide complexes as promising antibiotics to fight emerging drug resistance: new perspectives in tuberculosis. *Antibiotics* 9 (6): 337. https://www.mdpi.com/2079-6382/9/6/337.

23 Pandey, R., Mehta, N., Ghorui, A. et al. (2022). Bio-assisted preparation of nano zinc oxide and its behavior towards textile azo pollutants. *High Energy Chem.* 56 (2): 101–108. https://link.springer.com/10.1134/S0018143922020126.

24 Gavriely, S., Hadibrata, W., Nudelman, R. et al. (2021). One-pot bio-assisted synthesis of stable Ag–AgCl system using jellyfish-based scaffold for plasmonic photocatalysis applications. *Adv. Sustain. Syst.* 5 (7): 2100099. https://onlinelibrary.wiley.com/doi/10.1002/adsu.202100099.

25 Guo, Z., Song, Y., Wang, Y. et al. (2021). Macrochirality of self-assembled and co-assembled supramolecular structures of a pair of enantiomeric peptides. *Front. Mol. Biosci.* 8: 700964. https://www.frontiersin.org/articles/10.3389/fmolb.2021.700964/full.

26 Lv, M., Jan Cornel, E., Fan, Z., and Du, J. (2021). Advances and perspectives of peptide and polypeptide-based materials for biomedical imaging. *Adv. NanoBiomed. Res.* 1 (5): 2000109. https://onlinelibrary.wiley.com/doi/10.1002/anbr.202000109.

27 Schnitzer, T., Paenurk, E., Trapp, N. et al. (2021). Peptide–metal frameworks with metal strings guided by dispersion interactions. *J. Am. Chem. Soc.* 143 (2): 644–648. https://pubs.acs.org/doi/10.1021/jacs.0c11793.

28 Sun, T., Feng, Y., Peng, J. et al. (2022). Cofactors-like peptide self-assembly exhibiting the enhanced catalytic activity in the peptide-metal nanocatalysts. *J. Colloid Interface Sci.* 617: 511–524. https://linkinghub.elsevier.com/retrieve/pii/S0021979722003630.

29 Li, T., Lu, X.-M., Zhang, M.-R. et al. (2022). Peptide-based nanomaterials: self-assembly, properties and applications. *Bioact. Mater.* 11: 268–282. https://linkinghub.elsevier.com/retrieve/pii/S2452199X21004473.

30 Faikhruea, K., Choopara, I., Somboonna, N. et al. (2022). Enhancing peptide nucleic acid–nanomaterial interaction and performance improvement of peptide nucleic acid-based nucleic acid detection by using electrostatic effects. *ACS Appl. Bio Mater.* 5 (2): 789–800. https://pubs.acs.org/doi/10.1021/acsabm.1c01177.

31 Mishra, N.K., Kumar, V., and Joshi, K.B. (2015). Fabrication of gold nanoparticles on biotin-di-tryptophan scaffold for plausible biomedical applications. *RSC Adv.* 5 (79): 64387–64394. http://xlink.rsc.org/?DOI=C5RA11121H.

32 Jin, Z., Mantri, Y., Retout, M. et al. (2022). A charge-switchable zwitterionic peptide for rapid detection of SARS-CoV-2 main protease. *Angew. Chem. Int. Ed.* 61 (9): e202112995. https://onlinelibrary.wiley.com/doi/10.1002/anie.202112995.

33 Retout, M., Mantri, Y., Jin, Z. et al. (2022). Peptide-induced fractal assembly of silver nanoparticles for visual detection of disease biomarkers. *ACS Nano* 16 (4): 6165–6175. https://pubs.acs.org/doi/10.1021/acsnano.1c11643.

34 Mishra, N.K., Kumar, V., and Joshi, K.B. (2015). Thermoplasmonic effect of silver nanoparticles modulates peptide amphiphile fiber into nanowreath-like assembly. *Nanoscale* 7 (47): 20238–20248. http://xlink.rsc.org/?DOI=C5NR06577A.

35 Kumar, V., Mishra, N.K., Gupta, S., and Joshi, K.B. (2017). Short peptide amphiphile cage facilitate engineering of gold nanoparticles under the laser field. *ChemistrySelect* 2 (1): 211–218. https://onlinelibrary.wiley.com/doi/10.1002/slct.201601548.

36 Kumar, V., Gupta, S., Mishra, N.K. et al. (2017). Laser-induced fabrication of gold nanoparticles on shellac-driven peptide nanostructures. *Mater. Res. Express* 4 (3): 035036. https://iopscience.iop.org/article/10.1088/2053-1591/aa63e0.

37 Tan, H., Liu, S., He, Y. et al. (2021). Spider toxin peptide-induced NIR gold nanocluster fabrication for GSH-responsive cancer cell imaging and nuclei translocation. *Front. Bioeng. Biotechnol.* 9: 780223. https://www.frontiersin.org/articles/10.3389/fbioe.2021.780223/full.

38 Decandio, C.C., Silva, E.R., Hamley, I.W. et al. (2015). Self-assembly of a designed alternating arginine/phenylalanine oligopeptide. *Langmuir* 31 (15): 4513–4523. https://pubs.acs.org/doi/10.1021/acs.langmuir.5b00253.

39 Bhangu, S.K., Baral, A., Zhu, H. et al. (2021). Sound methods for the synthesis of nanoparticles from biological molecules. *Nanoscale Adv.* 3 (17): 4907–4917. http://xlink.rsc.org/?DOI=D1NA00496D.

40 Baral, A., Bhangu, S.K., Cimino, R. et al. (2020). Sono-Assembly of the [Arg-Phe]4 octapeptide into biofunctional nanoparticles. *Nanomaterials* 10(9): 1772. https://www.mdpi.com/2079-4991/10/9/1772.

41 Yan, L., Cai, Y., Zheng, B. et al. (2012). Microwave-assisted synthesis of BSA-stabilized and HSA-protected gold nanoclusters with red emission. *J. Mater. Chem.* 22 (3): 1000–1005. http://xlink.rsc.org/?DOI=C1JM13457D.

42 Chuang, K.-T. and Lin, Y.-W. (2017). Microwave-assisted formation of gold nanoclusters capped in bovine serum albumin and exhibiting red or blue emission. *J. Phys. Chem. C* 121 (48): 26997–27003. https://pubs.acs.org/doi/10.1021/acs.jpcc.7b09349.

43 Zhou, Y., Lv, Y., Dong, H. et al. (2021). Ultrasensitive assay of amyloid-beta oligomers using Au-vertical graphene/carbon cloth electrode based on poly(thymine)-templated copper nanoparticles as probes. *Sensors Actuators B Chem.* 331: 129429. https://linkinghub.elsevier.com/retrieve/pii/S092540052031769X.

44 Aladaghlo, Z., Javanbakht, S., Fakhari, A.R., and Shaabani, A. (2021). Gelatin microsphere coated Fe3O4@graphene quantum dots nanoparticles as a novel magnetic sorbent for ultrasound-assisted dispersive magnetic solid-phase extraction of tricyclic antidepressants in biological samples. *Microchim. Acta* 188 (3): 73. https://link.springer.com/10.1007/s00604-021-04727-y.

45 Eivazzadeh-Keihan, R., Bahreinizad, H., Amiri, Z. et al. (2021). Functionalized magnetic nanoparticles for the separation and purification of proteins and peptides. *TrAC Trends Anal. Chem.* 141: 11629. https://linkinghub.elsevier.com/retrieve/pii/S016599362100114X.

46 Zhou, H., Wang, Y., and Lu, H. (2021). Intracellular delivery of His-tagged proteins via a hybrid organic–inorganic nanoparticle. *Polym. J.* 53 (11): 1259–1267. https://www.nature.com/articles/s41428-021-00526-7.

47 Singh, R.K., Malosse, C., Davies, J. et al. (2021). Using gold nanoparticles for enhanced intradermal delivery of poorly soluble auto-antigenic peptides. *Nanomed. Nanotechnol. Biol. Med.* 32: 10232. https://linkinghub.elsevier.com/retrieve/pii/S1549963420301751.

48 Xie, M., Slocik, J.M., Kelley-Loughnane, N., and Knecht, M.R. (2022). Effect of a mixed peptide ligand layer on Au nanoparticles for optical control of catalysis. *ACS Appl. Nano Mater.* 5 (7): 9379–9388. https://pubs.acs.org/doi/10.1021/acsanm.2c01674.

49 Abbas, M., Susapto, H.H., and Hauser, C.A.E. (2022). Synthesis and organization of gold-peptide nanoparticles for catalytic activities. *ACS Omega* 7 (2): 2082–2090. https://pubs.acs.org/doi/10.1021/acsomega.1c05546.

50 Lea, M.C. (1889). Allotropic forms of silver. *Am. J. Sci.* s3–37 (222): 476–479. https://ajs.scholasticahq.com/article/62812.

51 Gurunathan, J. and Kang, K. (2019). Mitochondrial peptide humanin protects silver nanoparticles-induced neurotoxicity in human neuroblastoma cancer cells (SH-SY5Y). *Int. J. Mol. Sci.* 20 (18): 4439. https://www.mdpi.com/1422-0067/20/18/4439.

52 Pal, I., Bhattacharyya, D., Kar, R.K. et al. (2019). A peptide-nanoparticle system with improved efficacy against multidrug resistant bacteria. *Sci. Rep.* 9 (1): 4485. https://www.nature.com/articles/s41598-019-41005-7.

53 Ramírez-Acosta, C.M., Cifuentes, J., Castellanos, M.C. et al. (2020). PH-responsive, cell-penetrating, core/shell magnetite/silver nanoparticles for the delivery of plasmids: preparation, characterization, and preliminary in vitro evaluation. *Pharmaceutics* 12 (6): 56. https://www.mdpi.com/1999-4923/12/6/561.

54 Zharkova, M.S., Golubeva, O.Y., Orlov, D.S. et al. (2021). Silver nanoparticles functionalized with antimicrobial polypeptides: benefits and possible pitfalls of a novel anti-infective tool. *Front. Microbiol.* 12: 750556. https://www.frontiersin.org/articles/10.3389/fmicb.2021.750556/full.

55 Gupta, S., Kumar, V., and Joshi, K.B. (2017). Solvent mediated photo-induced morphological transformation of AgNPs-peptide hybrids in water-EtOH binary solvent mixture. *J. Mol. Liq.* 236: 266–277. https://linkinghub.elsevier.com/retrieve/pii/S016773221730452X..

56 Gupta, S., Kashyap, M., Kumar, V. et al. (2018). Peptide mediated facile fabrication of silver nanoparticles over living diatom surface and its application. *J. Mol. Liq.* 249: 600–608. https://linkinghub.elsevier.com/retrieve/pii/S0167732217338199.

57 Ottoni, C.A., Lima Neto, M.C., Léo, P. et al. (2020). Environmental impact of biogenic silver nanoparticles in soil and aquatic organisms. *Chemosphere* 239: 124698. https://linkinghub.elsevier.com/retrieve/pii/S0045653519319228.

58 Zhu, S., Wang, X., Li, S. et al. (2020). Near-infrared-light-assisted in situ reduction of antimicrobial peptide-protected gold nanoclusters for stepwise killing of bacteria and cancer cells. *ACS Appl. Mater. Interfaces* 12 (9): 11063–11067. https://pubs.acs.org/doi/10.1021/acsami.0c00310.

59 Mokashi-Punekar, S., Walsh, T.R., and Rosi, N.L. (2019). Tuning the structure and chiroptical properties of gold nanoparticle single helices via peptide sequence variation. *J. Am. Chem. Soc.* 141 (39): 15710–15716. https://pubs.acs.org/doi/10.1021/jacs.9b08798.

60 Jabir, M.S., Taha, A.A., Sahib, U.I. et al. (2019). Novel of nano delivery system for Linalool loaded on gold nanoparticles conjugated with CALNN peptide for application in drug uptake and induction of cell death on breast cancer cell line. *Mater. Sci. Eng. C* 94: 949–964. https://linkinghub.elsevier.com/retrieve/pii/S0928493118308853.

61 Li, Y., Yuan, M., Khan, A.J. et al. (2019). Peptide-gold nanocluster synthesis and intracellular Hg2+ sensing. *Colloids Surf. A Physicochem. Eng. Asp.* 579: 123666. https://linkinghub.elsevier.com/retrieve/pii/S0927775719306429.

62 Gao, P., Wu, S., Chang, X. et al. (2018). Aprotinin encapsulated gold nanoclusters: a fluorescent bioprobe with dynamic nuclear targeting and selective detection of trypsin and heavy metal. *Bioconjug. Chem.* 29 (12): 4140–4148. https://pubs.acs.org/doi/10.1021/acs.bioconjchem.8b00773.

63 Subair, R., Tripathi, B.P., Formanek, P. et al. (2016). Polydopamine modified membranes with in situ synthesized gold nanoparticles for catalytic and environmental applications. *Chem. Eng. J.* 295: 358–369. https://linkinghub.elsevier.com/retrieve/pii/S1385894716302145.

64 Wu, W., Tang, Z., Wang, K. et al. (2018). Peptide templated AuPt alloyed nanoparticles as highly efficient bi-functional electrocatalysts for both oxygen reduction reaction and hydrogen evolution reaction. *Electrochim. Acta* 260: 168–176. https://linkinghub.elsevier.com/retrieve/pii/S0013468617324088.

65 Wagner, J.T., Long, A.K., Sumner, M.B. et al. (2020). Peptide controlled assembly of palladium nanoparticles on high-aspect-ratio gold nanorods. *J. Phys. Chem. C* 124 (50): 27743–27753. https://pubs.acs.org/doi/10.1021/acs.jpcc.0c07653.

15

Pharmacotherapy Approach of Peptide-Assisted Nanoparticle

Shivani A. Kumar, Rimon Ranjit Das, and Surbhi Malik

Department of Physics, Amity Institute of Applied Sciences, Amity University, Noida, UP, India

Abstract

Peptide-nanoparticle conjugates (PNCs) have recently emerged as a promising medicinal tool. Synergy between the two potential kinds of accouterments allows for greater control over their natural actions, exceeding the separate paraphernalia's innate limitations. PNCs have been developed for a variety of functions over the years, including drug delivery, blockage of pathogenic biomolecular relationships, molecular imaging, and liquid vivisection. Here, in this template, we present a thorough overview of the essential sequences and structural features used to create polypeptide self-assembled nanoparticles, direct particles to their target and affect their distribution in cells and tissues, stimulate and regulate cellular response, and stabilize nano-constructs both biologically and physically. This study presents a detailed summary of recent developments in the broad topic of PNCs, as well as a guideline for scientists who are new to the field.

Keywords *peptide-nanoparticle conjugates; medicine transport; protein commerce asset; molecular tomography, liquid biopsy; vivo imaging; peptide therapy*

15.1 Introduction

Peptides have attracted much attention in the biomedical field as a novel material that is able to express the capabilities of proteins while maintaining a high degree of flexibility in molecular design. Current approaches to

Green Synthesis of Nanomaterials: Biological and Environmental Applications, First Edition.
Edited by Archana Chakravarty, Preeti Singh, Saiqa Ikram, and R.N. Yadava.

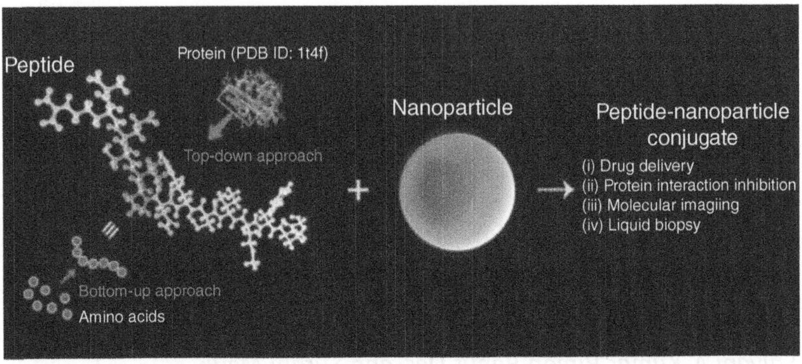

Figure 15.1 Finding synthetic bioactive peptides and fusing them with nanoparticles for medicinal uses.

the discovery of genetically manipulated bioactive peptides can be divided into two orders of magnitude (Figure 15.1): (i) construct and mesh peptide libraries from the arbitrary amino acid composition in a given macromolecular topology (peptide library bank, bottom-up) and (ii) isolation of biologically active sequences from natural proteins based on three-dimensional conformation (3D) of them (structure-based design, top-down approach) [1–4]. Membrane peptide libraries enable rapid production of effective binders for a wide range of target bacteria (e.g. peptides, DNA, RNA, and inorganic materials). On the other hand, the top-down strategy has an advantage over the bottom-up approach in that peptide sequences targeting a certain list point on biomolecules can be detected based on structural components of them.

A great number of studies have proved the use of synthesized bioactive peptides during the last few decades, and some of these compounds have been successfully commercialized. In the previous two decades, 28 noninsulin peptide medications have been authorized worldwide [5]. Furthermore, about 150 peptide therapies are in clinical trials, with promising outcomes for eventual commercialization [6]. Despite recent advances, most peptides have yet to be widely used because of (i) their poorer target binding affinity and specificity than proteins; (ii) susceptibility to protease digestion in biochemical environments [7]; (iii) limited circulatory half-lives necessitating repeated doses to maintain efficacy [8]; and (iv) inability to retain ingrain folding structures when isolated from proteins [9].

Numerous researchers have found that objectifying peptides using abiotic additives (e.g. small molecule composites, gasoline chelates) is a potential way to overcome these problems, inherent disadvantages of peptides [10, 11]. In particular, nanoparticles (NPs) have demonstrated the potential to act as conjugate seats not only enhancing peptide activity but also adopting abiotic properties, leading to synergistic effects (face 2). As a result, peptide-NP conjugates (PNCs) have been hailed as a catwalk for a number of medical applications. This study focuses on PNC, highlighting current advances in PNC-based technologies and their applications in personal, imaging, and therapeutic procedures. The benefits of using PNCs will first be briefly described (side 2), followed by a description of examples of their successful activities in the biomedical field, such as the distribution of manufactured drugs. On demand (side 3), blocking trafficking of pathogenic proteins (side 4), the molecules are largely susceptible imaging (face 5) and liquid biopsies (face 6). Finally, we will discuss exploratory operations that have been rapidly generated but still pose various obstacles in terms of clinical interpretation.

15.2 The Peptide-NP Conjugation

Nanomaterials (a few hundreds of nanometers in size) have unique physiochemical packages that are separate from those of bulk accessories. The development of finagled accessories that can uniquely interact nanobiomaterials is beneficial due to their ultra-small size and low volume-to-face-area rate [12]. Tone assembly is the most straightforward method for creating peptide-based nanostructures [13, 14]. Despite this, the naturalness of the thermodynamic process prevents the creation of nanoscale constructions with perfectly controlled shape, size, and composition. In contrast, peptide-NP conjugation allows for greater control over the structural components of nanostructures, enabling for quick changes to the overall form, dimension, and size of the conjugates by using NP pulpits that have been adapted for the necessary procedures.

Multivalent is another important feature that PNC can provide. Noncovalent interactions such as hydrogen bonds, ionic bonds, Van der Waals forces, mound forces, and hydrophobic relationships underlie the most fundamental relationships in natural systems. Although the individual bands are relatively weak, their combined action allows for strong list kinetics (the multivalent list effect causes a significant decrease in dissociation kinetics), based on the principle of rule that the strength of the collaborating list is proportional to the number of pairs of individual lists (Figure 15.2a) [15–17]. Multivalent relations, in addition to improving list strength, also improve selectivity by utilizing the viscosity of commerce modules on a face to fete target polyvalent shells (Figure 15.2b) [18].

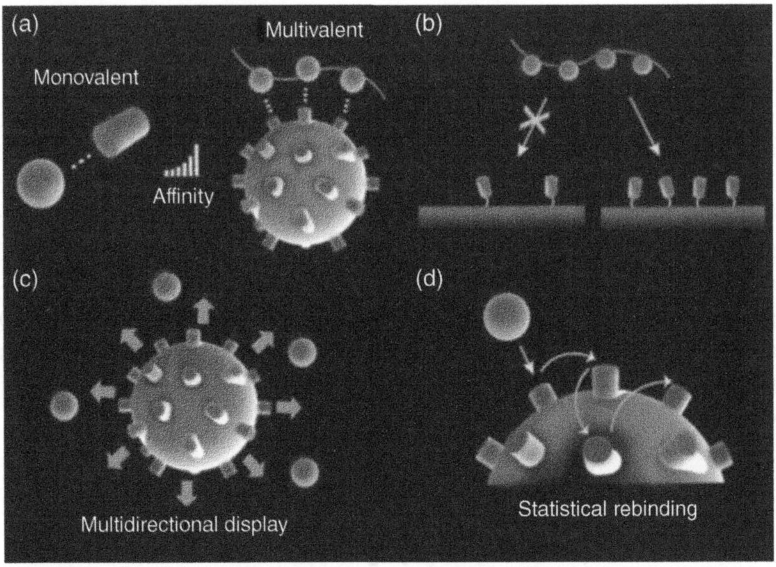

Figure 15.2 (a) Differentiation of interactions between monovalent and multivalent compounds. (b) Selectivity in multivalent interactions. (c) Multidirectional ligand display and (d) statistical rebinding on a multivalent compound.

Multiple list locations are important for allowing powerful multivalent tapes and providing statistical gaps for multiple monovalent list events to happen. The exhibit-sure of peptides in various directions, as shown in Figure 15.2c, leads in fewer opportunities to engage binding mates [19]. Peptides on NP pulpits display several re-binding areas during the dissociation process post list, which can improve the time frame of target accessories on the face, known as the statistical re-binding medium (Figure 15.2d) [20]. Co-conjugation with various types of peptides and/or other natural/non-biological accouterments, on the other hand, provides new functionalities for mongrel accouterments, such as vulnerable response elusion [21], theranostics [22], and multi-target directed treatment with a single material [23, 24]. As a result, peptides might possibly compete with or surpass natural proteins when shown on a nanostructure face, despite their low individual affinity, specificity, and selectivity [25, 26].

Non-biological properties of NPs introduce additional portions and functions that would otherwise be unavailable to their PNCs. For example, because of the advantages of deep imaging depth and excellent spatial resolution, NPs that absorb and emit near infrared (NIR, 700–1100 nm) light have been laboriously used in *in vivo* imaging [27]. When certain NPs come into contact with light energy, they form reactive oxygen species (ROS), which can damage memory macromolecules and eventually cause cell ablation (photodynamic remedy) [28]. Furthermore, employing photo-thermal and photoacoustic properties of NPs, absorbed light energy can be transformed to heat and sound energy, providing a non-invasive therapeutic alternative for illnesses like as cancer [29, 30]. Another intriguing type of nanoparticles is glamorous nanoparticles (MNPs), which can be used to treat illnesses remotely and actively.

MNPs can be extensively synthesized at a target site in natural systems and release guest particles in a diamond-controlled form in response to external glam stimuli [31, 32]. Several *in vitro* experiments have demonstrated that MNPs with multiple ligands on the list can effectively distinguish target biomaterials from the resulting mixing [33]. Likewise, upon exposure to the charm field, the arrangement of the MNPs on the surface can be tuned in a variety of ways, allowing the construction of novel cell culture platforms using MNPs [34].

15.3 Targeted Drug Delivery

Pharmaceutical materials have to be introduced to particular places with inside the body, which stays a massive difficulty. Peptides have lately emerged as an enormous mag that has the capacity to offer modular selectivity to medication transport structures, ensuing in progressed overall performance for the implicit remedy of plenty of most important fitness disorders, inclusive of most cancers and neurological ailments [35, 36]. Peptides engage with numerous herbal structures in precise ways, permitting them to be utilized in a huge variety of scripts for useful effects [37]. Nonetheless, vim-runs' brief *in vivo* half-existence and sub-top-quality biodistribution and pharmacokinetics have hampered their giant use in remedy transport [38]. Mixing peptides with NPs is a trustworthy manner to resolve the problems with the prevailing peptide-primarily based totally transport technology. In addition to their cap potential to synopsize and coat healing compounds, NPs may be finagled to preferentially supply medicines to the goal place after being functionalized with peptides as focused on agents, growing tube rotation time. As a result, researchers have coupled diverse focused on peptides to diverse forms of NPs with the intention to create greater powerful and adaptive remedy transport structures. The connection of load to the junction of cells is one of the precise moves of peptide-intermediated focused on. The diverse partitions that have to be conquered formerly in the cellular, tons on my own aiming to the right cellular withinside the first place, make transport to the nexus enormously difficult.

To keep away from being spoiled earlier than accomplishing the nexus, they have to have a few kind of endosomal break out mechanism [39]. When that is accomplished, the flyspeck has to recover from the nexus's safeguards. A bilayer phospholipid membrane protects the nexus, that's accessed usually via way of means of nuclear severance complexes (NSCs) with diameters starting from 20 to 150 nm [40]. The atoms or particles have to now no longer most effective be small sufficient to healthy thru the NSC, however it have to even have a corresponding nuclear localization signal (NLS), which fits as a key card to get entrance. Pan et al. determined a way to those problems *in vitro* via way of means of the use of mesoporous silica NPs conjugated with TAT peptide to supply doxorubicin (DOX) to the HeLa nexus. Their findings monitor that patches with a diameter of much less than 50 nm have been powerful in attaining TAT peptide-mediated nuclear absorption and uninterrupted DOX launch into the nexus throughout a 24-hour incubation period. Tkachenko et al. took a exceptional strategy, the use of a multipeptide conjugated gold NP(AuNP) grounded gadget for this purpose [41, 42]. They determined that the use of brief peptides, one for cell endocytosis and the alternative for nuclear focused on of the flyspeck, is greater powerful than the use of an unmarried lengthy sequence. When incubated for two hours at 37 °C, the 25 nm AuNP turned into capable of penetrate the nexus in 80% of HepG2 cells. Although Li and his colleague's intention turned into to supply siRNA for gene silencing, they used a 13 nm AuNP-primarily based totally tool connected with an NLS peptide [43]. They observed that their compound turned into powerful in inhibiting TK1 protein and TK1 mRNA frequency *in vitro* and decreasing excrescence boom via way of means of 250% in an *in vivo* mouse version while as compared to a control. Transdermal management for the remedy of most cancers is every other interesting operation for NSCs. The stratum corneum, the farthest subcaste of pores and skin, is the largest stumbling block to this operation's transport. Niu advanced an AuNP-primarily based totally approach for plasmid DNA (pDNA) transport that used conjugated TAT peptides [44]. TAT peptides enhance pores and skin penetration and NP gene transfection for a powerful topological transport gadget, in line with their findings. Patlolla et al. took use of TAT peptides' pores and skin saturation homes via way of means of conjugating them to 180 nm nanolipid crystal NPs (NLCNs) [45]. Compared with other complexes studied, they concluded that their combination penetrated up to 120 μm into the skin of mice, with improved tracking of the collected plaques in both the cornea and the dermal layer. For the use of Vemurafenib, Zou took a different approach, choosing a liposome NP coupled to the TD peptide [46]. Their findings suggest that TD peptides are capable of opening up paracellular pathways in the stratum corneum for the application of transdermal cancer patches. Peptides have also been shown to help NPs overcome other physiological barriers, such as the blood-brain barrier (BBB), which is a key barrier to efficient drug delivery to the brain. The BBB acts as a barrier, surrounding blood vessels that supply sugar to the brain, with the main purpose of preventing harmful substances from entering the delicate vessels behind it [47]. Peptides have been used by researchers to facilitate the transport of NPs across the BBB. For example, Georgieva et al. used G23 peptide-conjugated polymers to deliver drugs to the BBB *in vitro* and *in vivo* [48]. The NPs (165 nm) used the G23 peptide to bind to ganglioside receptors GM1 and GT1b of the navigator progeny induced on hCMEC/D3 cells (killer BBB model), leading to the possibility transcoding was reduced four-fold compared with the G23 peptide-deficient polymers. Yao et al. reported the delivery of pDNA through the BBB using dendrigraft NP poly-l-lysines (DGL) conjugated poly(ethylene glycol) (Cut) and the inferred transcellular peptide LIM Kinase 2 (LNP) [49]. Their innovative approach uses LNP, which helps additional BBB cells to absorb more cells. PNC-based techniques have shown several successful examples, including excellent targeting of diseased cells and perfusion across physiological barriers. However, various hurdles need to be addressed before this technique can be fully implemented, such as immunogenicity, long-term toxicity, and ultimate goals. If these concerns are addressed, the PNC approach will provide an essential system capable of delivering highly regulated drugs.

15.4 Pathogenic Protein Interaction Inhibition

One of the most difficult issues in pharmaceutical research is resolving the "unsolvable" targets [50]. Approximately 80% of proteins involved in fatal diseases have binding sites for tiny patch ligands [51]. Protein-based medicinals are one tactic that can be used to solve this problem. Despite this, their widespread use has been hampered by their low heat stability and the difficulty of medicating comparable proteins [52]. PNCs provide a fresh perspective on these insurmountable difficulties. The Lim group, for example, demonstrated inorganic NPs that act as an altar for stabilizing vim-drift folding structures, which are capable of improving target affinity and selectivity [53]. Figure 15.3a shows a spiral structure stabilized by reduced conformational entropy cost achieved by using cyclic peptides and trading with inorganic faces [54–60]. Bioactive α-spiral p53 vim-runs stabilized on AuNP shells successfully honored their target protein, MDM2, which is believed to block the p53-mediated apoptotic pathway, based on this concept. Inhibiting

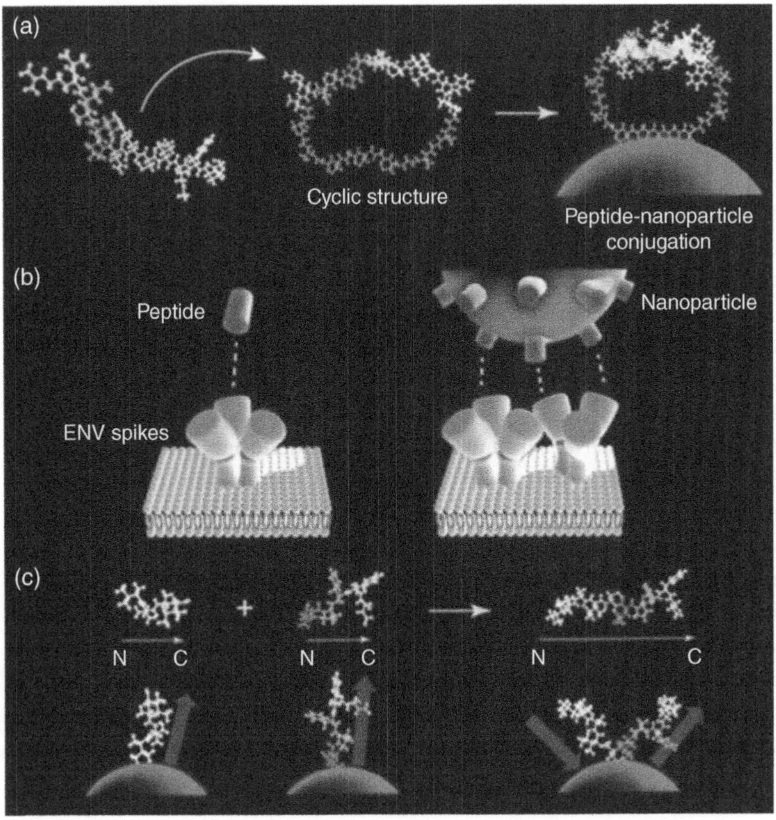

Figure 15.3 (a) Molecular simulations showing how the stability of the helical structure is caused by gold nanoparticle binding. (b) Free peptides and peptide-nanoparticle conjugates' associations with the HIV-1 spike proteins. (c) Gold nanoparticles with peptide hybrid functionalization that prevent amyloid formation.

the α-helix intermediated commerce between Rev protein and Rev response element (RRE) RNA, which regulates HIV-1 gene expression, proved the therapeutic potential of the cyclic peptide – nanomaterial conjugate system [24, 55].

The majority of problems in pharmaceutical research is solving "difficult" criteria [50]. Approximately 80% of proteins involved in life-threatening diseases have binding sites for minor patch ligands [51]. Protein-based drugs are one tactic that can be used to address this problem. Even so, their widespread use has been hampered by their poor thermal stability and difficulty in processing equivalent proteins [52]. The Crew Members bring a fresh perspective to these insurmountable difficulties. For example, Lim group demonstrated inorganic NPs to act as an altar to stabilize vim-drift folding structures, potentially enhancing target affinity and selectivity [53]. Figure 15.3a shows that the helical structure is stabilized by reducing the conformational entropy cost achieved by using cyclic peptides and swapping with inorganic faces [54]. The bioactive p53 α-helical viruses stable on the AuNP shell have successfully honored their target protein, MDM2, which is believed to block the p53-mediated apoptotic pathway, based on this concept. The inhibition of α-helix-mediated metabolism between Rev proteins and RNA by Rev response factor (RRE), which regulates HIV-1 gene expression, demonstrates the therapeutic potential of the cyclic peptide conjugate system – nanomaterials [24, 41, 55]; PNC function [61–68] can be significantly enhanced in certain activities by refining the peptide's valence. High-viscosity peptide-coated NPs were shown to have higher binding affinity [41]; however, binding motifs fine-tuned with some distance or ligand viscosity have been shown to further enhance controlled interactions with target microbes. Dendrimers are a method to modulate the ligand's valency [69–78]. Dendrimers are hyperactive polymers whose functional group size, surface properties, content and viscosity are carefully adjusted through relatively simple chemical reactions. Scientists and researchers recently demonstrated that polyglycerol dendrimers decorated with hemagglutinin (HA)-targeting peptides can suppress infection of influenza A virus (IAV) [25], which exploits multiple HAs to increase ability of host cells to attach to the skin surface. Surprisingly, despite the enhanced antiviral effort of PNC [79–87] using a multivalued list of HA-targeting peptides, inhibitory potency did not increase proportionally to peptide viscosity. In contrast, when inhibitory effort reached a specific level, enhanced chemotherapy lowered it, suggesting that optimization of facial engineering was necessary. PNC also has the advantage of being able to use multiple calibration routes by combining multiple types of spheres into a single nanoformulation system. Blancafort et al. recently conjugated polyglycidyl methacrylate NPs with peptides targeting Engrailed 1 (EN1), a non-drug [88–93] reuptake factor implicated in cell proliferation, metastasis, and chemoresistance in cancers basic cancers such as bone cancer. In the internal violation of this NP, an anticancer drug, docetaxel (DTX), was reintroduced. *In vitro* and *in vivo* studies have demonstrated that using PNC to combine peptides and chemotherapeutic drugs [94–103] induces more apoptosis in cancer cells than using DTX or EN1 peptides alone. On the other hand, Jeong et al. have shown that conjugation of several types of peptides on nanomaterials is an excellent strategy to maximize the restorative effects [24]. Two separate peptides were coupled to autobond nanotubes to suppress the Rev/RRE RNA and Rev/CRM1 relationship, resulting in a 150-fold increase in HIV resistance compared with leptomycin B.

As stated earlier, peptides have shown great promise in overcoming their inherent limitations when combined with NP shells [104–109]. PNCs have been shown to outperform single peptides and even proteins in terms of affinity, selectivity and, therefore, therapeutic efficacy. This PNC technique has also been shown to be effective in other procedures, such as molecular imaging, and individual/prognostic procedures, such as liquid organism anatomy, which will be discussed later.

15.5 Molecular Imaging

Molecular imaging [110–113] provides high-resolution optical intelligence about organic growth. It enables the apprehension of diseased cells and tissues, helping both preclinical researchers and physicians to determine disease progression and response to treatment. Advances in nanobiotechnology have accelerated the development of cell imaging by improving the focusing efficiency of imaging modalities. Peptides [114–120] have been effectively used as novel nanoprobes due to their long-term stability, target specificity, and rapid blood-duct competition, among many other agents that have been used to provide selectivity [37]. The modular design of the same peptides allows them to be integrated into a wide range of imaging modalities, yielding excellent results in animal models and preclinical studies. Because of their small size, peptides are consistently affected by low-list affinity, enzyme instability, and rapid renal competition, despite their benefits. These problems can be solved by combining them with NPs, which have long been used to improve the kinetics of target peptides [37]. Since NPs can be designed to accommodate a wide range of target areas and imaging modalities, they are an excellent means of transport. The potential of peptide/NP complexes to enhance the target background signal is a significant advantage. This can be achieved by coating various imaging tests on the surface of the NP or by increasing the viscosity of certain peptides on the surface of the NP. PNC can be used for multi-targeted nanotherapy if certain types of peptides are combined with restorative drugs. This section describes the recent improvements made possible by using the PNC as an imaging nanoprobe for a wide range of applications, including NIR luminescence imaging, computed tomography (CT), emission tomography. positron emission tomography (PET), magnetic resonance imaging (MRI), and multimodality imaging [121–126].

Imaging agents with a migration spread between 700 and 1100 nm were used in NIR luminescence imaging. NIR light penetrates deeper into the scarf than visible light, allowing for better scarf photography. PNC has recently been used to enable sensitive detection of abnormal tissues with high specificity. Addiction and associates generated fluorescent NPs containing cyclic peptides that were combined with Zn2 ions to generate strong NIR fluorescence signals. This imaging agent has been further enhanced using integrin v3-specific RGD peptides for more efficient development point scanning. This outgrowth-specific imaging agent is photochromic and has limited migration intensity, emitting a clear NIR fluorescence signal from the target. NIR imaging has also been used using fluorophores benzo bis(-thiadiazole). The fluorescent agents are conjugated with peptides specific for the gastrin-releasing peptide receptor (GRPR). According to *in vitro* and *in vivo* studies, these conjugates successfully synthesize into target tissues with high target background signal. A CT scan uses multiple X-ray axes to create cross-sectional images of the body's bones, blood vessels, and soft tissues. Because of their excellent X-ray attenuation, AuNPs are one of the most commonly used imaging agents for CT examination. AuNPs can be used in various applications due to their biocompatibility, stability, and adaptability [37]. AuNPs have been used as selective CT cleavage agents in combination with peptides. Zhu et al. used RGD peptides to decorate AuNP tangled dendrimers (AuDENP) and used these nanoprobes to image the CT growth. AuDENPs outperform Omnipaque in terms of X-ray attenuation. Hao and colleagues recently generated a poly(lactic-co-glycolic acid)-AuNP (PLGA)-AuNP NP shell-core structure. This, combined with Angiopep-2, a glioma-targeting peptide, enhances the cell's selective uptake of PLGA-AuNPs, resulting in superior developmental recognition and accuracy better CT image resolution. PNC-based methods have also been used to image cerebral vascular thrombosis in the brain. For direct CT imaging of cerebral vascular thrombosis, glycol-chitosan-coated AuNPs (GC-AuNPs) were conjugated with fibrin-specific peptides. Compared with peptide-free GC-AuNPs, this novel imaging agent extensively accumulates at the target site and is maintained in situ for longer periods (up to three weeks).

PET is recognized as an excellent non-invasive imaging method for detecting color situations. Conversion of target-specific peptides and positron emitters into NPs enables the detection of abnormal apkins with a high degree of particularity. To visualize the drug bioavailability of targeted NPs, CLPFFD peptides targeting protein fusion filaments were paired with 18F-labeled AuNPs. Cheng altered the appearance of the Au tripod with RGD and 64Cu peptides (64Cu-RGD-Au-tripods) to provide separate, binary integrin-specific targeting and PET imaging, in a single work, another thing. Compared with PNCs co-injected with unbound RGD peptides, these novel PNCs resulted in a threefold improvement in contrast of photoacoustic images (PAI) in animal models of carriers. PET scan also demonstrated that approximately eight corrected NP wounds accumulated and persisted at the target site after 24 hours. Angiogenesis strapping is also a possibility using biodegradable PET dendritic nanoprobes containing RGD peptides. Due to the multivalent relationship, the selectivity of the list of peptide-peptide compounds is 50 times higher than that of monovalent peptides. After labeling the conjugates with 125I and 76Br, the investigation was extended to *in vitro* and *in vivo* PET imaging, showing that the targeted nanoscans exhibited higher levels of cellular uptake against untargeted copies.

MRI uses radio waves and an enchanting field to create high-resolution three-dimensional images of organs and apkins. Charm NPs (MNPs) were used to scan specific organs and their complex formation with targeted peptides. According to Xie and his team, MNPs conjugated with RGD peptides primarily target cells expressing integrin v3. According to their *in vivo* MRI data, conjugation selectivity was maintained in mice carrying U87MG. To enhance targeting and traction detection, MNPs were combined with polymers. For example, RGD peptide-coupled superparamagnetic polymeric micelle nanotubes (SPPM) have been used to selectively detect integrin-overexpressing cells. On the other hand, Simberg combined fibrin-specific peptide CREKA and external growth with amino dextran-coated superparamagnetic iron oxide (SPIO) for imaging and targeted therapy. These conjugates cluster at the growth site, the timbre is amplified and allow for highly selective magnetic resonance imaging.

Despite the fact that PNC has improved the image quality of staining modalities, improved and more accurate resolution imaging is required for early diagnosis and appropriate treatment. Multimodal disparity agents have recently been created to help experimenters and physicians visualize two or more imaging modalities at the same time. For example, 64Cu-HAuNS was developed to combine the capabilities of CT and MRI in a single NP system. The RGD peptides were further paralyzed on the surface of 64Cu-HAuNS to promote precise targeting and better cellular NP uptake, resulting in a binary imaging agent for CT and MRI almost exactly defined. The combination of luminescence and magnetic resonance imaging is another typical method. Dendritic mestizo NPs were labeled with Cy5 and Gd for luminescence and MRI, respectively, after being functionalized with activatable transcellular peptides (ACPPs) on their clade. ACPP increased cell NP uptake by up to 15-fold, suggesting that this method can be used to detect the rapid growth of NIR and MR.

The examples given above clearly show that the PNC can be used as a visual differential in a wide range of modalities. In the field of biomedical imaging, color peptides have been effectively used for point-specific targeting, which can be further improved by using a binary PNC composite, which produces high-quality images, high amounts of apkins, and specific organs. PNC-based molecular imaging is extremely promising, despite the fact that their intrinsic toxicity and loss of natural safety must be taken into account for successful clinical reintroduction.

15.6 Liquid Biopsy

Liquid biology has great potential as a new method for the assessment and prognosis of life-threatening disorders. It refers to any method of examining, decoding, and analyzing biomarkers complained of in body fluids, especially blood. Since liquid ligation is less invasive than traditional solids, it reduces the risk of incidents while increasing case compliance, allowing for more frequent sampling, early detection and monitor conditions more precisely. As a result, liquid ligation provides more detailed information about complaints over time, allowing for faster and more effective treatment.

CTCs, cell-free DNA, and miRNA's have all been consistent as potential biomarkers of death. Several studies have shown that genetic or proteomic profiles of these biomarkers are associated with disease progression, proliferation, precipitation, chemotaxis, and metastatic potential. Due to the scant attention paid to biomarkers of fluid dissection in human fluid, reliable analysis and sensitivity discovery remain a challenge. In addition, the molecular diversity of biomarkers, as well as the phenotypic changes that often occur during remedial treatment and complaint development, make it difficult to separate biomarkers, further limiting downstream research.

This subsection describes several novel methods using PNC to detect biomarkers with excellent specificity and sensitivity. Due to its high selectivity and high list affinity for several facial receptors, antibodies are among the most commonly used chemoattractants for isolating biomarkers associated with complaints. According to recent research, antibodies can be translated into smaller peptides that target specific receptors on the face. Compared to antibodies, peptides are small, durable and simple to manufacture, and have the potential to replace whole antibodies by overcoming the efficiency and consistency issues that existing antibody-based techniques have encountered. Given these advantages, the main disadvantage of peptides is their low list affinity for specific target apkins. However, PNC techniques can help these companies. For example, the multivalent list effect mentioned above can be easily realized in stained PNCs, enhancing the ability to separate biomarkers based on binding of peptides to biomarkers learning target [37, 47].

To isolate CTCs, Yang and Wang's group used peptides honoring the epidermal cell adhesion patch (EpCAM) and lethal epidermal growth factor receptor 2 (HER2). For magnetic immunoassay, these peptides were attached to iron oxide charm NPs. Despite the fact that peptides have lower affinity than antibodies, the PNC-based strategy revealed more than 90 and 70–80% of EpCAM and HER2-positive cancer cells, respectively, due to the relationship multivalent system. Another well-known growth-specific antigen capable of targeting EpCAM-negative CTCs is the epidermal growth factor receptor (EGFR). Dinh and colleagues used reverse-phase evaporation to generate nanoparticles with EGFR-targeting GE11 peptides dispersed on their lipid membranes. Charming EGFR peptide vesicles (EPMVs) have been shown to bind to the SMMC-7221 hepatocellular carcinoma cell type. EPMV has recently outperformed both the EpCAM-based Cell Search and Magnetic Immunoassay cleavage methods in terms of CTC isolation. For the detection and characterization of CTCs, EGFR peptides were additionally combined with face-enhanced Raman scattering AuNPs (SERS) [115]. Test-tube studies have shown more than 90% cancer cell blocking efficiency and the uniqueness of the discovery. Further clinical studies by pilots demonstrated that EGFR-specific PNCs were able to detect up to 720 CTCs/ml in samples from head and neck cancer patients.

Exosomes are extracellular vesicles (EVs) produced by endosomes and play an important role in signaling between cells. Exosomes are known to carry proteins and genetic information from their stem cells. To protect the exosomes involved in the outgrowth of colored EVs in the deadly liquid of flesh, a lot of sweat was required. To trap these vesicles, tetraspanin, a facial protein overexpressed in lethal exosomes, has been widely used. Gao and colleagues recently published a study of a novel NP with CD63-targeting peptides on its surface. Unlike ultracentrifugation technology, this exosome-targeted NP has a capture rate of 54. Overexpression of growth-related proteins AFP and GPC-1 on collected EVs, which are well-defined markers of liver and pancreatic growth, has been established in clinical studies using lethal serum samples. EVs buried in growth blocks have also been identified using other growth-specific sensors. Heat shock protein 70 (Hsp70), a molecular convoy, is highly expressed in developing cells as they mature. Vn96, an Hsp70-specific peptide, was used by Ghosh et al. to protect tram derived from malignant cells. Peptide-NP conjugates effectively protected Hsp70-presenting EVs from lethal serum, demonstrating super-centrifugation-like prisoner efficiency.

Cyclic nucleic acids, including cfDNA and miRNA, are another diagnostic concern. Since the NA fractions produced by rapidly growing cells can contain all the genetic information of early growing cells, the lifespan of cyclic nucleic acids (NAs) has been studied for decades. Several examining groups have recently used pep-derived nucleic acids (PNAs) to identify specific abnormalities in circulating NAs. PNAs are naturally produced NA analogs and have better long-term stability and reciprocal sequence binding than native NA. Testing PNA in combination with PN allows for sensitive and selective measurement of cyclic NA. Colored nanoparticles, such as nanoscale pure organic tissues (NMOFs) and nanoscale graphene oxides (NGOs), have been used in the PNA experiments. The most widely used method evaluates changes in the fluorescence signal. For example, the strong binding between NMOF or NGO scanning and PNA causes luminescence quenching, which can be regained when the PNA sweep is released from the assembly through hybridization between miRNAs. NPs conjugated with NMOFs and NGOs can efficiently remove specific miRNAs using this technique, even at concentrations as low as 10 pM. The PNA exam is often combined with the AuNP. Adsorption of miRNA or ctDNA on the surface of PNA-AuNP conjugates alters the electrical, optical, and plasmonic properties of the conjugates. To detect specific mutations in the development and methylation of the PIK3CA gene, Nguyen used AuNP-conjugated peptides [127]. The local face plasmon resonance (LSPR) [128–130] peak was shifted from 4.3 to 11.4 nm after ctDNA adsorption on PNA-AuNP conjugates.

Due to the lower selectivity on the list of free peptides, the multivalent binding activity of PNCs allows these short-chain amino acid composites to be used as convict agents for organismal analysis. Liquid with transport phenomenon comparable to that of an antibody. However, the maturity of PNC-based liquid live surgery platforms is still in the early stages of development; only a few of these biases have shown clinical utility. Further investigation of the collected biomarkers, such as molecular characterization and functional testing, could help PNC-based liquid organismal anatomy systems to link with the clinical world.

15.7 Summary and Outlook

Peptides might possibly leverage the structural and functional benefits of the two primary accessories in pharmacological inquiry since they are molecularly situated between proteins and tiny molecular composites. As was already mentioned, numerous peptides combined with NPs have shown promise for use in a range of applications, including the delivery of medications, the suppression of harmful biomolecular interactions, liquid vivisection, etc. No matter the outcome, using PNCs for therapeutic purposes is still challenging for the reasons listed below. To begin with, little is known about PNC activity in pathologic settings including the circulatory system and the intracellular environment. One such efficient method for improving the therapeutic effects of parent antimicrobial peptides is nanoparticle tagging. Alarcon et al. [132] have demonstrated the significance of cysteine in the conjugation process, while Liu et al. [131] have explored the potency of the nano-conjugates from several perspectives, such as size and shape. Additionally, peptides on nanomaterial shells are prone to enzymatic degradation, demanding novel protective strategies to maintain their activities while minimizing the structural and compositional complexity of the conjugates. Third, as this is a typical impediment to *in vivo* and clinical deployment, the PNCs' inherent immunogenicity needs to be addressed. The inherent functions of peptides are often lost during covalent conjugation with NPs or other functional components.

The random spatial organization of peptides is evident during the early equilibration event (as shown by MD simulations), but over time, the peptides become dispersed and aligned to both leaflets. Such alignment also shows that the peptide interacts with water molecules and negatively charged phosphate head groups of lipid moieties. The core of the peptide bilayer in nanoconjugate forms, which contains the acyl chains of lipids, is anticipated to facilitate the early hydrophobic interactions. This main impetus, however, cannot be taken as definitive because most peptides interact with the outer membranous part.

Abbreviations

3D	Three dimensional
ACPP	Activatable transcellular peptides
BBB	Blood brain barrier
CT, PET and MRI	Computed Tomography, Positron Emission Tomography, Magnetic Resonance Imaging
DNA	Deoxyribonucleic acid
DOX	Doxorubicin
GRPR	Gastrin releasing peptide receptor
HA	Hemagglutin
HIV	Human immunodeficiency virus
MD simulations	Molecular dynamics simulations
MNP	Magnetic nanoparticles
NA	Nucleic acids
NIR	Near infrared
NLS	Nuclear localization signal

NLCN	Nanolipid crystal nanoparticles
nm	Nanometer
NP	Nanoparticles
NSC	Nuclear severance complexes
PNC	Peptide nanoparticle conjugates
RNA	Ribonucleic acid
ROS	Reactive oxygen species
RRE	Rev response element
SPIO	Superparamagnetic iron oxide

References

1 Vanhee, P., van der Sloot, A.M., Verschueren, E. et al. (2011). Computational design of peptide ligands. *Trends Biotechnol.* 29 (5): 231–239.

2 Wang, S.H. and Yu, J. (2018). Structure-based design for binding peptides in anticancer therapy. *Biomaterials* 156: 1–15.

3 Marasco, D., Perretta, G., Sabatella, M., and Ruvo, M. (2008). Past and future perspectives of synthetic peptide libraries. *Curr. Protein Pept. Sci.* 9 (5): 447–467.

4 Ryvkin, A., Ashkenazy, H., Weiss-Ottolenghi, Y. et al. (2018). Phage display peptide libraries: deviations from randomness and correctives. *Nucleic Acids Res.* 46 (9): e52.

5 Henninot, A., Collins, J.C., and Nuss, J.M. (2018). The current state of peptide drug discovery: back to the future? *J. Med. Chem.* 61 (4): 1382–1414.

6 Lau, J.L. and Dunn, M.K. (2018). Therapeutic peptides: historical perspectives, current development trends, and future directions. *Bioorg. Med. Chem.* 26 (10): 2700–2707.

7 Weinstock, M.T., Francis, J.N., Redman, J.S., and Kay, M.S. (2012). Protease-resistant peptide design-empowering nature's fragile warriors against HIV. *Biopolymers* 98 (5): 431–442.

8 Talmadge, J.E. (1998). Pharmacodynamic aspects of peptide administration biological response modifiers. *Adv. Drug Deliv. Rev.* 33 (3): 241–252.

9 Klein, M. (2017). Stabilized helical peptides: overview of the technologies and its impact on drug discovery. *Expert Opin. Drug Discovery* 12 (11): 1117–1125.

10 Shu, J.Y., Panganiban, B., and Xu, T. (2013). Peptide–polymer conjugates: from fundamental science to application. *Annu. Rev. Phys. Chem.* 64: 631–657.

11 Xiao, Y., Reis, L.A., Feric, N. et al. (2016). Diabetic wound regeneration using peptide-modified hydrogels to target re-epithelialization. *Proc. Natl. Acad. Sci. U. S. A.* 113 (40): E5792–E5801.

12 Rizvi, S.A.A. and Saleh, A.M. (2018). Applications of nanoparticle systems in drug delivery technology. *Saudi Pharm. J.* 26 (1): 64–70.

13 Habibi, N., Kamaly, N., Memic, A., and Shafee, H. (2016). Self-assembled peptide based nanostructures: smart nanomaterials toward targeted drug delivery. *Nano Today* 11 (1): 41–60.

14 Jeong, W.J., Kwon, S.H., and Lim, Y.B. (2018). Modular self-assembling peptide platform with a tunable thermoresponsiveness via a single amino acid substitution. *Adv. Funct. Mater.* 28: 35.

15 Kitov, P.I. and Bundle, D.R. (2003). On the nature of the multivalency effect: a thermodynamic model. *J. Am. Chem. Soc.* 125 (52): 16271–16284.

16 Gargano, J.M., Ngo, T., Kim, J.Y. et al. (2001). Multivalent inhibition of AB(5) toxins. *J. Am. Chem. Soc.* 123 (51): 12909–12910.

17 Hong, S., Leroueil, P.R., Majoros, I.J. et al. (2007). The binding avidity of a nanoparticle-based multivalent targeted drug delivery platform. *Chem. Biol.* 14 (1): 107–115.

18 Martinez-Veracoechea, F.J. and Frenkel, D. (2011). Designing super selectivity in multivalent nano-particle binding. *Proc. Natl. Acad. Sci. U. S. A.* 108 (27): 10963–10968.

19 Jeong, W.J., Choi, S.H., Jin, K.S., and Lim, Y.B. (2016). Tuning oligovalent biomacro-molecular interfaces using double-layered alpha-helical coiled-coil nanoassemblies from lariat-type building blocks. *ACS Macro Lett.* 5 (12): 1406–1410.

20 Vauquelin, G. and Charlton, S.J. (2010). Long-lasting target binding and rebinding as mechanisms to prolong in vivo drug action. *Br. J. Pharmacol.* 161 (3): 488–508.

21 Osman, G., Rodriguez, J., Chan, S.Y. et al. (2018). PEGylated enhanced cell penetrating peptide nanoparticles for lung gene therapy. *J. Control. Release* 285: 35–45.

22 Fang, C. and Zhang, M. (2010). Nanoparticle-based theragnostics: integrating diagnostic and therapeutic potentials in nanomedicine. *J. Control. Release* 146 (1): 2–5.

23 Borglin, J., Selegard, R., Aili, D., and Ericson, M.B. (2017). Peptide functionalized gold nanoparticles as a stimuli responsive contrast medium in multiphoton microscopy. *Nano Lett.* 17 (3): 2102–2108.

24 Jeong, W.J., Kye, M., Han, S.H. et al. (2017). Inhibition of multimolecular RNA–protein interactions using multitarget-directed nanohybrid system. *ACS Appl. Mater. Interfaces* 9 (13): 11537–11545.

25 Lauster, D., Glanz, M., Bardua, M. et al. (2017). Multi-valent peptide–nanoparticle conjugates for infuenza-virus inhibition. *Angew. Chem. Int. Edit.* 56 (21): 5931–5936.

26 Chan, J.W., Lewis, D.R., Petersen, L.K. et al. (2016). Amphiphilic macromolecule nanoassemblies suppress smooth muscle cell proliferation and platelet adhesion. *Biomaterials* 84: 219–229.

27 Liu, T.M., Conde, J., Lipinski, T. et al. (2016). Revisiting the classification of NIR-absorbing/emitting nanomaterials for in vivo bioapplications. *NPG Asia Mater.* 8: e295.

28 Lucky, S.S., Soo, K.C., and Zhang, Y. (2015). Nanoparticles in photodynamic therapy. *Chem. Rev.* 115 (4): 1990–2042.

29 Emelianov, S.Y., Li, P.C., and Odonnell, M. (2009). Photoacoustics for molecular imaging and therapy. *Phys. Today* 62 (5): 34–39.

30 Liu, Y., Bhattarai, P., Dai, Z., and Chen, X. (2018). Photothermal therapy and photoacoustic imaging via nanotheranostics in fighting cancer. *Chem. Soc. Rev.* https://doi.org/10.1039/c8cs00618k.

31 Kudr, J., Haddad, Y., Richtera, L. et al. (2017). Magnetic nanoparticles: from design and synthesis to real world applications. *Nanomaterials* 7: 9.

32 McBain, S.C., Yiu, H.H.P., and Dobson, J. (2008). Magnetic nanoparticles for gene and drug delivery. *Int. J. Nanomedicine* 3 (2): 169–180.

33 Zhou, S.L., Li, J., Hong, G.B., and Chang, C.T. (2013). Dendrimer modified magnetic nanoparticles as adsorbents for removal of dyes. *J. Nanosci. Nanotechnol.* 13 (10): 6814–6819.

34 Yun, H.M., Ahn, S.J., Park, K.R. et al. (2016). Magnetic nanocomposite scaffolds combined with static magnetic field in the stimulation of osteoblastic differentiation and bone formation. *Biomaterials* 85: 88–98.

35 Modi, D.A., Sunoqrot, S., Bugno, J. et al. (2014). Targeting of follicle stimulating hormone peptide-conjugated dendrimers to ovarian cancer cells. *Nanoscale* 6 (5): 2812–2820.

36 Jiang, X., Bugno, J., Hu, C. et al. (2015). Targeted treatment of FLT3-overexpressing acute myeloid leukemia with MiR-150 nanoparticles guided by conjugated FLT3 ligand peptides. *Blood* 126: 23.

37 Spicer, C.D., Jumeaux, C., Gupta, B., and Stevens, M.M. (2018). Peptide and protein nanoparticle conjugates: versatile platforms for biomedical applications. *Chem. Soc. Rev.* 47 (10): 3574–3620.

38 Patel, A., Patel, M., Yang, X., and Mitra, A.K. (2014). Recent advances in protein and peptide drug delivery: a special emphasis on polymeric nanoparticles. *Protein Pept. Lett.* 21 (11): 1102–1120.

39 Yang, N.J. and Hinner, M.J. (2015). Getting across the cell membrane: an overview for small molecules, peptides, and proteins. *Methods Mol. Biol.* 1266: 29–53.

40 Wente, S.R. and Rout, M.P. (2010). The nuclear pore complex and nuclear transport. *Cold Spring Harb. Perspect. Biol.* 2 (10): a000562.

41 Pan, L., He, Q., Liu, J. et al. (2012). Nuclear-targeted drug delivery of TAT peptide-conjugated monodisperse mesoporous silica nanoparticles. *J. Am. Chem. Soc.* 134 (13): 5722–5725.

42 Tkachenko, A.G., Xie, H., Coleman, D. et al. (2003). Multifunctional gold nanoparticle-peptide complexes for nuclear targeting. *J. Am. Chem. Soc.* 125 (16): 4700–4701.

43 Li, N., Yang, H., Yu, Z. et al. (2017). Nuclear-targeted siRNA delivery for long-term gene silencing. *Chem. Sci.* 8 (4): 2816–2822.

44 Niu, J., Chu, Y., Huang, Y.F. et al. (2017). Transdermal gene delivery by functional peptide-conjugated cationic gold nanoparticle reverses the progression and metastasis of cutaneous melanoma. *ACS Appl. Mater. Interfaces* 9 (11): 9388–9401.

45 Patlolla, R.R., Desai, P.R., Belay, K., and Singh, M.S. (2010). Translocation of cell penetrating peptide engrafted nanoparticles across skin layers. *Biomaterials* 31 (21): 5598–5607.

46 Zou, L., Ding, W., Zhang, Y. et al. (2018). Peptide modifed vemurafenib-loaded liposomes for targeted inhibition of melanoma via the skin. *Biomaterials* 182: 1–12.

47 Poellmann, M.J., Bu, J., and Hong, S. (2018). Would antioxidant-loaded nanoparticles present an effective treatment for ischemic stroke? *Nanomedicine* 13: 2327–2340.

48 Georgieva, J.V., Brinkhuis, R.P., Stojanov, K. et al. (2012). Peptide-mediated blood–brain barrier transport of polymersomes. *Angew. Chem. Int. Ed.* 51 (33): 8339–8342.

49 Yao, H., Wang, K., Wang, Y. et al. (2015). Enhanced blood–brain barrier penetration and glioma therapy mediated by a new peptide modified gene delivery system. *Biomaterials* 37: 345–352.

50 Dang, C.V., Reddy, E.P., Shokat, K.M., and Soucek, L. (2017). Drugging the 'undruggable' cancer targets. *Nat. Rev. Cancer* 17 (8): 502–508.

51 Crews, C.M. (2010). Targeting the undruggable proteome: the small molecules of my dreams. *Chem. Biol.* 17 (6): 551–555.

52 Leader, B., Baca, Q.J., and Golan, D.E. (2008). Protein therapeutics: a summary and pharmacological classification. *Nat. Rev. Drug Discov.* 7 (1): 21–39.

53 Azzarito, V., Long, K., Murphy, N.S., and Wilson, A.J. (2013). Inhibition of alpha-helixmediated protein–protein interactions using designed molecules. *Nat. Chem.* 5 (3): 161–173.

54 Kim, B., Choi, S.J., Han, S.H. et al. (2013). Stabilization of alphahelices by the self-assembly of macrocyclic peptides on the surface of gold nanoparticles for molecular recognition. *Chem. Commun.* 49 (69): 7617–7619.

55 Jeong, W.J., Choi, S.J., Choi, J.S., and Lim, Y.B. (2013). Chameleon-like self-assembling peptides for adaptable biorecognition nanohybrids. *ACS Nano* 7 (8): 6850–6857.

56 Emileh, A., Tuzer, F., Yeh, H. et al. (2013). A model of peptide triazole entry inhibitor binding to HIV-1 gp120 and the mechanism of bridging sheet disruption. *Biochemistry* 52 (13): 2245–2261.

57 Bastian, A.R., Nangarlia, A., Bailey, L.D. et al. (2015). Mechanism of multivalent nanoparticle encounter with HIV-1 for potency enhancement of peptide triazole virus inactivation. *J. Biol. Chem.* 290 (1): 529–543.

58 Hartl, F.U. (2017). Protein misfolding diseases. *Annu. Rev. Biochem.* 86: 21–26.

59 Xiong, N., Zhao, Y.J., Dong, X.Y. et al. (2017). Design of a molecular hybrid of dual peptide inhibitors coupled on AuNPs for enhanced inhibition of amyloid beta-protein aggregation and cytotoxicity. *Small* 13: 13.

60 Soto, C., Sigurdsson, E.M., Morelli, L. et al. (1998). Beta-sheet breaker peptides inhibit fibrillogenesis in a rat brain model of amyloidosis: implications for Alzheimer's therapy. *Nat. Med.* 4 (7): 822–826.

61 Fradinger, E.A., Monien, B.H., Urbanc, B. et al. (2008). C-terminal peptides coassemble into Abeta42 oligomers and protect neurons against Abeta42-induced neurotoxicity. *Proc. Natl. Acad. Sci. U. S. A.* 105 (37): 14175–14180.

62 Mammen, M., Choi, S.K., and Whitesides, G.M. (1998). Polyvalent interactions in biological systems: implications for design and use of multivalent ligands and inhibitors. *Angew. Chem. Int. Edit.* 37 (20): 2755–2794.

63 Kwon, S.J., Na, D.H., Kwak, J.H. et al. (2017). Nanostructured glycan architecture is important in the inhibition of infuenza A virus infection. *Nat. Nanotechnol.* 12 (1): 48–54.

64 Kim, H.W., Yang, K., Jeong, W.J. et al. (2016). Photoactivation of noncovalently assembled peptide ligands on carbon nanotubes enables the dynamic regulation of stem cell differentiation. *ACS Appl. Mater. Interfaces* 8 (40): 26470–26481.

65 Bugno, J., Hsu, H.J., Pearson, R.M. et al. (2016). Size and surface charge of engineered poly(amidoamine) dendrimers modulate tumor accumulation and penetration: a model study using multicellular tumor spheroids. *Mol. Pharm.* 13 (7): 2155–2163.

66 Yang, Y., Sunoqrot, S., Stowell, C. et al. (2012). Effect of size, surface charge, and hydrophobicity of poly(amidoamine) dendrimers on their skin penetration. *Biomacromolecules* 13 (7): 2154–2162.

67 Russell, R.J., Kerry, P.S., Stevens, D.J. et al. (2008). Structure of infuenza hemagglutinin in complex with an inhibitor of membrane fusion. *Proc. Natl. Acad. Sci. U. S. A.* 105 (46): 17736–17741.
68 Sorolla, A., Ho, D., Wang, E. et al. (2016). Sensitizing basal-like breast cancer to chemotherapy using nanoparticles conjugated with interference peptide. *Nanoscale* 8 (17): 9343–9353.
69 Beltran, A.S., Graves, L.M., and Blancafort, P. (2014). Novel role of Engrailed 1 as a prosurvival transcription factor in basal-like breast cancer and engineering of interference peptides block its oncogenic function. *Oncogene* 33 (39): 4767–4777.
70 Wolf, B., Sanglier, J.J., and Wang, Y. (1997). Leptomycin B is an inhibitor of nuclear export: inhibition of nucleo-cytoplasmic translocation of the human immunodeficiency virus type 1 (HIV-1) Rev protein and Rev-dependent mRNA. *Chem. Biol.* 4 (2): 139–147.
71 Cassidy, P.J. and Radda, G.K. (2005). Molecular imaging perspectives. *J. R. Soc. Interface* 2 (3): 133–144.
72 Morse, D.L. and Gillies, R.J. (2010). Molecular imaging and targeted therapies. *Biochem. Pharmacol.* 80 (5): 731–738.
73 Kim, D., Kim, J., Park, Y.I. et al. (2018). Recent development of inorganic nanoparticles for biomedical imaging. *ACS Cent. Sci.* 4 (3): 324–336.
74 Zhang, P., Cui, Y., Anderson, C.F. et al. (2018). Peptide-based nanoprobes for molecular imaging and disease diagnostics. *Chem. Soc. Rev.* 47 (10): 3490–3529.
75 An, L., Shah Gilani, M.R., and Liang, G. (2014). Peptide-based nanostructures for cancer diagnosis and therapy. *Curr. Med. Chem.* 21 (21): 2453–2466.
76 Chapman, S., Dobrovolskaia, M., Farahani, K. et al. (2013). Nanoparticles for cancer imaging: the good, the bad, and the promise. *Nano Today* 8 (5): 454–460.
77 Altınoğlu, E.İ. and Adair, J.H. (2010). Near infrared imaging with nanoparticles. *Nanomed. Nanobiotechnol.* 2 (5): 461–477.
78 Owens, E.A., Henary, M., El Fakhri, G., and Choi, H.S. (2016). Tissue-specific near-infrared fuorescence imaging. *Acc. Chem. Res.* 49 (9): 1731–1740.
79 Grahame, S. (2008). Cancer drug discovery and development: in vivo imaging of cancer therapy. *J. Med. Imaging Radiat. Sci.* 39 (2): 105.
80 Cao, C.Y., Chen, Y., Wu, F.Z. et al. (2011). Caspase-3 controlled assembly of nanoparticles for fluorescence turn on. *Chem. Commun.* 47 (37): 10320–10322.
81 Fan, Z., Chang, Y., Cui, C. et al. (2018). Near infrared fluorescent peptide nanoparticles for enhancing esophageal cancer therapeutic efficacy. *Nat. Commun.* 9 (1): 2605.

82 Fan, Z., Sun, L., Huang, Y. et al. (2016). Bioinspired fluorescent dipeptide nanoparticles for targeted cancer cell imaging and real-time monitoring of drug release. *Nat. Nanotechnol.* 11 (4): 388–394.

83 Sun, Y., Qu, C., Chen, H. et al. (2016). Novel benzo-bis(1,2,5-thiadiazole) fluorophores for in vivo NIR-II imaging of cancer. *Chem. Sci.* 7 (9): 6203–6207.

84 Zhu, J., Fu, F., Xiong, Z. et al. (2015). Dendrimer-entrapped gold nanoparticles modified with RGD peptide and alpha-tocopheryl succinate enable targeted theranostics of cancer cells. *Colloids Surf. B: Biointerfaces* 133: 36–42.

85 Hao, Y., Zhang, B., Zheng, C. et al. (2015). The tumor-targeting core-shell structured DTX-loaded PLGA@Au nanoparticles for chemo-photothermal therapy and X-ray imaging. *J. Control. Release* 220 (Pt A): 545–555.

86 Kwon, S.P., Jeon, S., Lee, S.H. et al. (2018). Thrombin-activatable fluorescent peptide incorporated gold nanoparticles for dual optical/computed tomography thrombus imaging. *Biomaterials* 150: 125–136.

87 Kim, J.Y., Ryu, J.H., Schellingerhout, D. et al. (2015). Direct imaging of cerebral thromboemboli using computed tomography and fibrin-targeted gold nanoparticles. *Theranostics* 5 (10): 1098–1114.

88 Stockhofe, K., Postema, J.M., Schieferstein, H., and Ross, T.L. (2014). Radiolabeling of nanoparticles and polymers for PET imaging. *Pharmaceuticals (Basel)* 7 (4): 392–418.

89 Guerrero, S., Araya, E., Fiedler, J.L. et al. (2010). Improving the brain delivery of gold nanoparticles by conjugation with an amphipathic peptide. *Nanomedicine (London)* 5 (6): 897–913.

90 Guerrero, S., Herance, J.R., Rojas, S. et al. (2012). Synthesis and in vivo evaluation of the biodistribution of a 18F-labeled conjugate gold-nanoparticle-peptide with potential biomedical application. *Bioconjug. Chem.* 23 (3): 399–408.

91 Cheng, K., Kothapalli, S.R., Liu, H. et al. (2014). Construction and validation of nano gold tripods for molecular imaging of living subjects. *J. Am. Chem. Soc.* 136 (9): 3560–3571.

92 Almutairi, A., Rossin, R., Shokeen, M. et al. (2009). Biodegradable dendritic positron-emitting nanoprobes for the noninvasive imaging of angiogenesis. *Proc. Natl. Acad. Sci. U. S. A.* 106 (3): 685–690.

93 Sosnovik, D.E. and Weissleder, R. (2007). Emerging concepts in molecular MRI. *Curr. Opin. Biotechnol.* 18 (1): 4–10.

94 Xie, J., Chen, K., Lee, H.-Y. et al. (2008). Ultrasmall c(RGDyK)-coated Fe(3)O(4) nanoparticles and their specific targeting to integrin α(v)β(3)-rich tumor cells. *J. Am. Chem. Soc.* 130 (24): 7542–7543.

95 Khemtong, C., Kessinger, C.W., Ren, J. et al. (2009). In vivo of-resonance saturation magnetic resonance imaging of αvβ3-targeted superparamagnetic nanoparticles. *Cancer Res.* 69 (4): 1651.

96 Kessinger, C.W., Togao, O., Khemtong, C. et al. (2011). Investigation of in vivo targeting kinetics of α(v)β(3)-specific superparamagnetic nanoprobes by time-resolved MRI. *Theranostics* 1: 263–273.

97 Simberg, D., Duza, T., Park, J.H. et al. (2007). Biomimetic amplification of nanoparticle homing to tumors. *Proc. Natl. Acad. Sci. U. S. A.* 104 (3): 932–936.

98 Tian, M., Lu, W., Zhang, R. et al. (2013). Tumor uptake of hollow gold nanospheres after intravenous and intra-arterial injection: pET/CT study in a rabbit VX2 liver cancer model. *Mol. Imaging Biol.* 15 (5): 614–624.

99 Olson, E.S., Jiang, T., Aguilera, T.A. et al. (2010). Activatable cell penetrating peptides linked to nanoparticles as dual probes for in vivo fluorescence and MR imaging of proteases. *Proc. Natl. Acad. Sci. U. S. A.* 107 (9): 4311–4316.

100 Siravegna, G., Marsoni, S., Siena, S., and Bardelli, A. (2017). Integrating liquid biopsies into the management of cancer. *Nat. Rev. Clin. Oncol.* 14 (9): 531–548.

101 Hong, S. and Wang, A.Z. (2018). Nanotechnology enabling the use of circulating tumor cells (CTCs) as reliable cancer biomarkers. *Adv. Drug Deliv. Rev.* 125: 1–2.

102 Bu, J., Kim, Y.J., Kang, Y.T. et al. (2017). Polyester fabric sheet layers functionalized with graphene oxide for sensitive isolation of circulating tumor cells. *Biomaterials* 125: 1–11.

103 Li, P., Kaslan, M., Lee, S.H. et al. (2017). Progress in exosome isolation techniques. *Theranostics* 7 (3): 789–804.

104 Volik, S., Alcaide, M., Morin, R.D., and Collins, C. (2016). Cell-free DNA (cfDNA): clinical significance and utility in cancer shaped by emerging technologies. *Mol. Cancer Res.* 14 (10): 898–908.

105 Endzeliņš, E., Berger, A., Melne, V. et al. (2017). Detection of circulating miRNAs: comparative analysis of extracellular vesicle-incorporated miRNAs and cell-free miRNAs in whole plasma of prostate cancer patients. *BMC Cancer* 17 (1): 730.

106 Bu, J., Kang, Y.T., Lee, Y.S. et al. (2017). Lab on a fabric: mass producible and low-cost fabric filters for the high-throughput viable isolation of circulating tumor cells. *Biosens. Bioelectron.* 91: 747–755.

107 Suh, D.H., Kim, M., Choi, J.Y. et al. (2017). Circulating tumor cells in the differential diagnosis of adnexal masses. *Oncotarget* 8 (44): 77195–77206.

108 Murlidhar, V., Rivera-Báez, L., and Nagrath, S. (2016). Affinity versus label-free isolation of circulating tumor cells: who wins? *Small* 12 (33): 4450–4463.

109 Bu, J., Cho, Y.-H., and Han, S.-W. (2017). Enhancement of isolation sensitivity for the viable heterogeneous circulating tumor cells swelled by hypo-osmotic pressure. *RSC Adv.* 7 (78): 49684–49693.

110 Nagrath, S., Sequist, L.V., Maheswaran, S. et al. (2007). Isolation of rare circulating tumour cells in cancer patients by microchip technology. *Nature* 450 (7173): 1235–1239.

111 Bai, L., Du, Y., Peng, J. et al. (2014). Peptide-based isolation of circulating tumor cells by magnetic nanoparticles. *J. Mater. Chem. B* 2 (26): 4080–4088.

112 Peng, J., Zhao, Q., Zheng, W. et al. (2017). Peptide-functionalized nanomaterials for the efficient isolation of HER2-positive circulating tumor cells. *ACS Appl. Mater. Interfaces* 9 (22): 18423–18428.

113 Kuo, W.T., Lin, W.C., Chang, K.C. et al. (2015). Quantitative analysis of ligand–EGFR interactions: a platform for screening targeting molecules. *PLoS One* 10 (2): e0116610.

114 Ding, J., Wang, K., Tang, W.J. et al. (2016). Construction of epidermal growth factor receptor peptide magnetic nanovesicles with lipid bilayers for enhanced capture of liver cancer circulating tumor cells. *Anal. Chem.* 88 (18): 8997–9003.

115 Wang, X., Qian, X., Beitler, J.J. et al. (2011). Detection of circulating tumor cells in human peripheral blood using surface-enhanced Raman scattering nanoparticles. *Cancer Res.* 71 (5): 1526–1532.

116 Kang, Y.T., Kim, Y.J., Bu, J. et al. (2017). High-purity capture and release of circulating exosomes using an exosome-specific dual-patterned immunofiltration (ExoDIF) device. *Nanoscale* 9 (36): 13495–13505.

117 Luan, X., Sansanaphongpricha, K., Myers, I. et al. (2017). Engineering exosomes as refined biological nanoplatforms for drug delivery. *Acta Pharmacol. Sin.* 38 (6): 754–763.

118 Gao, X., Ran, N., Dong, X. et al. (2018). Anchor peptide captures, targets, and loads exosomes of diverse origins for diagnostics and therapy. *Sci. Transl. Med.* 10: 444.

119 Bu, J., Lee, T.H., Kim, I.S., and Cho, Y.-H. (2017). Microfluidic-based mechanical phenotyping of cells for the validation of epithelial-to-mesenchymal-like transition caused by insufficient heat treatment. *Sens. Actuators B: Chem.* 244: 591–598.

120 Mayer, M.P. and Bukau, B. (2005). Hsp70 chaperones: cellular functions and molecular mechanism. *Cell. Mol. Life Sci.* 62 (6): 670–684.

121 Ghosh, A., Davey, M., Chute, I.C. et al. (2014). Rapid isolation of extracellular vesicles from cell culture and biological fluids using a synthetic peptide with specific affinity for heat shock proteins. *PLoS One* 9 (10): e110443.

122 Wan, J.C.M., Massie, C., Garcia-Corbacho, J. et al. (2017). Liquid biopsies come of age: towards implementation of circulating tumour DNA. *Nat. Rev. Cancer* 17 (4): 223–238.

123 Uhlmann, E., Peyman, A., Breipohl, G., and Will, D.W. (1998). PNA: synthetic polyamide nucleic acids with unusual binding properties. *Angew. Chem. Int. Ed. Eng.* 37 (20): 2796–2823.

124 Wu, Y., Han, J., Xue, P. et al. (2015). Nano metal-organic framework (NMOF)-based strategies for multiplexed microRNA detection in solution and living cancer cells. *Nanoscale* 7 (5): 1753–1759.

125 Ryoo, S.R., Lee, J., Yeo, J. et al. (2013). Quantitative and multiplexed microRNA sensing in living cells based on peptide nucleic acid and nano graphene oxide (PANGO). *ACS Nano* 7 (7): 5882–5891.

126 Cai, B., Huang, L., Zhang, H. et al. (2015). Gold nanoparticles-decorated graphene field-effect transistor biosensor for femtomolar MicroRNA detection. *Biosens. Bioelectron.* 74: 329–334.

127 Nguyen, A.H. and Sim, S.J. (2015). Nanoplasmonic biosensor: detection and amplification of dual bio-signatures of circulating tumor DNA. *Biosens. Bioelectron.* 67: 443–449.

128 Jeong, W.J., Lee, M.S., and Lim, Y.B. (2013). Helix stabilized, thermostable, and protease-resistant self-assembled peptide nanostructures as potential inhibitors of protein–protein interactions. *Biomacromolecules* 14 (8): 2684–2689.

129 Deehan, M., Garces, S., Kramer, D. et al. (2015). Managing unwanted immunogenicity of biologicals. *Autoimmun. Rev.* 14 (7): 569–574.

130 Jeong, W.J., Bu, J., Kubiatowicz, L.J. et al. (2018). Peptide–nanoparticle conjugates: a next generation of diagnostic and therapeutic platforms? *Nano Converg.* 5: 38. https://doi.org/10.1186/s40580-018-0170-1.

131 Liu, L. et al. (2013). Te potent antimicrobial properties of cell penetrating peptide-conjugated silver nanoparticles with excellent selectivity for gram-positive bacteria over erythrocytes. *Nanoscale* 5: 3834–3840.

132 Poblete, H. et al. (2015). New insights into peptide–silver nanoparticle interaction: deciphering the role of cysteine and lysine in the peptide sequence. *Langmuir* 32: 265–273.

16

Unleashing the Potential of Green-Synthesized Nanoparticles for Effective Biomedical Application

G.K. Prashanth[1], Manoj Gadewar[2], M. Mutthuraju[3], Srilatha Rao[4], A.S. Sowmyashree[4], K. Shwetha[4], Mithun Kumar Ghosh[5], B.R. Malini[6], and Vinita Chaturvedi[7]

[1] *Research and Development Centre, Department of Chemistry, Sir M. Visvesvaraya Institute of Technology, Affiliated to Visvesvaraya Technological University, Bengaluru, KA, India*
[2] *Department of Pharmacology, School of Medical and Allied Sciences, KR Mangalam University, Gurgaon, HR, India*
[3] *Department of Chemistry, Sai Vidya Institute of Technology, Affiliated to Visvesvaraya Technological University, Bengaluru, KA, India*
[4] *Department of Chemistry, Nitte Meenakshi Institute of Technology, Bengaluru, KA, India*
[5] *Department of chemistry, Government College Hatta, Damoh, India*
[6] *Department of Chemistry, Akshara First Grade College, Bengaluru, KA, India*
[7] *Biochemistry Division, Central Drug Research Institute, CSIR, Lucknow, UP, India*

Abstract

Scientific research on nanoparticles (NPs) has been trending for the last few years. NPs are developed by different methods like biological and chemical. Chemical approaches have several drawbacks like use of toxic and expensive capping agents, and solvents. Synthesis of NPs using a biochemical approach follows the green chemistry philosophy. As biological methods overcome most of the disadvantages of chemical processes, they have attracted the researchers. In biological approach, NPs are synthesized by using various bio-organisms such as plant materials, bacteria, fungi, etc. Green synthesis is a subset of green chemistry with the goal of developing safer chemical products and processes that reduce or eliminate the generation and use of elements hazardous for user health or environment. NPs derived from bio-organism are called "Green Nanoparticles" (GNPs). In this chapter, we discuss the process of synthesis, characterization, and application of various GNPs in drug delivery and as anti-microbials and anti-carcinogens.

Green Synthesis of Nanomaterials: Biological and Environmental Applications, First Edition.
Edited by Archana Chakravarty, Preeti Singh, Saiqa Ikram, and R.N. Yadava.

Keywords *green chemistry; green nanoparticles; synthesis; characterization; drug delivery; anti-microbials; anti-carcinogens*

16.1 Introduction

In present times, GNPs are playing a lead role in improving human health and lifestyle. Metal NPs are most commonly used in material applications with proven diversity of properties such as photo-catalysis, fuel cells, production of microelectronic circuits, sensors, heavy metal detection, H_2O_2 sensing, etc. [1–12]. Also, NPs have an important role in removing contaminants from wastewater so that it can be used for various purposes [4].

With the advancement of nanotechnology, NPs are now being explored for their widespread usage in the biological applications, like in drug delivery, bio-sensing, circulating tumor deoxyribonucleic acid (CT-DNA) binding, cytotoxicity, and also as anti-microbials, anti-tubercular agents, anti-carcinogens, etc. [4–7, 13–26].

GNPs are serving as drugs against human pathogens that cause different diseases [4, 16, 17, 20–22, 24, 25]. GNPs have improved the human lifestyle by acting as sensors, small electric devices, transformers, etc. GNPs are proven to be better drugs in the treatment of many human diseases due to the fact that they are less toxic, cause lesser side effects, cost effective, with greater stability, easy availability (or good bioavailability), etc. GNPs are specifically formulated to attract/bind diseased cells to provide targeted delivery allowing direct therapy to the infected cells, thereby, increasing efficacy, and reducing side effects, and improving overall human health.

A variety of NPs have been synthesized at high temperature, pressure, and using toxic chemicals, all of which increase the overall cost of the process, and also affect the environment [27, 28]. Green nanochemistry has attracted increasing interest over the past decade as a potential alternative to the production of metallic NPs with minimal chemical/toxic waste generation [29]. The development of a simple, environmentally friendly method for nanomaterial synthesis has become a popular research topic in recent years. The use of renewable resources and environmentally friendly molecules for the synthesis of eco-friendly, bio-compatible, and economical green NPs is a fundamental component of this sustainable approach [5–7, 30]. The external design and ultimate efficacy of GNPs are influenced by various natural capping or reducing agents such as alkaloids, polyphenols, proteins, and flavonoids [31]. Various microorganisms and bio-molecules, such as fungi, vitamins, bacteria, plant extracts, enzymes, yeasts, and biodegradable polymers have been successfully used to produce metallic NPs by green methods [32, 33]. However, employment of micro-organisms is still challenging because of the requirement of infrastructure to maintain their culture, low yield, time constraint, and higher cost [34]. Plant-based NP synthesis offers many advantages such as milder reaction conditions, higher feasibility, and safer reagents. In addition, NPs obtained from these routes exhibit myriads of biological applications [35–38].

Almost 10 million people will die from cancer worldwide in 2020, making it the top cause of death [39]. As far as new cancer cases go, the most prevalent in 2020 were: breast (2.26 million cases); colon and rectum (1.93 million cases); lung (2.21 million cases); prostate (1.41 million cases); stomach (1.09 million cases); and skin (non-melanoma) (1.20 million cases) [39]. Treatment of cancer generally like radiotherapy, surgery, and/or systemic therapy (hormonal treatments, targeted biological therapies, and chemotherapy). A treatment plan choice must take into account the patient as well as the malignancy. In most cases detected at early stage, chemotherapy is a general therapeutic procedure that uses a variety of cytotoxic chemicals to stop the growth of cancer cells. However, these treatments also kill the body's normal cells and have negative physiological effects [40]. By employing metallic NPs in controlled targeted medication delivery, the side effects can be reduced by minimizing damage in normal cells [41]. As a result, copper, gold, or silver nanoparticles (AgNPs, AuNPs, and CuNPs) have a positive impact on the effectiveness of anti-cancer medications. The green metal NPs can interact with the deoxyribonucleic acid (DNA) of cancerous cells, causing the arrest of mitosis and cell cycle. In recent years, researchers have focused to develop GNPs, which can easily bind to the DNA of cancerous cells [42]. To demonstrate the interaction of GNPs with DNA, UV-visible (UV-vis) spectroscopy and gel electrophoresis have proven themselves to be practical, specific, and incredibly sensitive techniques [43]. In terms of properties and transport, a nanoscale drug carrier acts as a single entity. The size distribution of these nanoclusters is narrow, with a minimum of 1D (one dimension) ranging from 1 to 10 nm. Nanopowders are ultra-fine particle agglomerates that are known as nanoclusters or NPs, whereas nanocrystals are nanoparticle-sized crystals [44].

In this chapter, we discuss the synthetic process, characterization, and applications of various GNPs in drug delivery and as anti-microbials and anti-carcinogens.

16.2 Synthesis and Characterization of NPs

Nanomaterials can be synthesized by two approaches. Top-down technique, where a large molecule is broken down into tiny pieces using biological, chemical, and physical energy. The bottom-up technique, where massive nanostructures are created from the atomic layer using a variety of biological,

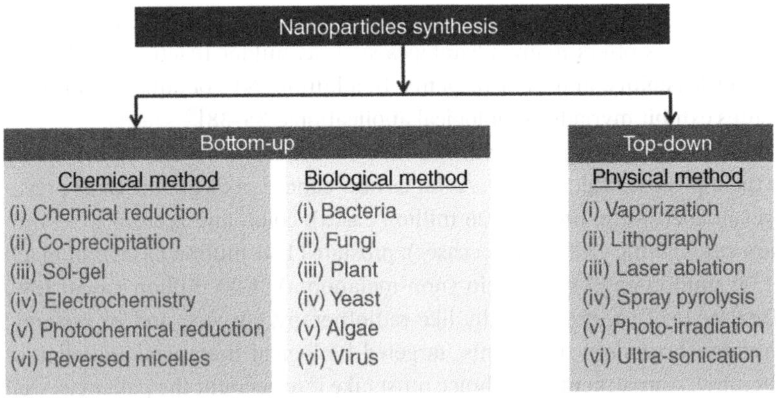

Figure 16.1 Approaches for NPs synthesis.

physical, and chemical reactions [45]. To construct nanostructured carriers (NC) using this technology, chemical and biological processes are commonly used (Figure 16.1). NPs are synthesized using both chemical and physical methods. The use of toxic substances in these methods can cause environmental toxicity, toxicity, and carcinogenicity [46]. To produce NPs, a reliable, clean, physiologically acceptable, and environmentally benign approach is required [47]. The NPs from biological methods could prove to be a viable option. It includes the adoption of both unicellular and multicellular microorganisms like bacteria [48], fungi [49], viruses [50], plants [51], yeasts [52], and actinomycetes [53]. Size, shape, physicochemical, and compositional properties of biologically produced NPs cover a wide range of topics [54].

Table 16.1 demonstrates the synthetic process of NPs following green chemistry. UV-vis spectroscopy is used to initially characterize the GNPs. X-ray diffraction (XRD) and Fourier-transform infrared (FT-IR) spectroscopy can also confirm the creation of NPs. The surface morphology and size of the GNPs are established by the transmission electron microscopy (TEM) and scanning electron microscopy (SEM).

16.3 GNPs as Anti-Carcinogens

Cancer is a general term that refers to a broad range of diseases characterized by uncontrolled cell division that can invade and damage healthy human tissue. Cancer has the potential to spread all over human body. Cardiovascular diseases are the leading cause of death globally (Kevin [63]). Cancer is the second most common cause of death in the world. However,

Table 16.1 NPs and green chemistry.

Serial number	NPs	Green chemistry approach applied	Reference
1	PLLA (poly-L-lactic acid)	Biodegradable	[55]
2	Silver (Ag) NPs	Plant extract	[22]
3	Dextrin–poly(lactide) nanogel	Easily degradable	[57]
4	Copper (Cu) NPs	Plant extract	[3]
5	Magnesium (Mg) stents	Non-toxic solvent	[58]
6	Cerium oxide (CeO_2) NPs	Plant extract	[10]
7	Poly-esteramine and poly(amido amine) NPs	Renewable	[59]
8	Iron (Fe) NPs	Plant extract	[60]
9	Chitosan-graphene oxide composite	Non-toxic solvent	[61]
10	Zinc (Zn) NPs	Plant extract	[62]

due to advancement in diagnosis and treatment, mortality rate is reduced to a great extent. Nanotechnology is an advanced tool for preventing the spread of cancer in the human body, diagnosis, and treatment of cancer with minimal side effects. Plants play an important role in treating various diseases, especially cancer, due to their wide variety of phytochemical and medicinal properties [64]. Scientific studies have shown that less than 80% of the world's population relies on herbal remedies for side effects [65, 66]. Secondary metabolites are low molecular weight molecules that have medicinal properties in many plants. Secondary metabolites are divided into three categories: polyphenols, terpenoids, and alkaloids [67–71].

Certain species of plants may play an important role in inhibiting or stimulating various signaling pathways in functional cells by producing anti-carcinogenic and melanoma secondary metabolites [72]. Plant-based anti-cancer drugs may contain quercetin [73], kaempferol [74], rutin [75], vincristine [76], colchicine [77], paclitaxel [78], linalool [79], and imodine [80].

Metal NPs prepared/synthesized by incorporating plant constituents have higher anti-cancer activity than the drugs prepared with respective plant constituents alone. Ag (Silver), Au (gold), and Cu (Copper)-based NPs have significant anti-cancer activity compared to other metals. Cu-based NPs are not much successful as cancer drugs compared to Ag- or Au-based NPs because of higher stability and lesser reactivity of Ag and Au to the environment than Cu. Among Au and Ag, Au-based NPs act as more effective anti-cancer drugs due to their higher conductivity and huge surface area. Gardea-Torresdey et al. [81] published the first research on AuNPs production using *Medicago sativa* (Alfaalfa or purple medic) plant extract [81]. Figure 16.2 displays the treatment of breast cancer by regulated and targeted delivery systems of nano-based drugs [82].

Green AuNPs are of significant interest in cancer diagnostics because of their unique advantages like high stability, high permeability, non-immunogenicity, and minimal cytotoxicity [83]. As green AuNPs can accumulate in cancer at the cell level, they are now being used in the diagnosis of cancer based on spectroscopic imaging applications, including NIR (Near Infrared) imaging, fluorescence imaging, photoacoustic imaging, and X-ray imaging [84]. Since early detection of cancer is crucial, it is important to develop strategies to reduce the potential impact of the disease progression and to increase the survival time [85]. Using imaging and biosensor technology, green AuNPs can detect specific cancer cell types or bio-markers as shown in Figure 16.3 [86]. Further, several Cu and Ag GNPs have shown significant anti-cancer properties [87, 88]. As a result, several scientists have been researching the cytotoxicity of nanoparticles. Different nanoparticles exhibit various levels of *in vitro*-cytotoxicity [89]. Ag NPs' ability to induce apoptosis is mediated by a reactive oxygen species (ROS)- and NIH3T3 cells have a JNK (c-Jun N-terminal kinase)-dependent mechanism involving the mitochondrial system [90]. Curcumin (Cur) in PLGA (poly lactic-coglycolic acid) based nano-spheres (NCur) were synthesized by NCur were employed as drug against prostate cancer Azandeh et al., [91]. Li et al. [92] introduced the cheapest Molybdenum disulfide (MoS_2) nano-sheet, which was used as photo-thermal therapy for breast cancer. Van et al. [93] modified PGG-PTX (poly(L-γ-glutamyl-glutamine)-paclit axel) nano-conjugation from PGA-PTX (poly(L-glutamic acid)-paclitaxel). They found higher anti-cancer activity in the PGG-PTX on H460 cell line (human lung cancer cell) than PGA-PTX. CuO NPs' plausible anti-cancer mechanism is depicted in Figure 16.4, and it was as hypothesized by Letchumanan et al. [94].

16.4 Green NPs as Anti-Microbials

In recent years, green synthesized metal and metal-oxide NPs have a lot of interest because of the environmental concern. Various plant components, like roots, leaves, fruits, stems, seeds, etc. have been used in the biosynthesis of NPs. Anti-septic effects of metal ions are widely recognized, and several plant extracts have anti-bacterial properties as well. Metal (M)-based NPs and M^{n+} released by them can enter the plasma membrane and cell wall of

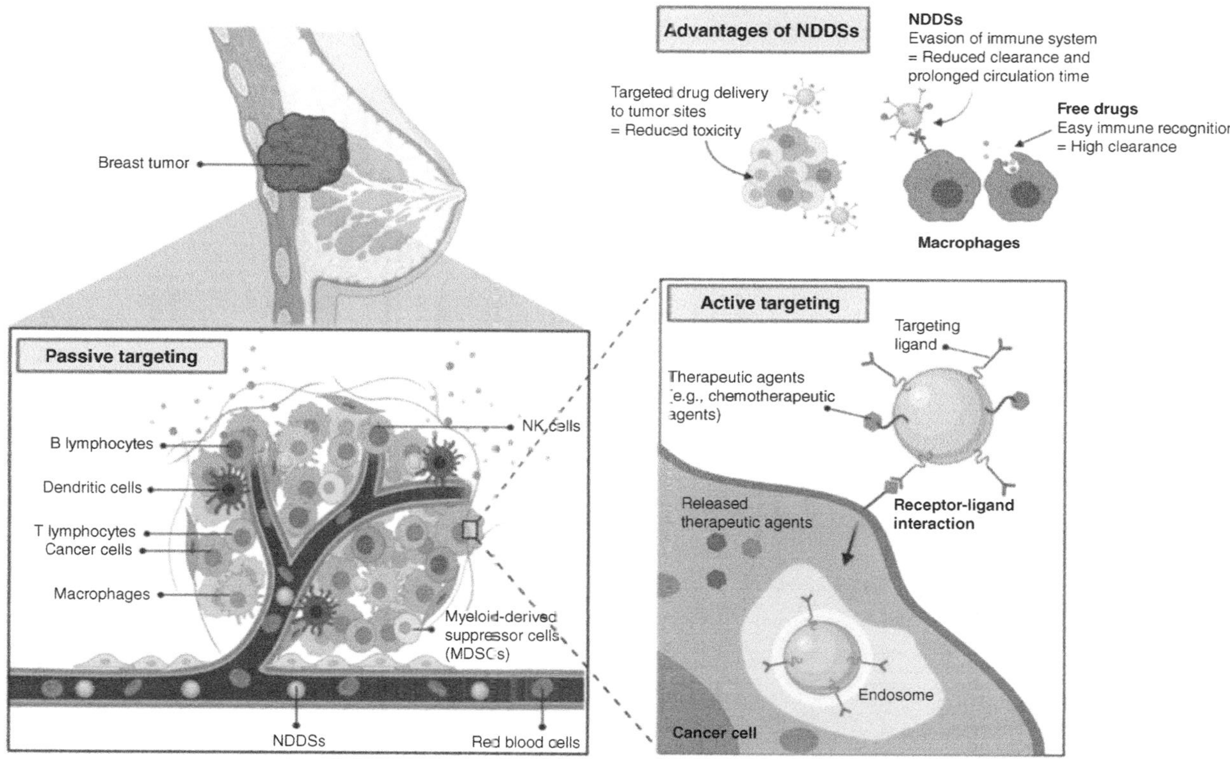

Figure 16.2 Schematic diagram for breast cancer treatment by nano-based drugs [82].
Source: Yap/Dove Press/CC BY-NC 4.0.

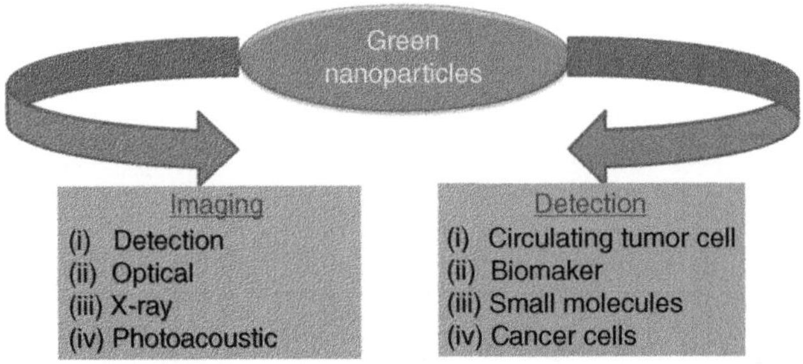

Figure 16.3 Detection of cancer by green AuNPs.

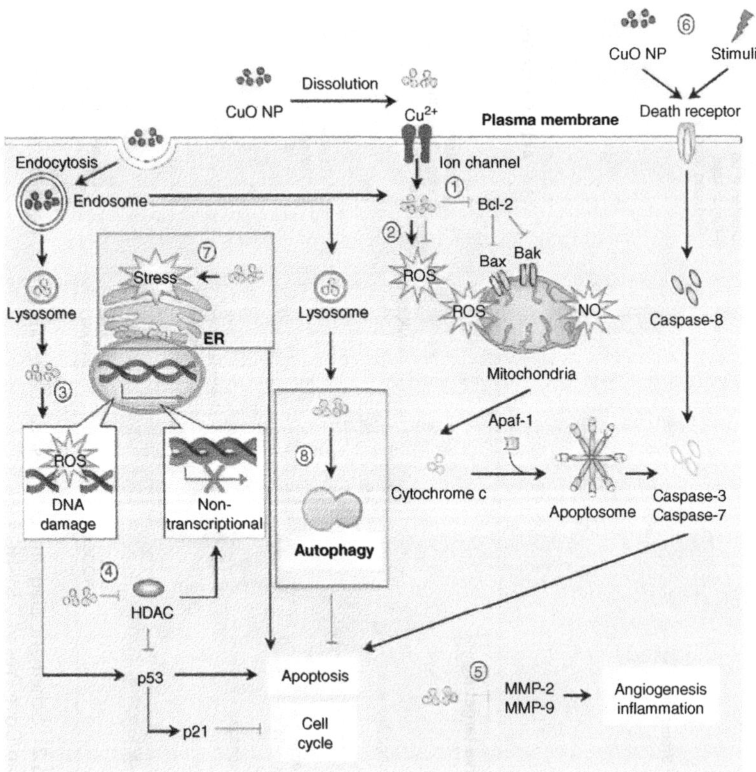

Figure 16.4 Possible mechanism of anti-cancer activity of CuO nanoparticles [94]. *Source:* Letchumanan/MDPI/CC by 4.0.

the bacteria, inactivate a wide range of enzymes, and inhibit adenosine triphosphate (ATP) formation. Green metal-based NPs inhibit the process of electron transport and produce ROS [95]. The produced ROS binds with the DNA of microorganisms and inhibits its replication and ultimately causes cell death as shown in Figure 16.5.

AuNPs synthesized by aqueous extract of the seeds of *Abelmoschus esculentus* (ladies finger or Okra) were used to inhibit *Candida albicans* and *Puccinia graminis tritci* [97]. The green AuNPs were synthesized by Abdel-Raouf et al. [98] using leaf extract of *Galaxaura elongate* (thalloid red algae). These AuNPs showed an impressive anti-bacterial activity against *Klebsiella pneumoniae* and *Escherichia coli* [98] (Table 16.2).

Khaled et al. [101] synthesized silver NPs (AgNPs) through herbal plant *Ipomoea asarifolia* (Convolvulaceae, ginger-leaf morning-glory). The NPs showed potential anti-bacterial activity against bacteria [101]. Similarly, green AgNPs were synthesized using *Croton Sparsifloras morang* (Croton), and these AgNPs exhibited potential anti-fungal activity [100].

Green CuNPs were synthesized by Punniyakotti et al. [102]) using leaf extract of *Cardiospermum halicacabum* (balloon vine exhibited significant anti-bacterial activity [102]). Green CuNPs synthesized by using leaf extract

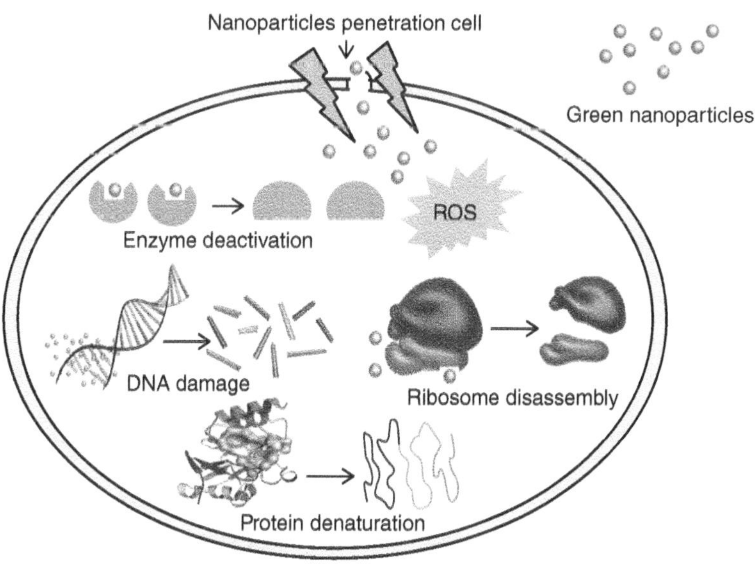

Figure 16.5 Schematic diagram of anti-microbial activity of GNPs [96] synthesized platinum NPs (PtNPs) by using plant extract of *Taraxacum laevigatum (the rock dandelion or red-seeded dandelion).* The NPs had anti-bacterial activity against *Pseudomonas aeruginosa* (*P. aeruginosa*) and *Bacillus subtilis* (*B. subtilis*) [96].

Table 16.2 Anti-microbial activity by GNPs.

Serial number	NPs	Plant extract	Anti-microbial	References
1	PtNPs	*Taraxacum laevigatum*	Bacteria	[96]
2	AuNPs	*Abelmoschus esculentus*	Bacteria	[99]
3	AuNPs	*Galaxaura elongate*	Bacteria	[100]
4	AgNPs	*Ipomoea asarifolia*	Bacteria	[101]
5	AgNPs	*Croton Sparsifloras morang*	Fungi	[97]
6	CuNPs	*Cardiospermum halicacabum*	Bacteria	[102]
7	CuNPs	*Celastrus paniculatus*	Fungi	[99]
8	FeNPs	*Platanus orientalis*	Fungi	[104]
9	FeNPs	*Lawsonia inermis* and *Gardenia jasminoides*	Bacteria	[103]
10	ZnNPs	*Lippia adoensi*	Bacteria	[105]
11	ZnONPs	*Salvia officinalis*	Fungal	[106]
12	ZnONPs	*Punica granatum,* Tamarind	Bacteria	[21]
13	AgNPs	Neam	Anti-microbial	[107]
14	AgNPs	*Bryophyllum pinnatum* and *Sonchus arvensi*	Anti-fungal	[13]
15	ZnONPs	Lemon juice	Bacteria, fungi, anti-tubercular	[17, 22]
16	CuONPs	Lemon juice	Anti-mycobacterial activity	[108]
17	CuONPs, Ag/CuO nanocomposites	Lemon juice	Anti-mycobacterial	[4]

of *Celastrus paniculatus* (black oil plant) showed significant anti-fungal activity against *Fusarium oxysporum* (plant pathogenic) [99].

Iron oxide NPs (FeNPs) were synthesized using leaf extract of *Platanus orientalis* (Chinar tree). The anti-fungal activity of FeNPs against *Mucor piriformis and Aspergillus niger* was remarkable [103]. Naseem et al. (2015)

synthesized green iron NPs (INPs) and evaluated their anti-bacterial activity against *Staphylococcus aureus*, *Escherichia coli*, *Proteus mirabilis*, and *Salmonella enterica* [104].

Zinc NPs (ZnNPs) synthesized by using leaf extract of *Lippia adoensi* (Koseret) showed anti-bacterial activity against Gram-positive and Gram-negative bacteria [105]. The synthesis of ZnO NPs (zinc oxide NPs) has been described by Abomuti et al. (2021) using the leaf extract of *Salvia officinalis* (culinary sage, common garden sage). ZnO NPs exhibited long-term potential in developing new anti-fungal drugs with many pharmacological targets [106]. Green synthesized Ag and CuO NPs were used as potential anti-microbials [107]. AgNPs from plant extract were used as drugs against **fungal infection** [56]. Sathyananda et al. [108] synthesized CuO and cobalt-doped CuO NPs using lemon juice and determined their anti-mycobacterial response. Green synthesized ZnONPs exhibited anti-tubercular activity against Mycobacterium tuberculosis H37Ra strain at concentrations as low as 12.5 μg/ml. Figure 16.6 depicts a possible anti-tubercular mechanism that was put up by Krishna et al. [22, 24, 25]. Antimicrobial activity of GNPs is provided in Table 16.2.

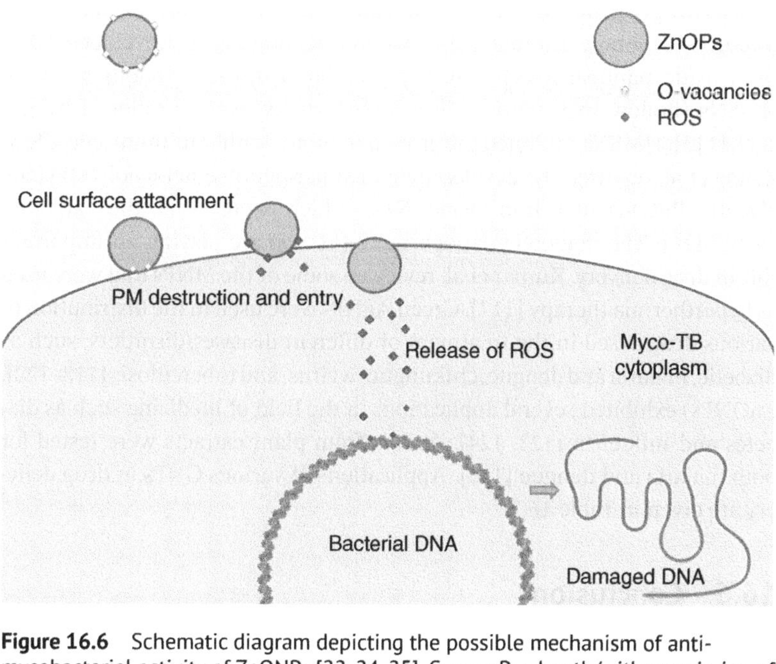

Figure 16.6 Schematic diagram depicting the possible mechanism of anti-mycobacterial activity of ZnONPs [22, 24, 25]. *Source:* Prashanth/with permission of Elsevier.

16.5 Applications of Green NPs in Another Drug Delivery

The higher surface area of NPs due to smaller size in proportion to their volume allows them to reach and penetrate cell barriers, make effective delivery of drugs possible to specific parts of the body. As a result, numerous NPs have been prepared, manipulated, and used as therapeutic transporters throughout the body. This has enabled early detection and treatment of a variety of disorders. NPs have been found to be applied in medicine as agents against pathogenic bacteria or as drug nano-carriers in specific areas of infection.

Pozdnyakov et al. (2016) synthesized AgNPs by using co-polymers such as N-vinylpyrrolidone and 1-vinyl-1,2,4-triazole under non-toxic water solvent. The non-toxic AgNPs exhibited potential anti-bacterial activity against Gram-positive and Gram-negative bacteria. The nobel non-toxic green AgNPs were used as anti-septic drugs [109]. Fatima et al. reported the synthesis of AuNPs and established their drug delivery efficacy against fungus and cancer [110]. Obayemi et al. [111] introduced multi-functional, biodegradable, and injectable anti-cancer drug D,L-lactide-co-glycolide NPs. These NPs were used as breast cancer drugs [111]. AgNPs from plant extract were used as drugs against several diseases, anti-bacterial and anti-oxidant [112], tumor [113], brain disorder [114], wound healing [115]. Developed IONs (iron oxide nanoparticles) were used for brain disorder treatment [114]. HMSNs (hollow mesoporous *silica nanoparticles*) were developed by Liu et al. [113]. HMSNs inhibited the growth of tumor with minimum side effect. Zhang et al. reported the development of super-paramagnetic poly(ethylene glycol) (PEG)-coated iron oxide NPs, which were used the targeting agent [116]. The magnetic nanoparticles (MNPs) are playing an important role in drug delivery. Kumar et al. reviewed some of the MNPs that were used as hyperthermia therapy [117]. Green AgNPs were used in the distribution of various drugs used in the treatment of different diseases/disorders, such as diabetic, malaria and dengue, chikungunya virus, and tuberculosis [118–122]. ZnONPs) exhibited several applications in the field of medicine such as diabetes and influenza [123, 124]. AuNPs from plant extracts were tested for both malaria and dengue [125]. Applications of various GNPs in drug delivery are given in Table 16.3.

16.6 Conclusion

Plant-based green-produced metallic NPs have significant physicochemical properties that make them ideal agents in nanobiotechnology research. Applications of plant and metal-based NPs as anti-carcinogens,

Table 16.3 Applications of GNPs in another drug delivery.

Serial Number	Name of NPs	Applications in drug delivery	References
1	Poly(D,L-lactide-co-glycolide)	Breast cancer cells	[111]
2	Molybdenum disulfide nanosheet	Photothermal agent	[92]
3	AgNPs in a polymer	Anti-septics	[109]
4	AgNPs	Wound healing	[115]
5	SiNPs-β	Inhibited tumor growth	[113]
6	AgNPs	Diabetic	[121]
7	AgNPs	Malaria	[120]
8	AgNPs	Dengue	[119]
9	AgNPs	Chikungunya virus	[122]
10	AgNPs	Tuberculosis	[118]
11	AuNPs	Immunomodulatory, antimicrobial, anti-cancer	[110]
12	AuNPs	Malaria and dengue	[125]
13	Iron oxide (PEG coating)	Targeting agent	[116]
14	Iron oxide NPs	Brain disorder	[114]
15	MNNPs	Hyperthermia therapy	[117]
16	ZnONPs	Diabetic	[123]
17	ZnONPs	Against influenza virus	[124]

anti-malarials, anti-bacterials, anti-oxidants, drug delivery agents, and as biosensors have been extensively investigated. Researchers are focusing on the plant-mediated green synthesis protocol, a one-step method that is safe, simple, energy efficient, reliable, fast, eco-friendly, and cost-effective. This periodic review summarizes the results of recent studies on green metal-based NPs in terms of their biological applications toward betterment of human health. Due to the presence of various phytochemicals in different plant extracts, significant differences in the size, shape, and applications of metal-based NPs have been discovered. This report will help the researchers to discover innovative green metal-based nanomaterials and explore their applications in drug discovery research.

Acknowledgments

The author Dr. G.K. Prashanth thanks the management of Sri KET Bengaluru for the support and encouragement extended toward the research activities.

References

1 Alabdallah, N.M. and Hasan, M.M. (2021). Plant-based green synthesis of silver nanoparticles and its effective role in abiotic stress tolerance in crop plants. *Saudi J. Biol. Sci.* 28 (10): 5631–5639. https://doi.org/10.1016/j.sjbs.2021.05.081.

2 Chandraker, S.K., Ghosh, M.K., Lal, M. et al. (2019). Colorimetric sensing of Fe^{3+} and Hg^{2+} and photocatalytic activity of green synthesized silver nanoparticles from the leaf extract of *Sonchus arvensis* L. *New J. Chem.* 43 (46): 18175–18183. https://doi.org/10.1039/C9NJ01338E.

3 Chandraker, S.K., Lal, M., Ghosh, M.K. et al. (2020). Green synthesis of copper nanoparticles using leaf extract of *Ageratum houstonianum* Mill. and study of their photocatalytic and antibacterial activities. *Nano Express* 1 (1): 010033. https://doi.org/10.1088/2632-959X/ab8e99.

4 Prashanth, G.K., Sathyananda, H.M., Prashanth, P.A. et al. (2022). Controlled synthesis of Ag/CuO nanocomposites: evaluation of their antimycobacterial, antioxidant, and anticancer activities. *Appl. Phys. A* 128:614. https://doi.org/10.1007/s00339-022-05748-x.

5 Ghosh, M.K., Jain, K., Khan, S. et al. (2020). New dual-functional and reusable bimetallic Y_2ZnO_4 nanocatalyst for organic transformation under microwave/green conditions. *ACS Omega* 5 (10): 4973–4981. https://doi.org/10.1021/acsomega.9b03875.

6 Ghosh, M.K., Sahu, S., Gupta, I., and Ghorai, T.K. (2020). Green synthesis of copper nanoparticles from an extract of *Jatropha curcas* leaves: characterization, optical properties, CT-DNA binding and photocatalytic activity. *RSC Adv.* 10 (37): 22027–22035. https://doi.org/10.1039/D0RA03186K.

7 Ghosh, M.K., Sahu, S.K., Sahoo, D., and Ghorai, T.K. (2020). Green synthesis, characterization and photocatalytic study of Cu based ZrO. *J. Indian Chem. Soc.* 97 (9b): 1507–1513.

8 Ghotekar, S., Pansambal, S., Bilal, M. et al. (2021). Environmentally friendly synthesis of Cr_2O_3 nanoparticles: characterization, applications and future perspective – a review. *Case Stud. Chem. Environ. Eng.* 3: 100089. https://doi.org/10.1016/j.cscee.2021.100089.

9 Kamel, S. and Khattab, T.A. (2021). Recent advances in cellulose supported metal nanoparticles as green and sustainable catalysis for organic synthesis. *Cellulose* 28 (8): 4545–4574. https://doi.org/10.1007/s10570-021-03839-1.

10 Naidi, S.N., Harunsani, M.H., Tan, A.L. et al. (2021). Green-synthesized CeO_2 nanoparticles for photocatalytic, antimicrobial, antioxidant and cytotoxicity activities. *J. Mater. Chem. B* 9 (28): 5599–5620. https://doi.org/10.1039/D1TB00248A.

11 Nas, M.S. and Calimli, M.H. (2021). Recent development of nanoparticle by green-conventional methods and applications for corrosion and fuel cells. *Curr. Nanosci.* 17 (4): 525–539. https://doi.org/10.2174/1573413716999200925163316.

12 Zamarchi, F. and Vieira, I.C. (2021). Determination of paracetamol using a sensor based on green synthesis of silver nanoparticles in plant extract. *J. Pharm. Biomed. Anal.* 196: 113912. https://doi.org/10.1016/j.jpba.2021.113912.

13 Chandraker, S.K., Lal, M., Dhruve, P. et al. (2021). Cytotoxic, antimitotic, DNA binding, photocatalytic, H_2O_2 sensing, and antioxidant properties of biofabricated silver nanoparticles using leaf extract of *Bryophyllum pinnatum* (Lam.) Oken. *Front. Mol. Biosci.* 465. https://doi.org/10.3389/fmolb.2020.593040.

14 Chandraker, S.K., Lal, M., Dhruve, P. et al. (2022). Bioengineered and biocompatible silver nanoparticles from *Thalictrum foliolosum* DC and their biomedical applications. *Clean Technol. Environ. Policy* 24: 2479–2494. https://doi.org/10.21203/rs.3.rs-1273479/v1.

15 Chandraker, S.K., Lal, M., Khanam, F. et al. (2022). Therapeutic potential of biogenic and optimized silver nanoparticles using *Rubia cordifolia* L. leaf extract. *Sci. Rep.* 12 (1): 1–15. https://doi.org/10.1038/s41598-022-12878-y.

16 Prashanth, G.K., Prashanth, P.A., Nagabhushana, B.M. et al. (2018). Comparison of anticancer activity of biocompatible ZnO nanoparticles prepared by solution combustion synthesis using aqueous leaf extracts of Abutilon indicum, Melia azedarach and Indigofera tinctoria as biofuels. *Artif. Cells Nanomed. Biotechnol.*, Taylor & Francis 46: 968–979. https://doi.org/10.1080/21691401.2017.1351982.

17 Prashanth, G.K., Prashanth, P.A., Ramani, M. et al. (2019). Comparison of antimicrobial, antioxidant and anticancer activities of ZnO nanoparticles prepared by lemon juice and citric acid fueled solution combustion synthesis. *BioNanoScience* 9 (4): 799–812. https://doi.org/10.1007/s12668-019-00670-8.

18 Sathyananda, H.M., Prashanth, P.A., Prashanth, G.K. et al. (2021). Evaluation of antimicrobial, antioxidant, and cytotoxicity activities of CuO nanopellets synthesized by surfactant free hydrothermal method. *J. Test. Eval.* 49: 2021. https://doi.org/10.1520/JTE20200538.

19 Noah, N.M. and Ndangili, P.M. (2022). Green synthesis of nanomaterials from sustainable materials for biosensors and drug delivery. *Sens. Int. 3*:100166. https://doi.org/10.1016/j.sintl.2022.100166.

20 Prashanth, G.K., Prashanth, P.A., Singh, P. et al. (2020). Effect of doping (with cobalt or nickel) and UV exposure on the antibacterial, anticancer, and ROS generation activities of zinc oxide nanoparticles. *J. Asian Ceram. Soc.* 8 (4): 1175–1187.

21 Prashanth, G.K., Prashanth, P.A., Bora, U. et al. (2015). In vitro antibacterial and cytotoxicity studies of ZnO nanopowders prepared by combustion assisted facile green synthesis. *Karbala Int. J. Mod. Sci.*, Elsevier 1: 67–77. https://doi.org/10.1016/j.kijoms.2015.10.007.

22 Krishna, P.G., Ananthaswamy, P.P., Yadavalli, T. et al. (2017). Antitubercular activity of ZnO nanoparticles prepared by solution combustion synthesis using lemon juice as bio-fuel. *Mater. Sci. Eng. C* 75: 1026–1033. Doi: https://doi.org/10.1016/j.msec.2017.02.093.

23 Krishna, P.G., Ananthaswamy, P.P., Yadavalli, T. et al. (2016). ZnO nanopellets have selective anticancer activity. *Mater. Sci. Eng. C* 62: 919–926.

24 Krishna, P.G., Ananthaswamy, P.P., Bora, U. et al. (2017). In vitro antibacterial and anticancer studies of ZnO nanoparticles prepared by sugar fueled combustion synthesis. *Adv. Mater. Lett.* 8: 24–29. https://doi.org/10.5185/amlett.2017.6424.

25 Krishna, P.G., Ananthaswamy, P.P., Mutta, N.B. et al. (2017). Comparison of antimicrobial and anticancer activity of ZnO nanoparticles prepared using different precursors by hydrothermal synthesis. *J. Chem. Pharm. Sci.* 10 (10): 192–197.

26 Ramesh, P., Saravanan, K., Manogar, P. et al. (2021). Green synthesis and characterization of biocompatible zinc oxide nanoparticles and evaluation of its antibacterial potential. *Sens. Bio-Sens. Res.* 31: 100399. https://doi.org/10.1016/j.sbsr.2021.100399.

27 Berenguel-Alonso, M., Ortiz-Gómez, I., Fernández, B. et al. (2019). An LTCC monolithic microreactor for the synthesis of carbon dots with photoluminescence imaging of the reaction progress. *Sensors Actuators B Chem.* 296: 126613.

28 Harada, M., Kuwa, M., Sato, R. et al. (2020). Cation distribution in monodispersed MFe_2O_4 (M= Mn, Fe, Co, Ni, and Zn) nanoparticles investigated by X-ray absorption fine structure spectroscopy: implications for magnetic data storage, catalysts, sensors, and ferrofluids. *ACS Appl. Nano Mater.* 3 (8): 8389–8402. https://doi.org/10.1021/acsanm.0c01810.

29 Polshettiwar, V. and Varma, R.S. (2010). Green chemistry by nano-catalysis. *Green Chem.* 12 (5): 743–754. https://doi.org/10.1039/B921171C.

30 Chung, I.M., Park, I., Seung-Hyun, K. et al. (2016). Plant-mediated synthesis of silver nanoparticles: their characteristic properties and therapeutic applications. *Nanoscale Res. Lett.* 11 (1): 1–14. https://doi.org/10.1186/s11671-016-1257-4.

31 Liang, W., Church, T.L., and Harris, A.T. (2012). Biogenic synthesis of photocatalytically active Ag/TiO_2 and Au/TiO_2 composites. *Green Chem.* 14 (4): 968–975. https://doi.org/10.1039/C2GC16082J.

32 Mohammadlou, M., Maghsoudi, H., and Jafarizadeh-Malmiri, H. (2016). A review on green silver nanoparticles based on plants: synthesis, potential applications and eco-friendly approach. *Int. Food Res. J.* 23 (2): 446–463.

33 Xin-gang, Z., Yuan-feng, Z., and Yan-bin, L. (2019). The spillovers of foreign direct investment and the convergence of energy intensity. *J. Clean. Prod.* 206: 611–621.

34 Gul, A.R., Shaheen, F., Rafique, R. et al. (2021). Grass-mediated biogenic synthesis of silver nanoparticles and their drug delivery evaluation: a biocompatible anti-cancer therapy. *Chem. Eng. J.* 407: 127202. https://doi.org/10.1016/J.CEJ.2020.127202.

35 Abdel-Halim, E.S., El-Rafie, M.H., and Al-Deyab, S.S. (2011). Polyacrylamide/guar gum graft copolymer for preparation of silver nanoparticles. *Carbohydr. Polym.* 85 (3): 692–697. https://doi.org/10.1016/j.carbpol.2012.12.035.

36 Huang, X., Wu, H., Pu, S. et al. (2011). One-step room-temperature synthesis of Au@Pd core–shell nanoparticles with tunable structure using plant tannin as reductant and stabilizer. *Green Chem.* 13 (4): 950–957. https://doi.org/10.1039/C0GC00724B.

37 Iravani, S. and Varma, R.S. (2019). Biofactories: engineered nanoparticles via genetically engineered organisms. *Green Chem.* 21 (17): 4583–4603. https://doi.org/10.1039/C9GC01759C.

38 Nasrollahzadeh, M., Sajadi, S.M., and Maham, M. (2015). *Tamarix gallica* leaf extract mediated novel route for green synthesis of CuO nanoparticles and their application for N-arylation of nitrogen-containing heterocycles under ligand-free conditions. *RSC Adv.* 5 (51): 40628–40635. https://doi.org/10.1039/C5RA04012D.

39 Max Roser and Hannah Ritchie (2015). "*Cancer*". OurWorldInData.org. https://ourworldindata.org/cancer'.

40 Iqbal, J., Abbasi, B.A., Mahmood, T. et al. (2017). Plant-derived anticancer agents: a green anticancer approach. *Asian Pac. J. Trop. Biomed.* 7 (12): 1129–1150. https://doi.org/10.1016/j.apjtb.2017.10.016.

41 Chandraker, S.K. and Shukla, R.G. (2021). Green synthesis of silver nanoparticles (AgNPs) from plants of Amarkantak region and assessment of their photocatalytic and biological activity (Doctoral dissertation, Indira Gandhi National Tribal University, Amarkantak, Madhya Pradesh-484886).

42 Ribeiro, A.P.C., Anbu, S., Alegria, E.C.B.A. et al. (2018). Evaluation of cell toxicity and DNA and protein binding of green synthesized silver nanoparticles. *Biomed. Pharmacother.* 101: 137–144. https://doi.org/10.1016/j.biopha.2018.02.069.

43 Kukreti, S. and Kaushik, M. (2019). Exploring the potential of environment friendly silver nanoparticles for DNA interaction: physicochemical approach. *J. Photochem. Photobiol. B Biol.* 194: 158–165. https://doi.org/10.1016/j.jphotobiol.2019.03.022.

44 Zhang, D., Ma, X.L., Gu, Y. et al. (2020). Green synthesis of metallic nanoparticles and their potential applications to treat cancer. *Front. Chem.* 799. https://doi.org/10.3389/fchem.2020.00799.

45 Das, R.K., Pachapur, V.L., Lonappan, L. et al. (2017). Biological synthesis of metallic nanoparticles: plants, animals and microbial aspects. *Nanotechnol. Environ. Eng.* 2 (1): 1–21. https://doi.org/10.1007/s41204-017-0029-4.

46 Gupta, R. and Xie, H. (2018). Nanoparticles in daily life: applications, toxicity and regulations. *J. Environ. Pathol. Toxicol. Oncol.* 37 (3): https://doi.org/10.1615/JEnvironPatholToxicolOncol.2018026009.

47 Jain, N., Bhargava, A., Majumdar, S. et al. (2011). Extracellular biosynthesis and characterization of silver nanoparticles using *Aspergillus flavus* NJP08: a mechanism perspective. *Nanoscale* 3 (2): 635–641.

48 Hamdiani, S. and Shih, Y.F. (2022). The antibacterial effect of silver nanoparticle (Ag0) and silver oxide (Ag_2O) in PLAPBS-D (polylactic acid-polybuthylene succinate-diatomite) green composites. *Mod. Phys. Lett. B* 2150598. https://doi.org/10.1142/S0217984921505989.

49 Jeevanandam, J., Kiew, S.F., Boakye-Ansah, S. et al. (2022). Green approaches for the synthesis of metal and metal oxide nanoparticles using microbial and plant extracts. *Nanoscale* 14 (7): 2534–2571. https://doi.org/10.1039/D1NR08144F.

50 Tortella, G.R., Pieretti, J.C., Rubilar, O. et al. (2022). Silver, copper and copper oxide nanoparticles in the fight against human viruses: progress and perspectives. *Crit. Rev. Biotechnol.* 42 (3): 431–449. https://doi.org/10.1080/07388551.2021.1939260.

51 Habeeb Rahuman, H.B., Dhandapani, R., Narayanan, S. et al. (2022). Medicinal plants mediated the green synthesis of silver nanoparticles and their biomedical applications. *IET Nanobiotechnol.* https://doi.org/10.1049/nbt2.12078.

52 Sharma, R. and Tripathi, A. (2022). Green synthesis of nanoparticles and its key applications in various sectors. *Mater. Today: Proc.* 48: 1626–1632.

53 Bedlovičová, Z. (2022). Green synthesis of silver nanoparticles using actinomycetes. In: *Green Synthesis of Silver Nanomaterials* (ed. K.A. Abd-Elsalam), 547–569. Elsevier https://doi.org/10.1016/C2020-0-00288-4.

54 Mohanpuria, P., Rana, N.K., and Yadav, S.K. (2008). Biosynthesis of nanoparticles: technological concepts and future applications. *J. Nanopart. Res.* 10 (3): 507–517. https://doi.org/10.1007/s11051-007-9275-x.

55 Tamai, H., Igaki, K., Kyo, E. et al. (2000). Initial and 6-month results of biodegradable poly-l-lactic acid coronary stents in humans. *Circulation* 102 (4): 399–404. https://doi.org/10.1161/01.cir.102.4.399.

56 Chandraker, S.K., Ghosh, M.K., Lal, M. et al. (2021). A review on plant-mediated synthesis of silver nanoparticles, their characterization and applications. *Nano Express* 2 (2): 022008. https://doi.org/10.1088/2632-959X/ac0355.

57 Das, D., Patra, P., Ghosh, P. et al. (2016). Dextrin and poly (lactide)-based biocompatible and biodegradable nanogel for cancer targeted delivery of doxorubicin hydrochloride. *Polym. Chem.* 7 (17): 2965–2975. https://doi.org/10.1039/C6PY00213G.

58 Peeters, P., Bosiers, M., Verbist, J. et al. (2005). Preliminary results after application of absorbable metal stents in patients with critical limb ischemia. *J. Endovasc. Ther.* 12 (1): 1–5. https://doi.org/10.1583/04-1349R.1.

59 Withey, A.B., Chen, G., Nguyen, T.U., and Stenzel, M.H. (2009). Macromolecular cobalt carbonyl complexes encapsulated in a click-cross-linked micelle structure as a nanoparticle to deliver cobalt pharmaceuticals. *Biomacromolecules* 10 (12): 3215–3226.

60 Ahmouda, K., Benhaoua, B., Laouini, S.E. et al. (2022). Plant extract FRAP effect on cation vacancies formation in greenly synthesized wüstite (FexO) nanoparticles: a new contribution. *Sustain. Chem. Pharm.* 25: 100563. https://doi.org/10.1016/j.scp.2021.100563.

61 Patel, M., Patel, J., and Dayaramani, R. (2022). Green-nanotechnology-driven drug delivery systems. In: *Sustainable Nanotechnology: Strategies, Products, and Applications*, 117–133. Wiley.

62 Riaz, T., Assey, N., Javed, M. et al. (2022). Biogenic plant mediated synthesis of monometallic zinc and bimetallic Copper/Zinc nanoparticles and their dye adsorption and antioxidant studies. *Inorg. Chem. Commun.* 140: 109449.

63 Mc Namara, K., Alzubaidi, H., and Jackson, J.K. (2019). Cardiovascular disease as a leading cause of death: how are pharmacists getting involved? *Integr. Pharm. Res. Pract.* 8: 1–11.

64 Kuruppu, A.I., Paranagama, P., and Goonasekara, C.L. (2019). Medicinal plants commonly used against cancer in traditional medicine formulae in Sri Lanka. *Saudi Pharm. J.* 27 (4): 565–573. https://doi.org/10.1016/j.jsps.2019.02.004.

65 Evin, A., Bouby, L., Bonhomme, V. et al. (2022). Archaeophenomics of ancient domestic plants and animals using geometric morphometrics: a review. *Peer Community Journal* 2: e27. doi : 10.24072/pcjournal.126. https://peercommunityjournal.org/articles/10.24072/pcjournal.126/.

66 Schaal, B. (2019). Plants and people: our shared history and future. *Plants People Planet* 1 (1): 14–19. https://doi.org/10.1002/ppp3.12.

67 Kabera, J.N., Semana, E., Mussa, A.R., and He, X. (2014). Plant secondary metabolites: biosynthesis, classification, function and pharmacological properties. *J. Pharm. Pharmacol.* 2 (7): 377–392.

68 Prashanth, G.K. and Krishnaiah, G.M. (2014). Chemical composition of the leaves of *Azadiracta indica* Linn (Neem). *Int. J. Adv. Eng. Technol. Manag. Appl. Sci.* 1 (5): 21–31.

69 Prashanth, G.K. and Krishnaiah, G.M. (2014). Phytochemical screening and GC-MS analysis of the leaves of *Pongamia Pinnata* Linn. *Int. J. Innov. Res. Sci. Eng. Technol.* 3 (11): 17329–17334.

70 Prashanth, G.K., Krishnaiah, G.M., and Sathyananda, H.M. (2015). Qualitative phytochemical screening and GCMS analysis of the leaves of *Indigofera tinctoria*. *Int. J. Innov. Res. Sci. Eng. Technol.* 4: 7544–7547. https://doi.org/10.15680/IJIRSET.2015.0408080.

71 Seca, A.M. and Pinto, D.C. (2018). Plant secondary metabolites as anticancer agents: successes in clinical trials and therapeutic application. *Int. J. Mol. Sci.* 19 (1): 263. https://doi.org/10.3390/ijms19010263.

72 Fulda, S. and Efferth, T. (2015). Selected secondary plant metabolites for cancer therapy. *World J. Tradit. Chin. Med.* 1 (1): 24. https://doi.org/10.15806/j.issn.2311-8571.2014.0005.

73 Almatroodi, S.A., Alsahli, M.A., Almatroudi, A. et al. (2021). Potential therapeutic targets of quercetin, a plant flavonol, and its role in the therapy of various types of cancer through the modulation of various cell signaling pathways. *Molecules* 26 (5): 1315. https://doi.org/10.3390/molecules26051315.

74 Imran, M., Salehi, B., Sharifi-Rad, J. et al. (2019). Kaempferol: a key emphasis to its anticancer potential. *Molecules* 24 (12): 2277. https://doi.org/10.3390/molecules24122277.

75 Satari, A., Ghasemi, S., Habtemariam, S. et al. (2021). Rutin: a flavonoid as an effective sensitizer for anticancer therapy; insights into multifaceted mechanisms and applicability for combination therapy. *Evid. Based Complement. Alternat. Med.* 2021: https://doi.org/10.1155/2021/9913179.

76 Qu, Y., Safonova, O., and De Luca, V. (2019). Completion of the canonical pathway for assembly of anticancer drugs vincristine/vinblastine in *Catharanthus roseus*. *Plant J.* 97 (2): 257–266. https://doi.org/10.1111/tpj.14111.

77 Balkrishna, A., Das, S.K., Pokhrel, S. et al. (2019). Colchicine: isolation, LC–MS QT of screening, and anticancer activity study of *Gloriosa superba* seeds. *Molecules* 24 (15): 2772. https://doi.org/10.3390/molecules24152772.

78 Zein, N., Aziz, S.W., El-Sayed, A.S., and Sitohy, B. (2019). Comparative cytotoxic and anticancer effect of taxol derived from *Aspergillus terreus* and *Taxus brevifolia*. *Biosci. Res.* 16 (2): 1500–1509.

79 Rodenak-Kladniew, B., Castro, M.A., Crespo, R. et al. (2020). Anti-cancer mechanisms of linalool and 1, 8-cineole in non-small cell lung cancer A549 cells. *Heliyon* 6 (12): e05639. https://doi.org/10.1016/j.heliyon.2020.e05639.

80 Zhou, R.S., Wang, X.W., Sun, Q.F. et al. (2019). Anticancer effects of emodin on HepG2 cell: evidence from bioinformatic analysis. *Biomed. Res. Int.* https://doi.org/10.1155/2019/3065818.

81 Gardea-Torresdey, J.L., Gomez, E., Peralta-Videa, J.R. et al. (2003). Alfalfa sprouts: a natural source for the synthesis of silver nanoparticles. *Langmuir* 19 (4): 1357–1361. https://doi.org/10.1021/la020835i.

82 Yap, K.M., Sekar, M., Fuloria, S. et al. (2021). Drug delivery of natural products through nanocarriers for effective breast cancer therapy: a comprehensive review of literature. *Int. J. Nanomedicine* 16: 7891. https://doi.org/10.2147/IJN.S328135.

83 Gong, F., Yang, N., Wang, X. et al. (2020). Tumor microenvironment-responsive intelligent nanoplatforms for cancer theranostics. *Nano Today* 32: 100851. https://doi.org/10.1016/j.nantod.2020.100851.

84 Luo, D., Wang, X., Burda, C., and Basilion, J.P. (2021). Recent development of gold nanoparticles as contrast agents for cancer diagnosis. *Cancer* 13 (8): 1825. https://doi.org/10.3390/cancers13081825.

85 Zhang, Q., Yang, M., Zhu, Y., and Mao, C. (2018). Metallic nanoclusters for cancer imaging and therapy. *Curr. Med. Chem.* *25* (12): 1379–1396. https://doi.org/10.2174/0929867324666170331122757.

86 Khan, M., Shaik, M.R., Adil, S.F. et al. (2018). Plant extracts as green reductants for the synthesis of silver nanoparticles: lessons from chemical synthesis. *Dalton Trans.* 47 (35): 11988–12010. https://doi.org/10.1039/C8DT01152D.

87 Hasanin, M., Al Abboud, M.A., Alawlaqi, M.M. et al. (2022). Ecofriendly synthesis of biosynthesized copper nanoparticles with starch-based nanocomposite: antimicrobial, antioxidant, and anticancer activities. *Biol. Trace Elem. Res.* 200 (5): 2099–2112. https://doi.org/10.1007/s12011-021-02812-0. Epub 2021 Jul 20.

88 Oves, M., Rauf, M.A., Aslam, M. et al. (2022). Green synthesis of silver nanoparticles by *Conocarpus lancifolius* plant extract and their antimicrobial and anticancer activities. *Saudi J. Biol. Sci.* 29 (1): 460–471. https://doi.org/10.1016/j.sjbs.2021.09.007.

89 Hsin, Y.H., Chen, C.F., Huang, S. et al. (2008). The apoptotic effect of nanosilver is mediated by a ROS- and JNK dependent mechanism involving the mitochondrial pathway in NIH3T3 cells. *Toxicol. Lett.* 2008 (179): 130–139. https://doi.org/10.1016/j.toxlet.2008.04.015.

90 Park, S., Lee, Y.K., Jung, M. et al. (2007, 2007). Cellular toxicity of various inhalable metal nanoparticles on human alveolar epithelial cells. *Inhal. Toxicol.* 19: 59–65. https://doi.org/10.1080/08958370701493282.

91 Azandeh, S.S., Abbaspour, M., Khodadadi, A. et al. (2017). Anticancer activity of curcumin-loaded PLGA nanoparticles on PC3 prostate cancer cells. *Iran. J. Pharm. Res.* 16 (3): 868.

92 Li, X., Gong, Y., Zhou, X. et al. (2016). Facile synthesis of soybean phospholipid-encapsulated MoS2 nanosheets for efficient in vitro and in vivo photothermal regression of breast tumor. *Int. J. Nanomedicine* 11: 1819. https://doi.org/10.2147/IJN.S104198.

93 Van, S., Das, S.K., Wang, X. et al. (2010). Synthesis, characterization, and biological evaluation of poly (L-γ-glutamyl-glutamine)-paclitaxel nanoconjugate. *Int. J. Nanomedicine* 5: 825. https://doi.org/10.2147/IJN.S13482.

94 Letchumanan, D., Sok, S.P.M., Ibrahim, S. et al. (2021). Plant-based biosynthesis of copper/copper oxide nanoparticles: an update on their applications in biomedicine, mechanisms, and toxicity. *Biomolecules* 11 (4): 564. Published 2021 Apr 12. https://doi.org/10.3390/biom11040564.

95 Abdal Dayem, A., Hossain, M.K., Lee, S.B. et al. (2017). The role of reactive oxygen species (ROS) in the biological activities of metallic nanoparticles. *Int. J. Mol. Sci.* 18: 120. https://doi.org/10.3390/ijms18010120.

96 Tahir, K., Nazir, S., Ahmad, A. et al. (2017). Facile and green synthesis of phytochemicals capped platinum nanoparticles and in vitro their superior antibacterial activity. *J. Photochem. Photobiol. B Biol.* 166: 246–251. https://doi.org/10.1016/j.jphotobiol.2016.12.016.

97 Jayaseelan, C., Ramkumar, R., Rahuman, A.A. et al. (2013). Green synthesis of gold nanoparticles using seed aqueous extract of *Abelmoschus esculentus* and its antifungal activity. *Ind. Crop. Prod.* 45:423–429. https://doi.org/10.1016/j.indcrop.2012.12.019.

98 Abdel-Raouf, N., Al-Enazi, N.M., and Ibraheem, I.B. (2017). Green biosynthesis of gold nanoparticles using *Galaxaura elongata* and characterization of their antibacterial activity. *Arab. J. Chem.* 10: S3029–S3039. https://doi.org/10.1016/j.arabjc.2013.11.044.

99 Mali, S.C., Dhaka, A., Githala, C.K., and Trivedi, R. (2020). Green synthesis of copper nanoparticles using *Celastrus paniculatus* Willd. leaf extract and their photocatalytic and antifungal properties. *Biotechnol. Rep.* 27: e00518. https://doi.org/10.1016/j.btre.2020.e00518.

100 Kathiravan, V., Ravi, S., Ashokkumar, S. et al. (2015). Green synthesis of silver nanoparticles using *Croton sparsiflorus* morong leaf extract and their antibacterial and antifungal activities. *Spectrochim. Acta A Mol. Biomol. Spectrosc.* 139: 200–205. https://doi.org/10.1016/j.saa.2014.12.022.

101 Khaled, J.M., Alharbi, N.S., Kadaikunnan, S. et al. (2017). Green synthesis of Ag nanoparticles with anti-bacterial activity using the leaf extract of an African medicinal plant, *Ipomoea asarifolia* (Convolvulaceae). *J. Clust. Sci.* 28 (5): 3009–3019.

102 Punniyakotti, P., Panneerselvam, P., Perumal, D. et al. (2020). Anti-bacterial and anti-biofilm properties of green synthesized copper nanoparticles from *Cardiospermum halicacabum* leaf extract. *Bioprocess Biosyst. Eng.* 43 (9): 1649–1657. https://doi.org/10.1007/s00449-020-02357-x.

103 Devi, H.S., Boda, M.A., Shah, M.A. et al. (2019). Green synthesis of iron oxide nanoparticles using *Platanus orientalis* leaf extract for antifungal activity. *Green Process. Synth.* 8 (1): 38–45. https://doi.org/10.1515/gps-2017-0145.

104 Naseem, T. and Farrukh, M.A. (2015). Antibacterial activity of green synthesis of iron nanoparticles using *Lawsonia inermis* and *Gardenia jasminoides* leaves extract. *J. Chem.* 2015: https://doi.org/10.1155/2015/912342.

105 Demissie, M.G., Sabir, F.K., Edossa, G.D., and Gonfa, B.A. (2020). Synthesis of zinc oxide nanoparticles using leaf extract of *Lippia adoensis* (koseret) and evaluation of its antibacterial activity. *J. Chem.* 2020: https://doi.org/10.1155/2020/7459042.

106 Abomuti, M.A., Danish, E.Y., Firoz, A. et al. (2021). Green synthesis of zinc oxide nanoparticles using *Salvia officinalis* leaf extract and their photocatalytic and antifungal activities. *Biology* 10 (11): 1075. https://doi.org/10.3390/biology10111075.

107 Verma, A. and Mehata, M.S. (2016). Controllable synthesis of silver nanoparticles using neem leaves and their antimicrobial activity. *J. Radiat. Res. Appl. Sci.* 9 (1): 109–115. https://doi.org/10.1016/j.jrras.2015.11.001.

108 Sathyananda, H.M., Prashanth, P.A., Prashanth, G.K. et al. (2022). Evaluation of antimycobacterial, antioxidant, and anticancer activities of CuO nanoparticles through cobalt doping. *Appl. Nanosci.* 12: 79–86. https://doi.org/10.1007/s13204-021-02156-0.

109 Pozdnyakov, A.S., Emel'yanov, A.I., Kuznetsova, N.P. et al. (2016). Nontoxic hydrophilic polymeric nanocomposites containing silver nanoparticles with strong antimicrobial activity. *Int. J. Nanomedicine* 11:1295. https://doi.org/10.2147/IJN.S98995.

110 Fatima, F., Bajpai, P., Pathak, N. et al. (2015). Antimicrobial and immunomodulatory efficacy of extracellularly synthesized silver and gold nanoparticles by a novel phosphate solubilizing fungus *Bipolaris tetramera*. *BMC Microbiol.* 15 (1): 1–10. https://doi.org/10.1186/s12866-015-0391-y.

111 Obayemi, J.D., Malatesta, K.A., Odusanya, O.S. et al. (2016). Abstract C60: injectable, biodegradable micro-and nano-particles loaded with prodigiosin-based drug for localized anticancer drug delivery. *Cancer Epidemiol. Biomark. Prev.* 25: C60.

112 Mitiku, A.A. and Yilma, B. (2017). Antibacterial and antioxidant activity of silver nanoparticles synthesized using aqueous extract of *Moringa stenopetala* leaves. *Afr. J. Biotechnol.* 16 (32): 1705–1716. https://doi.org/10.5897/AJB2017.16010.

113 Liu, J., Luo, Z., Zhang, J. et al. (2016). Hollow mesoporous silica nanoparticles facilitated drug delivery via cascade pH stimuli in tumor microenvironment for tumor therapy. *Biomaterials* 83: 51–65. https://doi.org/10.1016/j.biomaterials.2016.01.008.

114 Sun, Z., Worden, M., Thliveris, J.A. et al. (2016). Biodistribution of negatively charged iron oxide nanoparticles (IONPs) in mice and enhanced brain delivery using lysophosphatidic acid (LPA). *Nanomedicine* 12 (7): 1775–1784. https://doi.org/10.1016/j.nano.2016.04.008.

115 Kaler, A., Mittal, A.K., Katariya, M. et al. (2014). An investigation of in vivo wound healing activity of biologically synthesized silver nanoparticles. *J. Nanopart. Res.* 16 (9): 1–10. https://doi.org/10.1007/s11051-014-2605-x.

116 Zhang, Y., Sun, C., Kohler, N., and Zhang, M. (2004). Self-assembled coatings on individual monodisperse magnetite nanoparticles for efficient intracellular uptake. *Biomed. Microdevices* 6 (1): 33–40. https://doi.org/10.1023/b:bmmd.0000013363.77466.63.

117 Kumar, C.S. and Mohammad, F. (2011). Magnetic nanomaterials for hyperthermia-based therapy and controlled drug delivery. *Adv. Drug Deliv. Rev.* 63 (9): 789–808. https://doi.org/10.1016/j.addr.2011.03.008.

118 Agarwal, P., Mehta, A., Kachhwaha, S. et al. (2013). Green synthesis of silver nanoparticles and their activity against *Mycobacterium tuberculosis*. *Adv. Sci. Eng. Med.* 5 (7): 709–714. https://doi.org/10.1166/asem.2013.1307.

119 Arjunan, N.K., Murugan, K., Rejeeth, C. et al. (2012). Green synthesis of silver nanoparticles for the control of mosquito vectors of malaria, filariasis, and dengue. *Vector Borne Zoonotic Dis.* 12 (3): 262–268.

120 Dinesh, D., Murugan, K., Madhiyazhagan, P. et al. (2015). Mosquitocidal and antibacterial activity of green-synthesized silver nanoparticles from *Aloe vera* extracts: towards an effective tool against the malaria vector *Anopheles stephensi*? *Parasitol. Res.* 114 (4): 1519–1529. https://doi.org/10.1007/s00436-015-4336-z.

121 Jini, D. and Sharmila, S. (2020). Green synthesis of silver nanoparticles from *Allium cepa* and its in vitro antidiabetic activity. *Mater. Today: Proc.* 22: 432–438. https://doi.org/10.1016/j.matpr.2019.07.672.

122 Sharma, V., Kaushik, S., Pandit, P. et al. (2019). Green synthesis of silver nanoparticles from medicinal plants and evaluation of their antiviral potential against chikungunya virus. *Appl. Microbiol. Biotechnol.* 103 (2): 881–891. https://doi.org/10.1007/s00253-018-9488-1.

123 Bala, N., Saha, S., Chakraborty, M. et al. (2015). Green synthesis of zinc oxide nanoparticles using *Hibiscus subdariffa* leaf extract: effect of temperature on synthesis, anti-bacterial activity and anti-diabetic activity. *RSC Adv.* 5 (7): 4993–5003. https://doi.org/10.1039/C4RA12784F.

124 Rabiee, N., Bagherzadeh, M., Ghadiri, A.M. et al. (2020). Green synthesis of ZnO NPs via *Salvia hispanica*: evaluation of potential antioxidant, antibacterial, mammalian cell viability, H1N1 influenza virus inhibition and photocatalytic activities. *J. Biomed. Nanotechnol.* 16 (4): 456–466. https://doi.org/10.1166/jbn.2020.2916.

125 Murugan, K., Benelli, G., Panneerselvam, C. et al. (2015). Cymbopogon citratus-synthesized gold nanoparticles boost the predation efficiency of copepod *Mesocyclops aspericornis* against malaria and dengue mosquitoes. *Exp. Parasitol.* 153: 129–138. https://doi.org/10.1016/j.exppara.2015.03.017.

126 Liu, Z., Robinson, J.T., Sun, X., and Dai, H. (2008). PEGylated nanographene oxide for delivery of water-insoluble cancer drugs. *J. Am. Chem. Soc.* 130 (33): 10876–10877. https://doi.org/10.1021/ja803688x.

127 Krishna, P.G., Mishra, P.C., Naika, M.M. et al. (2021). Photocatalytic activity induced by metal nanoparticles synthesized by sustainable approaches: a comprehensive review, rising stars in catalysis and photocatalysis. *Front. Chem. Photocatal. Photochem.* 2022: 10. https://doi.org/10.3389/fchem.2022.917831.

128 Robinson, J.T., Perkins, F.K., Snow, E.S. et al. (2008). Reduced graphene oxide molecular sensors. *Nano Lett.* 8 (10): 3137–3140.

Index

Green Synthesis of Nanomaterials: Biological and Environmental Applications, First Edition.
Edited by Archana Chakravarty, Preeti Singh, Saiqa Ikram, and R.N. Yadava.

o

p

Printed and bound by CPI Group (UK) Ltd, Croydon, CR0 4YY
17/09/2026
14964255-0002